Signals and Communication Technology

For further volumes:
http://www.springer.com/series/4748

Arturas Medeisis · Oliver Holland
Editors

Cognitive Radio Policy and Regulation

Techno-Economic Studies to Facilitate Dynamic Spectrum Access

Springer

Editors
Arturas Medeisis
Department of Telecommunication Engineering
Vilnius Gediminas Technical University
Vilnius
Lithuania

Oliver Holland
Centre for Telecommunications Research
King's College London
London
UK

ISSN 1860-4862 ISSN 1860-4870 (electronic)
ISBN 978-3-319-04021-9 ISBN 978-3-319-04022-6 (eBook)
DOI 10.1007/978-3-319-04022-6
Springer Cham Heidelberg New York Dordrecht London

Library of Congress Control Number: 2014930579

The companion website to this book can be found at www.cost-terra.org/CR-policy-book.

Printed on acid-free paper

Springer is part of Springer Science+Business Media (www.springer.com)

Preface

Dear Reader,

The book at hand was authored by a large team of highly dedicated people and represents a consolidated summary of results of nearly 4 years of their work. Its aim is to investigate the highly complex policy and regulatory aspects that govern the ecosystem of wireless services and innovation. It seeks to identify policy changes that can be instantiated to become effective enablers for the development and proliferation of advanced wireless communications systems, particularly those based on Cognitive Radio (CR) technologies. A distinguishing feature of this book is its consideration of the prospects of CR from two diverging standpoints: technological development and economic market reality. This book therefore provides a broad survey of various techno-economic and policy aspects of CR development, and offers the reader an understanding of the intricate complexities involved in such aspects, as well as providing a toolbox of possible solutions to enable the evolutionary leap towards successful implementation of CR technology.

This book might thus be seen as a quite unique survey giving a holistic techno-economic treatise on the subject of CR policy and regulation. This is particularly significant given the importance of the current radio spectrum governing framework and its adaptations needed to pave the way for CR and Dynamic Spectrum Access (DSA) applications to develop and flourish in real-world deployment environments. It should also be understood that the motivation and basis for any regulatory advancements would in turn rely on the ability to prove the soundness and economic benefits of proposed CR development scenarios.

It is sincerely hoped that this book will become a source of subject reference material and new ideas of value to academic researchers in the field of wireless communications, especially those working on CR and DSA and their interaction with regulatory and policy issues. Moreover, this book should be equally useful for the industry and regulatory professionals concerned with radio spectrum management and the general development of wireless communications. This is particularly the case noting the breadth of considered regulatory and strategic issues covered in the book. Together, these provide a solid basis upon which academic or professional work can flourish.

The intention is that this book will also become a helpful reading reference for advanced postgraduate studies on burgeoning subjects of wireless technologies.

This book will provide learned knowledge and inspiration to study novel wireless technologies such as CR and their promotion at the intersection among policy, communications technology and economic interests. Possible study topics and directions that directly relate to the subjects covered in this book include:

- Cognitive Radio and Software-defined Radio (SDR);
- Dynamic Spectrum Access and "White Space" Technologies;
- Radio Spectrum Management;
- Advanced Wireless Communications;
- Telecommunications Policy and Strategy;
- Telecommunications Business.

The overall structure of this book is organised so as to carefully lead the reader through the key constituent elements in consideration of CR policy and regulation. First, the introduction together with Chap. 1 gives an insight into the role and modern structures of wireless policy and radio spectrum management. This includes discussion of the state-of-the-art approaches to standardisation and regulation of emerging CR and DSA technologies and applications.

Chapter 2 presents an overview and analysis of CR deployment scenarios, thus setting the foundation and common terminology for further analysis.

Chapters 3 and 4 cover, respectively, technical and business considerations around the process of bringing CR to reality. This analysis is capped by the impact assessment of policy developments discussed in Chap. 5.

Chapter 6 presents the reader with a set of example case studies that describe several practical scenarios for applying and developing CR technologies in different contexts: TV White Spaces, ISM Bands and in challenging applications to medical environments.

Finally, Chap. 7 wraps up the preceding analysis by offering some forward-looking insights as well as several examples of quite concrete and specific proposals that may be conducive for the realisation of CR and DSA technologies.

On this occasion we would like to thank our colleagues, chapter editors and each and every author who has contributed their time and work towards the drafting of this book.

Arturas Medeisis
Oliver Holland

Contents

Abbreviations

2G	2nd Generation mobile communications system, e.g. GSM
3G	3rd Generation mobile communications system, e.g. UMTS
3GPP	3rd Generation Partnership Project
ABM	Agent-Based Modelling
AP	Access Point (usually in WLAN, e.g. Wi-Fi)
ARPU	Average Revenue Per User
ASA	Authorised Shared Access (similar to LSA)
BS	Base Station
BSO	Beneficial Sharing Opportunities
CAGR	Compound Annual Growth Rate
CAPEX	Capital Expenditure
CEPT	European Conference of Postal and Telecommunications administrations
CoC	Certificate of Conformity
COST	European Cooperation framework in Science and Technology
COST-TERRA	COST Action IC0905 TERRA
CPC	Cognitive Pilot Channel
CR	Cognitive Radio
CRAHN	Cognitive Radio Ad Hoc Network
CRN	Cognitive Radio Network
CRNMS	Cognitive Radio Network Management System
CRS	Cognitive Radio System
CSMA	Carrier Sense Multiple Access
CSS	Cooperative Spectrum Sensing
DAB	Digital Audio Broadcasting
DFS	Dynamic Frequency Selection
DoC	Declaration of Conformity
DSA	Dynamic Spectrum Access
DSA II	Dynamic Spectrum Access Information Infrastructure
DTT	Digital Terrestrial Television (same as DVB-T)
DVB-T	Digital Video Broadcasting—Terrestrial (same as DTT)
EC	European Commission of the European Union

ECC	Electronic Communications Committee of CEPT
EIRP	Effective Isotropically Radiated Power
ENG	Electronic News Gathering (part of PMSE)
ETSI	European Telecommunications Standards Institute
FBMC	Filter Bank Multi-Carrier modulation
FCC	Federal Communications Commission, USA regulatory agency
GDB	Geolocation Data Base
GSM	Global System for Mobile Communications
HSPA	High Speed Packet Access, an UMTS operating mode
IA	Impact Assessment
IETF	Internet Engineering Task Force
IL	Intuitive Logics, a scenario planning method
IMT	International Mobile Telecommunications (an umbrella term used by ITU to describe 3rd and 4th generation of mobile technologies)
ISM	Industrial, Scientific, Medical (frequency band, e.g. 2400–2483.5 MHz)
ITS	Intelligent Transport Systems
ITU	International Telecommunication Union
ITU-R	Radiocommunications Sector of ITU
LSA	Licensed Shared Access (similar to ASA)
LTE	Long Term Evolution (a 4th Generation mobile communications standard)
LTE-A	LTE-Advanced
M2M	Machine-to-Machine communication
MAC	Medium Access Control layer (ref. Open System Interconnection model)
MBBA	Mobile Broadband Access
MNO	Mobile Network Operator
MS	Mobile Station
MVNO	Mobile Virtual Network Operator
NRA	National Regulatory Authority (in charge of telecommunications)
O&M	Operation and Maintenance
OFDM	Orthogonal Frequency Division Modulation
OOB	Out-Of-Band
OPEX	Operational Expenditure
OSA	Opportunistic Spectrum Access
OSS	Open-Source Software
PHY	Physical layer (lowest level of Open System Interconnection model)

PMSE	Programme Making and Special Events (wireless microphones and other similar equipment used by broadcasters/event organisers)
PMT	Probabilistic Modified Trends, a scenario planning method
PSD	Power Spectral Density
PU	Primary User (of spectrum)
QoS	Quality of Service
R&D	Research and Development
R&TTE	Radio and Telecommunications Terminal Equipment Directive
RA	Radiocommunications Assembly of the ITU
RAN	Radio Access Network (i.e. an access front-end part of mobile network)
RAT	Radio Access Technology
RBS	Reconfigurable Base Station
REM	Radio Environment Map
RLAN	Radio Local Area Network (same as WLAN)
RSC	Radio Spectrum Committee (within EU regulatory structure)
RSPG	Radio Spectrum Policy Group, an EU committee
RSSI	Received Signal Strength Indicator
SDO	Standards Development Organisation
SDR	Software Defined Radio
SINR	Signal-to-Interference-and-Noise Ration
SLA	Service Level Agreement
SNR	Signal-to-Noise Ratio
SU	Secondary User (of spectrum)
TC RRS	ETSI's Technical Committee on Reconfigurable Radio Systems
TVWS	White Spaces in TV Bands (e.g. 470–862 MHz in Europe)
UHF	Ultra High Frequency Band (300–3000 MHz)
UMTS	Universal Mobile Telecommunications System
US	United States
VHF	Very High Frequency Band (30–300 MHz)
WAPECS	Wireless Access Platform for Electronic Communications Services, a European policy initiative
WLAN	Wireless Local Area Network (same as RLAN)
WRC	World Radiocommunications Conference of ITU
WS	White Space (a geo-temporal gap in utilisation of radio spectrum)
WSD	White Space Device
WSDB	White Space Database (same as GDB)

Introduction

As wireless communications continue their explosive growth and diffusion across all spheres of public and private life, the problem of scarcity of suitable radio spectrum resource becomes ever more pressing. For example, recent history shows us several examples where large amounts of money had been paid by operators to secure access to only moderate portions of spectrum in internationally harmonised bands.

It is commonly acknowledged that spectrum scarcity is not a physical phenomenon, but rather an effect of the inherent inefficiencies of current spectrum assignment and usage practices. The traditional "command and control" school of spectrum management provides for an orderly use of resources and avoidance of interference through permanent and exclusive allocations of large portions of spectrum to monopolistic or oligopolistic sets of users. As a result, the spectrum is often hoarded and used rather uneconomically by incumbent users, limiting opportunities for access by new users. Moves towards the use of market-based policies for spectrum management, such as those advocated by the European Commission, provide a partial solution to these problems by giving users financial incentives to use spectrum more efficiently. However, these policies are up to now built rather superficially around the historic framework of traditional spectrum access rights.

This status quo is now challenged by the growing prominence of two new complementary technologies: CR and SDR. The former will enable radio devices to detect the environment, both internal such as the user and application context and external such as the spectrum characteristics, learn about such context, and take decisions on appropriate wireless access in the most optimal configurations based on that detected context and the learned experience. The latter technology will facilitate radio equipment dynamically adapting transmission characteristics so that transmissions can fit within identified spectrum spaces, connect to available networks or generally be optimised collaboratively with receivers to connect in a point-to-point fashion or more generally in a network. In the context of this book, SDR might therefore be considered as a technical facilitator of the CR concept. Moreover, whenever speaking of CR technologies in this book, it is to be inherently understood that SDR can be an integral capability to achieve associated flexibility although not a prerequisite—especially considering that various other adaptive radio capabilities outside of SDR might be used to realise forms of radio flexibility in CR.

CR technologies might create significant opportunity for truly open access to the vast wealth of natural radio spectrum resource. More intensive and efficient use of spectrum without the need to migrate incumbent users offers the possibility for large economic and social benefit to all humanity. However, the disruptive paradigms of self-managed spectrum access and associated capabilities such as DSA are still in its conceptual development phase, and require coordinated international research effort.

To bring DSA and the many other potential benefits of CR/SDR from concept to realisation in products, industry and other stakeholders need to agree with regulators on the common sets of technical standards as well as rules and conditions for CR/SDR operation and associated spectrum access. In Europe, regulatory and standardisation bodies, such as the European Conference of Postal and Telecommunications administrations (CEPT) and the European Telecommunications Standards Institute (ETSI), have already begun deliberating on these subjects; however, this early start has been hampered by conflicting interests and a lack of coordinated industry contributions.

While the technical R&D aspects of CR/SDR technologies are being currently addressed by several major research initiatives, the subject of an optimised regulatory framework and rules for CR/SDR have been thus far largely overlooked. It is the realisation of this gap that gave the initial impetus for the start of COST Action IC0905 "COST-TERRA" in early 2010. In turn, this built the basis for cooperation of a team of authors that led to the writing of this book.

COST[1] is an intergovernmental framework for European Cooperation in Science and Technology, which strives to support multidisciplinary research and networking of researchers across the European continent and beyond. As one of its projects (known as "Actions"), COST-TERRA has run for 4 years from May 2010 to April 2014, with the objective of studying means for creating a comprehensive techno-economic regulatory framework for radio spectrum access rules for CR/SDR. Over its duration, COST-TERRA has become a "think-tank" involving participants from more than 20 countries, hailing from academia as well as regulators and industry members.

Jointly, the team of experts in COST-TERRA have reviewed and progressed to great length the analysis of CR/SDR technologies, as well as their potential to be used in DSA scenarios. The results of COST-TERRA activities, the many scientific inputs and ideas discussed in COST-TERRA meetings, are hereby laid down in front of you. Besides the content in the pages of the book, some complementary material from COST-TERRA activities, such as summaries of contributions, copies of presentations, etc., may be found at the website: http://www.cost-terra.org/CR-policy-book.

Given all of the above, we would like to extend our thanks to the COST ICT Domain Committee and the COST Office for facilitating the activities in COST-TERRA.

[1] For more information about COST, please visit their website at: http://www.cost.eu/

Chapter 1
State-of-the-Art in Policy and Regulation of Radio Spectrum

Arturas Medeisis and Oliver Holland

Abstract This chapter sets the stage for the rest of book by presenting the current state of affairs in the management of radio spectrum and related standardisation and regulatory initiatives pertaining to the emerging fields of CR and DSA. Section 1.1 discusses the international structure of spectrum management from the global ITU level down to the regional and national level. It also outlines how the ITU has started approaching the consideration of DSA challenges. The next Sect. 1.2 looks at how these efforts have been matched by the European regulators. Next, two Sects. 1.3 and 1.4 examine the complex issue of standardisation of CR/DSA technologies, starting from the general overview of work in global and regional standardisation bodies, followed by the analysis of drivers and obstacles. Standardisation is particularly covered because it has intrinsic interactions with regulation. For example, a well-targeted standardisation initiative involving strong industry players and perhaps an industry association can provide significant motivation for regulators to adapt regulations to support that initiative, in support of economic/industrial and national interests. The chapter is concluded by Sect. 1.5 that takes a closer look at the developments in two countries that have been particular champions of CR technologies: namely the United States and the United Kingdom. This is done with a viewpoint on the status of TV White Space access implementation—currently a key driver of DSA.

A. Medeisis (✉)
Department of Telecommunication Engineering, Vilnius Gediminas Technical University, Vilnius, Lithuania
e-mail: arturas.medeisis@vgtu.lt

O. Holland
Centre for Telecommunications Research, King's College London, London, UK
e-mail: oliver.holland@kcl.ac.uk

A. Medeisis and O. Holland (eds.), *Cognitive Radio Policy and Regulation*,
Signals and Communication Technology, DOI: 10.1007/978-3-319-04022-6_1,

1.1 International Regulations and DSA

Peter Anker

Ministry of Economic Affairs and Delft University of Technology, Delft, The Netherlands

1.1.1 Introduction

Radio waves are used to deliver a broad range of services and applications, for instance, mobile telephony, radio and television broadcasting, maritime radio, research into the (birth of) the universe, and even for heating food in a microwave oven. However, it is not possible for users to use this resource without limitations. The use of radio waves at a particular frequency by one user will influence the use of the same, or nearby frequencies, by other users at the same time. Radio receivers will have difficulties to distinguish the intended signal from all other signals it receives. This phenomenon is called interference. Hence, coordination is needed in the use of radio waves between the various users to manage the problems associated with interference. As the propagation of radio waves is not hindered by national borders, this coordination will need to be performed on an international level.

Particularly for users, it is also often important that services and the related equipment are standardized, i.e. these services can operate with similar equipment in various countries in the same frequency band. As a result of this harmonisation of allocations, the spectrum can be used more efficiently and the equipment can be used over much wider geographical areas, increasing the size of the market for such equipment and reducing production costs. In the case of a number of applications, international harmonisation is even necessary owing to the nature of the application.

Historical developments have led to a situation in which governments have taken the role of 'supreme coordinator' in the use of the radio spectrum. Spectrum management has become based on the avoidance of interference and technically efficient use of spectrum. This section gives an overview of the international regulatory framework for spectrum regulations and their ability to support the introduction of CR. The section focuses thereby on the general framework for Europe. Detailed European regulations on CR and standardisation of CR are dealt with separately in other sections of this chapter.

1.1.2 Different Levels of Spectrum Management: European Case

1.1.2.1 ITU

Spectrum is globally governed by the International Telecommunications Union (ITU), a specialized agency of the United Nations. The Radiocommunication Sector of the ITU (ITU-R) develops and adopts the Radio Regulations, a binding international treaty, with a voluminous set of rules, recommendations and procedures for the regulation of radiocommunications. The Radio Regulations are based on avoidance of radio interference through the division of spectrum in bands which are allocated to one or more services out of some 40 different radio services. These radio services include services such as fixed, mobile, satellite, amateur, radio navigation and radio astronomy. Most bands are shared among primary and secondary services. Primary services have priority in case of conflicts resulting in harmful interference. Harmful interference is defined as *Interference* which endangers the functioning of a *radionavigation service* or of other *safety services* or seriously degrades, obstructs, or repeatedly interrupts a *radiocommunication service* operating in accordance with Radio Regulations ([1], article 1.169)

A wide range of regulatory, operational, and technical provisions ensure that radio services are compatible with one another and harmful interference among services of different countries is avoided. The Radio Regulations are regularly updated in response to changes in needs and to new demands at World Radiocommunication Conferences (WRC), which are held every three to four years [2].

The Radio Regulations are an international treaty between countries. This means that it only concerns the relations between countries. Individual countries can adopt some or all of the allocated services of each band and they are allowed to deviate from the Radio Regulations as long as no harmful interference is caused to the recognised services in other countries.

1.1.2.2 CEPT/ECC

The Electronic Communications Committee (ECC) of the European Conference of Postal and Telecommunications Administrations (CEPT) brings together 48 countries to develop common policies and regulations in electronic communications and related applications for Europe. Its primary objective is to harmonize efficient use of the radio spectrum, satellite orbits and numbering resources across Europe. It takes an active role at the international level, preparing common European proposals to represent European interests in the ITU and other international organisations. The ECC work is carried out in partnership with all stakeholders including the EC and ETSI.

There are four different regulatory deliverables developed by the ECC:

- *ECC Decisions* are regulatory texts providing measures on significant harmonisation matters, which CEPT member administrations are strongly urged to follow. ECC Decisions are not obligatory legislative documents, as any other CEPT deliverable; however, they are normally implemented by many CEPT administrations.
- *ECC Recommendations* are measures which national administrations are encouraged to apply. They are principally intended as harmonisation measures for those matters where ECC Decisions are not yet relevant, or as guidance to CEPT member administrations.
- *ECC Reports* are the result of studies by the ECC normally in support of a harmonisation measure.
- *CEPT Reports* are the final results of studies developed in order to support responses to EC mandates. In many cases the results in the report form the basis for future EC Decisions on harmonized technical conditions of use (see the following section on the European Union).

As noted above, CEPT deliverables are non-binding. This gives the National Regulatory Authorities (NRA) a large degree of flexibility when it comes to adapting these to country specific conditions, legacy usages and circumstances.

1.1.2.3 European Union (EU)

Throughout the 1990s the EC gradually increased its involvement in spectrum issues, as the RF spectrum use started to affect the 'internal market'. The first intervention was related to the creation of a single European (internal) market for equipment. On the 9th of March 1999 the European Commission published the R&TTE Directive 1999/5/EC [3]. This Directive covers most products which use the radio frequency spectrum, including unlicensed devices. All equipment that is placed on the market must comply with a set of essential requirements, covering the protection of health and safety, electromagnetic emission and immunity of the equipment and effective use of the radio spectrum so as to avoid harmful interference.

Equipment manufactured in accordance with a "Harmonised Standard" may be placed on the market within the whole European Union (EU) (see also the following Sects. 1.3 and 1.4 on standardisation). However, certain restrictions may apply to the use of radio equipment if the frequencies are not harmonised in the European Union (EU). If a Harmonised Standard is used, the manufacturer has to perform some specific radio tests and can make its own declaration of conformity (self-declaration) which states that the product satisfies the essential requirements. There is no need for an external body to perform the testing. When a Harmonised Standard is not available or not appropriate, a manufacturer needs to demonstrate more extensively how the requirements of the Directive are being met through testing, to be documented in a 'technical construction file'. This file has to be reviewed and approved by a notified body.

Involvement of the European Union with radio spectrum management came with the introduction of the new regulatory framework. This framework was aimed at further liberalisation, harmonisation and simplification of the regulations in the telecommunications sector. The Framework Directive (2002/21/EC), *on a common regulatory framework for electronic communications networks and services*, states that the allocation and assignment of radio frequencies by national regulatory authorities are to be based on objective, transparent, non-discriminatory and proportionate criteria [4]. The related Authorisation Directive (2002/20/EC) specifies the circumstances under which the granting of an individual license is being allowed [5]. The Directive states that granting of an individual license is only allowed to ensure efficient use of radio frequencies. The Directive also limits the conditions that may be attached to the rights of use for radio frequencies. The licensing and the formulation of the conditions under which the radio frequencies may be used are left to the Member States.

Under this new regime harmonisation of spectrum is still left to CEPT. However, the associated Radio Spectrum Decision by the European Commission (2002/676/EC) created the possibility to impose technical harmonisation measures upon the Member States [6]. This Decision created a legal framework for 'the harmonised availability and efficient use of radio spectrum in the European Union (EU) for the establishment and functioning of the internal market in Community policy areas, such as electronic communications, broadcasting and transport'. In the implementation of the Decision the European Commission is assisted by the newly formed Radio Spectrum Committee (RSC). The RSC is composed of experts from the Member States.

The European Commission can issue mandates to CEPT to solicit advice on technical harmonisation measures. The RSC approves the CEPT Report and associated technical implementation measures prepared by the Commission. The implementation of these measures is mandatory for the EU Member States.

Next to the RSC, the Radio Spectrum Policy Group (RSPG) was set up to facilitate consultation and to develop and support radio spectrum policy. The Radio Spectrum Policy Group (RSPG) is a group of high-level representatives of the Member States which advises the European Commission on radio spectrum policy at a strategic level.

The revision of the regulatory framework in 2009 introduces two governing principles that will have implications on the future regulation. Firstly, general authorisation should be the general rule when authorizing access to spectrum. Individual licensing can still be used but such deviations from the general principle must be justified. Secondly, the principles of technology and service neutrality should be the general rule for both general and individual authorisation of access to spectrum. Deviations from this principle are still allowed but must be justified. As the allocation of spectrum to specific technologies or services is an exception to the principles of technology and service neutrality and reduces the freedom to choose the service provided or technology used, any proposal for such allocation should be transparent and subject to public consultation [7].

1.1.2.4 ETSI

The European Telecommunications Standards Institute (ETSI) is an independent, non-profit organisation, whose mission is to produce globally applicable standards for Information & Communications Technologies including fixed, mobile, radio and TV broadcasting, internet and several other areas. ETSI plays a major role in developing a wide range of standards and other technical documentation as Europe's contribution to world-wide ICT standardisation. This activity is supplemented by other activities such as interoperability testing services. ETSI's prime objective is to support global harmonisation by providing a forum in which all key players can contribute actively.

ETSI is recognised as an official European standards organisation by the European Commission and works under mandates from the Commission to prepare Harmonised Standards under the provisions of the R&TTE Directive. Membership is open to all interested parties. Harmonised Standards are standards adopted by the European Standards Organisations (ETSI, CEN and CENELEC), prepared in accordance with the General Guidelines agreed by them with the EC, and in response to a mandate issued by the Commission after consultation with EU Member States. The reference of a Harmonised Standard must be published in the Official Journal (OJEU) in order to give a presumption of conformity to the essential requirements of the R&TTE Directive.

ETSI is an officially recognised partner of the ECC, which is reflected in a Memorandum of Understanding (MoU). The cooperation between ETSI and the ECC plays an important role to ensure the objective of harmonised and efficient use of the radio spectrum across Europe.

1.1.2.5 National Spectrum Management Authority

Based on the international allocations and regulatory provisions the national spectrum regulator-NRA-grants access to spectrum for users. An EU Member State has the right to set conditions on the use of spectrum under the Framework Directive. These conditions can include appropriate limits that aim to avoid harmful interference to other radio services. These conditions can be harmonised on a European wide basis either through a European Commission Spectrum Decision (which is mandatory for EU Member states to implement) or by an ECC Decision or Recommendation. Alternatively, if no mandatory harmonised guidance is available, a regulatory deliverable can be developed on a national basis.

Usually a license gives an exclusive right to operate in a specific frequency range, in a specific location or geographic area and under specific technical conditions (e.g., power level, antenna height, antenna location etc.) and other conditions such as service obligations and (network) build-out requirements. The compliance of spectrum users with the license obligations is monitored and enforced by the NRAs.

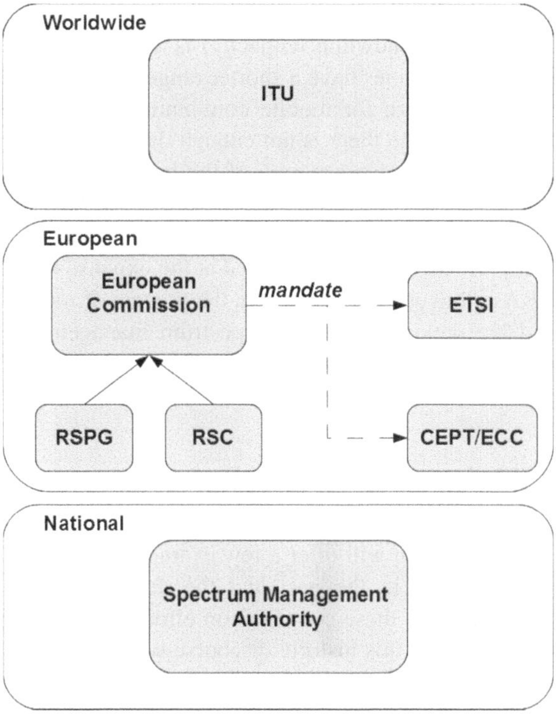

Fig. 1.1 The international and national regulatory framework for spectrum regulations in Europe

If the demand for spectrum within a particular band is considered to be significantly less than the supply, licenses are usually granted on a first come first served basis. When spectrum demand exceeds the supply, the spectrum regulator has to use another mechanism to award the licenses. Increasingly, regulators have turned to comparative hearings or "beauty contests" and more recently to spectrum auctions [8].

In summary, the current spectrum management model operates on both a national and international level as depicted in Fig. 1.1.

In the current paradigm all decisions are made by the spectrum regulator. Therefore, this traditional spectrum management model is commonly referred to as Command & Control. This Command and Control model has its limitations. The two most eminent are: all (usable) spectrum is allocated but some of the portions of the spectrum are hardly used, and the method to allocate and assign spectrum is slow in responding to changes in market and technology.

In the past, the inefficiencies in spectrum utilisation introduced by this bureaucratic command and control spectrum management model were tolerable. As demand grew, advancing technology ensured that new frequency bands were available, and there was no need to deal with inefficiently used spectrum. More recently, demand has grown very rapidly and technology has delivered new services and devices to serve that demand. However, the opening up of even higher frequency bands is not progressing at the same pace and not all frequencies are alike. More bandwidth (capacity) is available in the higher frequency range, but higher frequencies have a shorter range, ceteris paribus. To give an example, ideal frequency range for mobile communications is roughly 0.1–3 GHz. Below this frequency range there is not enough data throughput capacity available and above this range the coverage area of the base stations becomes too small.

This means that the NRAs more or less have run out of useable spectrum to assign for new services and technologies. Hence, services based on new technologies can only be introduced at the expense of existing services. Consequently, NRAs all over the world are in the process of modernising their spectrum policies, and are seeking alternative spectrum management models which allow a much more efficient and flexible utilisation of the spectrum [8, 9].

1.1.3 Lessons from the Past

This subsection will offer a few example case studies of the coordination of radio spectrum use in the past and the development of radio spectrum regulations resulting from these coordination efforts. Besides offering very revealing lessons from the past, this historic discourse will provide the foundation for proposing an actor-centric approach to analyse the links between different stake-holders involved in coordination of spectrum use. It may be seen that until now, most of the advances that have been made in the coordination of radio spectrum usage were triggered by problems with a specific service. This will be illustrated in the three cases to be discussed in the rest of this subsection. Each case is concluded with an assessment that places the observed coordination efforts in an actor-centric perspective on alignment. This discussion will be then continued in Sect. 5.2 of this book.

1.1.3.1 Marconi and the Birth of Spectrum Management

At the time of Marconi, spectrum was like an open and untouched pasture. Marconi was the first to enter this pasture to exploit this common resource. He started his business by selling wireless stations for use on-board ships. As others also started to enter the business, he changed his strategy. He decided to sell not only the equipment but also wireless telegraphy as a service. For that purpose he set up a new company, the Marconi International Marine Communications Company in 1900. He built his own land based radio stations along the sea-trade routes on the shores of Britain, Ireland, Belgium, Italy, Canada and New Foundland. He trained his own radio telegraphists and placed them on all ships he equipped with a wireless radio station. These radio telegraphists, or marconists as they were called, were only allowed to communicate with Marconi wireless stations both land based and onboard other ships [10]. By doing so, he created a very successful private business using a public resource, radio waves.

The behavior of the Marconi Company led to governmental involvement in the use of radio waves. Kaiser Wilhelm of Germany convened an international conference on the use of radio telegraphy in 1903. Representatives of nine countries gathered in Berlin for the *Preliminary Conference on Wireless Telegraphy* [11]. Complete agreement was not reached, but the Conference drafted a protocol that served as the basis for a future international agreement on the use of wireless telegraphy. Among the articles of the protocol was the requirement that all coastal stations were required to exchange messages with all ships without distinction as to the radio system being used [12].

This preliminary Conference was followed in 1906 by the first Radio Telegraph Conference of Berlin. Twenty-nine countries adopted the first *International Radiotelegraph Convention*. Two important provisions of the Convention were firstly, a requirement to accept all messages from coastal stations and ships regardless of the system used and secondly, priority for distress calls. The annex to this Convention contained the first regulations governing wireless telegraphy. It was decided to use two wavelengths corresponding to 1000 kHz and 500 kHz for public correspondence.

The interconnection among radio operators was considered to be of public interest to support the safety of the man at sea, and the continuous availability of the service should be assured at all times. This need for rules of engagement and international coordination was strengthened at the next Radio Telegraph Conference which took place in London, shortly after the Titanic disaster in 1912 [10, 13].

To conclude, it was not the introduction of new technology—radio—as such that made it necessary to coordinate the use of the radio frequency spectrum and design new regulations. It was the use of this new technology by Marconi which triggered it. Marconi used this new technology in such a way that a conflict became apparent between his efforts of realizing private objectives and the realization of the newly identified public objectives.

Regulations were put in place to safeguard the public interests in the use of maritime communications. The regulations allowed for as much (business case) freedom as possible for the maritime service with the exception of a few standardized channels for the exchange of public messages and as an emergency signalling frequency. The outcome of the coordination efforts provided the support for a public service using a commercial incentive scheme, i.e. combining the public and private interests in a creative new combination.

1.1.3.2 Spectrum Auctions

In 1959 economist Ronald Coase posed that the allocation of spectrum should be determined by the forces of the market rather than as a result of government decisions. Radio licenses should be bought and sold like any other scarce resource in our economy, such as land or labour. Rights should be assigned to individual users via an auction with the provision that these rights can subsequently be traded in an open market. The market should not only decide who will own the license, but also what services will be provided. If a business model would fail, the right to use the radio spectrum could be bought by another operator with a different, more successful, business model or by a new entrant. The problem of interference could be solved by delimiting the rights. These delimitations should not only come from strict regulations, but also as a result of transactions on the market [14].[1]

At that time, Coase's idea was taken as a big joke by the FCC [16]. Nonetheless, the idea of a model based on trading of the property rights has since been discussed among economists,[2] but a property rights model was only considered seriously by spectrum management authorities in the early 1990s. At that time a broad consensus in political thinking had emerged in support of deregulation; the introduction of market forces was considered for a number of infrastructures that had been heavily regulated in the past, including mobile telephony [16].

Deregulation changed the set of objectives pursued by the government. One of the new objectives pertaining to mobile communications became the creation of a market for radio spectrum usage rights for mobile communications. The institutional change that was already proposed in the late 1950s by Coase perfectly fitted the newly defined objectives. Hence, various countries chose to auction the spectrum rights for mobile telephony [18].[3]

1.1.3.3 Wi-Fi and License Exempt Use of Spectrum

In 1985 the FCC decided—for the purpose of deregulation—to allow the use of spread spectrum for communication purposes in three bands designated for Industrial, Scientific and Medical (ISM) applications (900 MHz, 2.4 GHz and 5.8 GHz). These where bands that could be used without the need for a license but applications had to be limited in output power and had to tolerate interference from other users, including other ISM applications.

The new (for civil applications) technology of spread spectrum and the introduction of regulations to support it, triggered NCR Corporation to use spread spectrum for a nagging issue from their sales force: the lack of 'mobility' in their cash register product portfolio. Through their involvement in IEEE, as a leading standards developing organization, NCR became the de facto leader in the IEEE 802.11 Working Group resulting in a highly successful Wireless-LAN standard [20].

[1] Coase generalized this idea in his Noble prize winning essay "The Problem of Social Cost" [15].

[2] See note 6 of Baumol and Robyn for an overview of references [17].

[3] New Zealand was probably the first country that experimented with the definition of long-term, tradable property rights to radio channels, and the first country to auction these rights to the highest bidder [19].

This case shows that the introduction of new technology will also need associated institutional arrangements supportive of this technology. Before 1985, the institutional arrangements were based on exclusive rights. In such a setting there was no need to use a new technology (spread spectrum) that facilitated shared access.

It further shows that alignment between technology and the institutional arrangements is necessary but is by itself not enough for successful introduction of this new technology. In the institutional arrangements that were set up by the FCC, it is up to the radio equipment manufacturers to coordinate the efficient use of the radio spectrum, including graceful degradation of service levels under increasing load conditions and avoiding interference. The coordination activities necessary to develop new technology to achieve alignment between this new institutional arrangement and technology were only realized after a private actor (NCR) had a private objective that materialized in a compelling business case. This private objective of NCR was compatible with the public objectives of the FCC.

1.1.3.4 Conclusions

This review of historical cases has provided evidence of the value in applying an actor-centric approach to the process of alignment. Each of the cases described above were triggered by problems related to private actors on the one hand and public actors on the other hand pursuing the realization of their private, respectively public objectives. A successful outcome can be concluded when private and public actors can realize their objectives simultaneously, by designing a business opportunity in theory and allowing it to be transformed into a viable business case in practice.

1.1.4 The New Regulatory Paradigm of DSA

DSA solutions have to address the lack of available (accessible) spectrum in the current static model. In the current spectrum management model, radio spectrum is divided into fixed and non-overlapping blocks, sometimes separated by so-called guard bands, and exclusively assigned to different services and wireless technologies, while a lot of spectrum usage is only local and limited in time. In an economic sense, there appears to be a paradox whereby the rights to the radio spectrum are fully assigned, but a lot of radio spectrum remains unused in practice when considered on a time or geographical basis. Under the current command and control model it is very difficult to make this unused spectrum available.

There are two basic alternative regimes considered, a regime based on exclusive property rights and a regime based on spectrum commons with strict general rules on the use of spectrum without the need for individual licenses [8, 21]. In these discussions, CR has been closely linked to the commons. Advocates of the commons see CR technology as an enabler to realise true radio spectrum commons

[22]. However, technologies such as CR do not favour one regime over another. CR can be used in both spectrum management regimes, as it can also be used to facilitate an efficient market-based regime based on property rights [8]. CR, as a technology, is an enabling tool to realise this goal of increased flexibility in access to spectrum.

The key feature of such a CR is its ability to recognise unused parts of spectrum that are assigned to conventional users and adapt its communication strategy to use these parts while minimising the interference that it causes to the conventional users. An important consequence is that CR can be an enabling technology to facilitate a paradigm shift for spectrum management from a regime based on static spectrum assignments to a regime based on more dynamic forms of spectrum access [8, 23].

1.1.4.1 Adapting the Regulatory Framework for DSA

The first question is if there is any international regulation in place that prohibits DSA through the use of CR. The short answer to that question is: No. Administrations that wish to implement CR have two different alternatives to do so [24].

Firstly, CR can be used under any service defined in the Radio Regulations, i.e., if the CR is used to deliver mobile communications, the CR can be treated in the same way as an ordinary mobile radio, and will be allowed to operate under the provisions for the mobile service. This means that the CR can use bands that are allocated to the mobile service as far as the (international) regulations on interference and sharing conditions are met.

A second option is to implement CR on a so-called non-interference basis ([1]: article 4.4). This means that the CR is allowed to operate as long as it doesn't cause harmful interference to, and shall not claim protection from, harmful interference caused by a station operating in accordance with the provisions of the Radio Regulations. These provisions only apply for cross-border communications (and interference), since the Radio Regulations are an international treaty between countries. Hence, individual countries are allowed to deviate from the Radio Regulations as long as no harmful interference is caused to the services in other countries.

However, to realise the full potential of CR, the radio will need to have dynamic access to a wide range of spectrum bands, which might currently be divided in a number of frequency bands designated for different radio services. Introduction of DSA is only possible if these exclusively designated frequency bands are opened up for other services and technologies. Hence, there is a need to enhance the regulatory framework to allow for more flexibility in the use of radio spectrum.

1.1.4.2 Activities Within the ITU

At the World Radio Conference 2007 (WRC-07) it was decided to put Software Defined Radio and CR on the agenda for the World Radio Conference of 2012 under agenda item 1.19. Study Group 1 (Spectrum management) of the ITU-R was responsible for the studies needed in preparation of this agenda item of the WRC-12. As part of these studies, the following definition was developed [25]:

> Cognitive Radio System: A radio system employing technology that allows the system to obtain knowledge of its operational and geographical environment, established policies and its internal state; to dynamically and autonomously adjust its operational parameters and protocols according to its obtained knowledge in order to achieve predefined objectives; and to learn from the results obtained.

The World Radio Conference of 2012 (WRC-12) came to the conclusion that SDR and CR are related technologies which can be used in any radio service within the Radio Regulations. There is no need to incorporate the definitions of SDR and CR in the Radio Regulations. However, WRC-12 reiterated that any radio system implementing CR technology needs to operate in accordance with the provisions of the Radio Regulations.

In other words, WRC-12 confirmed that CR can be used under any of the services defined in the Radio Regulations. Administrations that wish to implement CR already can do so. However, it was also noted that there remain questions around the deployment and use of CR. A common concern was expressed within the ITU-R about how the protection of existing services from potential interference from the services implementing CR technology, especially from the DSA capability of CR, could be realised. ITU-R and the WRC-12 came to the conclusion that there is need for further studies within ITU-R on the implementation of CR technologies within a radiocommunication service and on sharing among different radiocommunication services with regard to the capabilities of CR, in particular dynamic access to frequency bands.

ITU-R came to aforementioned conclusion that there is a need for further studies on CR during the discussions on the future work programme of the ITU-R at the Radiocommunications Assembly (RA). This need for further studies is expressed in ITU-R Resolution 58. The RA was held in January 2012, in the week prior to the WRC-12. The WRC-12 confirmed this need for further studies in WRC-12 Recommendation 76.

ITU-R Study Group 5 (Terrestrial services) already started work on the possibilities for the introduction of CR in the mobile service and the operational implications of this introduction. ITU-R Report M.2225 provides a general description of CR systems and describes a set of deployment scenarios for the introduction of CR systems in the land mobile service (excluding international mobile telecommunications (IMT)). ITU-R Report M.2242 describes how introduction of CR in the IMT systems may be used for more dynamic and flexible radio resource management and optimisation.

Working Party 5A of Study Group 5 is now working on a second report on CR systems in the land mobile service (excluding IMT). This report aims to present existing, emerging and potential applications of CR systems in the land mobile service from a technical perspective, including the impact on the use of spectrum.

It is now up to the other study groups to study possibilities for the introduction of CR technology for the radio services under their purview.

1.2 European Regulatory Developments Related to CR

Arturas Medeisis

Vilnius Gediminas Technical University, Vilnius, Lithuania

1.2.1 Introduction

This section will review the European regulatory developments pertinent to CR. However it must be clarified from the beginning that the current state of thinking in European regulatory circles does not really consider the CR as the subject of regulatory policy, but rather as pure technological innovation phenomenon that should find its way to fit into the existing regulatory service definitions and spectrum access rules. Moreover, the CR term is generally understood in current European (and elsewhere) regulatory context as a moniker for DSA-enabling solution, and in that sense the White Space Devices (WSD), as may be deployed in the traditional TV Bands, are seen as proxy CR systems to be encountered in the near future. Thus in the rest of this section the terms WSD and CR are used interchangeably and should be seen as being synonymous.[4]

1.2.2 Historical Background

As was presented in Sect. 1.1, the Europeans firmly believe that management of radio spectrum should follow the road of broadest possible international harmonisation and therefore all European regulations in the area of spectrum management are initiated on the international (regional) level through co-operation arrangements between CEPT and EU. So also the formal consideration of the subject of CR, first embodied as WSDs in the TV Bands, started when the European Commission issued to CEPT in 2007 a Mandate on the "Technical considerations regarding harmonisation options for the Digital Dividend". The principal goal of this mandate was the most flexible spectrum usage in the band 470–862 MHz while allowing the widest possible range of uses and technologies.

---

[4] The author would like to acknowledge assistance of Dr. Alexandre Kholod, the chairman of CEPT PTSE43, in compiling the material for this section.

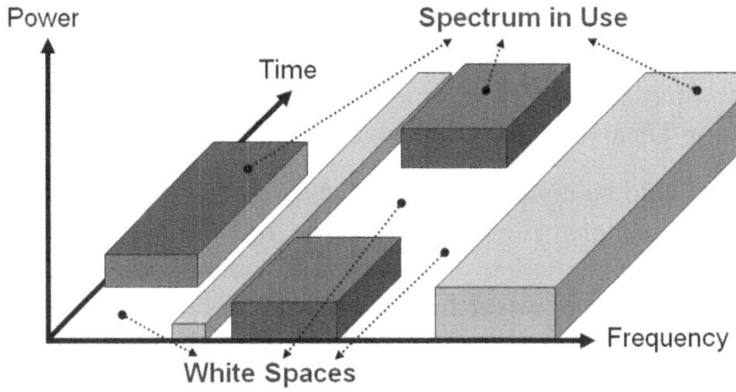

Fig. 1.2 Illustration of the concept of white spaces

As far as the practicability of implementation of new/future applications within the white space spectrum in the band 470–862 MHz is concerned, the CEPT provided its preliminary review in CEPT Report 24 [26]. This seminal high level policy document was important as it set the ground for all further European developments in this field. Most notably, it formalised the concept of White Spaces, see Fig. 1.2, as "*a part of the spectrum, which is available for a radiocommunication application (service, system) at a given time in a given geographical area on a non-interfering/non-protected basis with regard to primary services and other services with a higher priority on a national basis* [26]".

Note the two critical points that were laid solid into the definition of WS and, accordingly, became the ground rules for authorising access to them (i.e. the DSA):

- the WSDs would be operated on a strictly non-protected non-interfering basis, and
- the incumbent would always retain higher priority.

These principles were clearly well intentioned and aimed at protecting the safe operation of incumbent services, which was especially critical in the case of high political importance of TV broadcasting deployed within the initial candidate band of 470–862 MHz. They also contributed to shaping the following considerations of WSDs in order to establish their operational requirements so as to ensure strict compliance with the above principles.

The main burden of establishing those technical conditions for co-existence was entrusted to a specially created Project Team Spectrum Engineering #43 (PT SE43), which was active from 2009 until January 2013. The PT SE43 was heavily attended by national regulators and industry alike, and over its lifetime produced three important technical reports, that form the basis of current European regulatory regime for deployment of WSD/DSA systems:

- ECC Report 159 establishing key technical principles for possible operation of CR in WS of the frequency band 470–790 MHz [27];

- ECC Report 185 that further defined the technical and operational requirements for the operation of WSDs in 470–790 MHz [28], and
- ECC Report 186 describing requirements for WSD operation under the concept of Geo-location Database (GDB) managed operation [29].

First of all it should be noted that the reports focused only on the portion of the TV Bands IV&V, namely 470–790 MHz. The remaining band 790–862 MHz (800 MHz band) was earmarked as "Digital Dividend" to be freed from TV Broadcasting owing to switch-over from analogue transmission to more spectrally efficient digital transmission based on DVB-T standard. That Digital Dividend band would be licensed to broadband IMT systems, such as implemented by the LTE standard. Ironically, by the time of completion of PTSE43 work, after the World Radiocommunications Conference that took place in 2012, it was already becoming obvious that the TV operations might be soon squeezed further due to the extension of the mobile allocation into the 700 MHz band (694–790 MHz). The important corollary of this is that the reduction of band available to TV means increasing density of TV transmitters, which in turn minimizes the potential availability of WS.

Secondly, it is important to note that ECC Report 185 was developed as complementary to ECC Report 159, whereas ECC Report 186, though complementing the discussions in ECC Report 159, may be also considered on its own, as the principles set in this report could be easily extended to other frequency bands.

The Report 159 reviewed various possible mechanisms for ensuring coexistence (sensing, GDB and beacons) and essentially concluded that in terms of practical feasibility for ensuring protection of incumbent users from WSDs interference, the GDB approach would be the prime (and only) solution. This led to re-focusing the bulk of subsequent CEPT work towards consideration of that approach. Nevertheless, the Report 185 still provided certain consideration of sensing, especially cooperative sensing, as well as beacons, e.g. for protection of radio microphones. This report also presented further additional studies for co-existence and established more explicit conditions for WSDs such as transmit power limits.

Then the Report 186 is solely focusing on practical implementation of GDB approach, such as requirements for database operation, its interfacing with master and slave WSDs as well as what type of data should be stored in the GDB. More detailed analysis of European GDB approach is presented in Sect. 3.1.

By the time PTSE43 was finishing its work, in September 2012 CEPT has kicked-off a complementary regulatory analysis by creating a new Project Team Frequency Management #53. Its mandate was to review the technical reports produced by PTSE43 and based on them to formulate the overall regulatory framework for WSDs under GDB approach, including guidelines for national implementation. By that time also another concept for DSA was starting to emerge under the name of Licensed Shared Access, which was also added to the PTFM53 mandate. The target dates for completion of PTFM53 considerations were set to

2014/2015, which means that by the time of writing this book it was too early to sum up conclusions of this project team.

Another important element of the European regulatory landscape is the standardisation aspect which is being addressed through the work of ETSI. The CR-related aspects are being addressed in its Technical Committees on Reconfigurable Radio Systems (TC RRS) and on Electromagnetic compatibility and Radio Matters (TC ERM). Their key objective is the development of the standards for WSD operation. Please refer to Sects. 1.3 and 1.4 for more detailed analysis of standardisation activities.

1.2.3 Current State-of-the-Art of European CR/WSD Regulatory Policies

With departure in above described historical background, the current European philosophy that will define the CR/WSD regulatory policies may be summarised by the following key points:

- Spectrum sensing alone is not feasible approach [27], however collaborative sensing may provide certain added value as complementary feature to GDB approach [28];
- Geo-location database (also in combination with sensing) is the only viable option for controlled deployment of WSDs [29], as further discussed in Sect. 3.1;
- The WSD deployment is expected to cover three typical use cases, see Fig. 1.3 [28]:
  - Indoor wireless access;
  - Outdoor wireless access;
  - Machine-to-Machine connectivity;
- The WSD maximum transmit powers for all above use cases can be set either by the geo-location database or at the hardware level by WSD manufacturers, or by an algorithm implemented in the firmware [28];
- Protection of different services/systems remains of paramount importance. As far as the band 470–790 MHz is concerned, this covers such incumbent services and systems as Broadcasting Service, wireless microphones (aka Programme Making Special Events—PMSE applications), Radio Astronomy Service, Aeronautical Radionavigation Service as well as Mobile Services in adjacent bands [27, 28]. This might even include the need to protect operation of cable TV head-ends, which was left to be considered as mostly national issue [28]. Different incumbent users of the band 470–790 MHz are schematically shown in Fig. 1.4.

Regardless of these well matured and solidified European views on the WSD regulation, the practical enactment of these policies had been somewhat lagging, with obvious reason that national administrations did not (and still often do not)

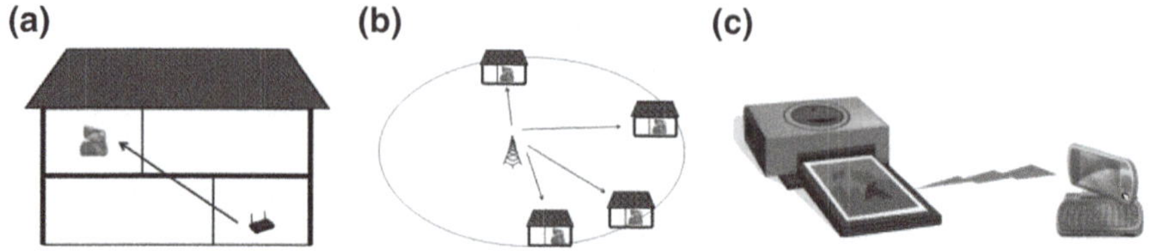

Fig. 1.3 Three classes of WSDs considered in current European regulations: **a** indoor, **b** outdoor, **c** Machine-to-Machine [28]

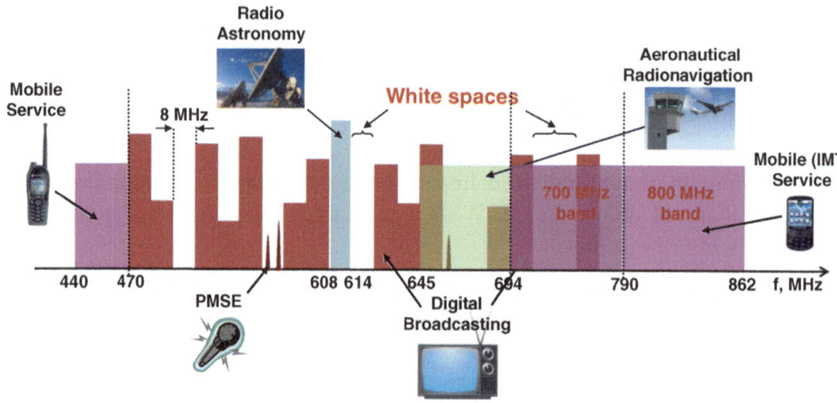

Fig. 1.4 Incumbent users of the band 470–790 MHz to be protected from WSD interference

perceive the market demand for such applications. One European administration that was steadily progressing with the practical implementation of WSD regulation is the United Kingdom. Its case study is offered in separate Sect. 1.5.2 later in this chapter.

1.2.4 The Emerging Value of CR as Part of Spectrum Sharing Paradigm

As was discussed above, up to now the European regulatory focus was mostly on allowing the deployment of WSDs in TV Bands as a certain niche application. However as of lately, the CR started gaining wider recognition as important element of regulatory toolbox for enabling the further growth of spectrum sharing as dominant means for radio spectrum access. This development may be linked to the political push by the EU, which in March 2012 approved the Radio Spectrum Policy Programme (RSPP) [30].

The key objectives and concrete envisioned actions of the RSPP may be summarised as follows:

- providing at least 1200 MHz of total spectrum to accommodate the growth of wireless data traffic;
- assessing the need for harmonised bands as part of the above spectrum portfolio;
- authorising spectrum trading in all harmonised bands that allow flexible use;
- fostering different modes of spectrum sharing in Europe, with the aim of ensuring its most efficient use and allowing access for innovative products.

This high-level political programme will definitely set a renewed focus on possibilities for shared spectrum use, including DSA option. The first obvious implementation, designed to address the spectrum need of mobile operators and hence adding to the balance of those 1200 MHz of spectrum, would be the Licensed Shared Access mechanism (see discussion in Sects. 2.5 and 2.6). But also other interesting and more general approaches are appearing, such as the concept of Beneficial Sharing Opportunity (BSO) [31].

BSO is a concept that allows judging whether sharing is appropriate and desirable for a given band:

- the BSO may exist in both licensed and licence-exempt frequency bands;
- the BSO condition is deemed fulfilled if the net benefit of application A (incumbent) is less than the combined net benefit of applications (A + B + C − (cost of sharing)).

So in effect the EC is seeing the BSO as a twofold concept: one being the most efficient use of licence-exempt bands (commons, such as 2.4 GHz ISM band), the other being the licensed sharing, namely the LSA concept.

What is however most important with new approach, compared with the original European focus on WSDs in TV bands, is the emergence of recognition of the general value of DSA in fostering spectrum sharing, and its corollary that identification of white spaces may be pertaining to any band, where the radio spectrum usage can be reliably identified in geographic and time domains. In that sense also the value of GDB was re-affirmed as the key infrastructure for enabling dynamic spectrum management.

1.3 An Overview on CR Standardisation

Markus Mueck

ETSI Reconfigurable Radio Systems Technical Committee, Sophia Antipolis, France

It may be noted that the strength and breadth of standardisation efforts and interest in TV White Space and related areas has helped to drive forward the regulatory case for CR. Moreover, standardisation of new technologies provides a level of stability that can assist regulators in planning for the allowance of them, and can yield dependable and consistent results. Therefore, this section provides a summary overview of standardisation activities in various standardisation bodies pertaining to the field of CR.

1.3.1 IEEE

The IEEE has developed one of the first CR Standards (IEEE 802.22 for wireless regional area network (WRAN) on secondary usage of TVWS) and currently has a number of projects ongoing for short-term (e.g., IEEE 802.11af enabling Wi-Fi in TVWS) and longer-term (e.g., IEEE DySPAN-SC developing solutions for DSA) requirements.

IEEE standards for secondary usage of TVWS include in particular:

- **IEEE 802.22** [32]: This standard was aimed at bringing broadband access to hard-to-reach, low population density areas, typical of rural environments and developing countries, and is based on a point to multi-point network topology. It provides a solution for a CR-based PHY/MAC air interface for use by license-exempt devices on a non-interfering basis in spectrum that is allocated to the TV Broadcast Service.
- **IEEE 802.11af** [33]: This standard provides modifications to both the 802.11 physical layers (PHY) and the 802.11 Medium Access Control Layer (MAC), to meet the legal requirements for channel access and coexistence in the TV White Space. It is based on Wi-Fi technology, appropriately modified to meet the regulatory requirements, database access, out of band emissions and channel bandwidths.
- **IEEE 802.19** [34]: The IEEE 802.19 Wireless Coexistence Working Group (WG) develops standards for coexistence between wireless standards of unlicensed devices. IEEE 802.19 Task Group 1 develops corresponding solutions for Wireless Coexistence in the TV White Space.

The IEEE Dynamic Spectrum Access Networks Standards Committee (DySPAN-SC) is developing standards in the areas of DSA, CR, interference management, coordination of wireless systems, advanced spectrum management, and policy languages for next generation radio systems [35]:

- **IEEE 1900.1 Working Group on Definitions and Concepts for Dynamic Spectrum Access : Terminology Relating to Emerging Wireless Networks, System Functionality, and Spectrum Management**: This standard provides terms and definitions in the field of DSA and related technologies;
- **IEEE 1900.2 Working Group on Recommended Practice for the Analysis of In-Band and Adjacent Band Interference and Coexistence Between Radio System**: This standard provides guidance for the analysis of coexistence and interference between various radio services in the specific context of spectrum management, policy-defined radio, adaptive radio, and software-defined radio;

- **IEEE 1900.3 Working Group on Recommended Practice for Conformance Evaluation of Software Defined Radio (SDR) Software Modules**: IEEE 1900.3 WG has been disbanded;
- **IEEE 1900.4 Working Group on Architectural Building Blocks Enabling Network-Device Distributed Decision Making for Optimized Radio Resource Usage in Heterogeneous Wireless Access Networks**: This standard provides solutions for Architectural Building Blocks Enabling Network-Device Distributed Decision Making for Optimized Radio Resource Usage in Heterogeneous Wireless Access Networks; IEEE 1900.4a is an amendment providing an architecture and interfaces for DSA in White Space Frequency bands; IEEE 1900.4.1 is a Standard for Interfaces and Protocols Enabling Distributed Decision Making for Optimized Radio Resource Usage in Heterogeneous Wireless Networks;
- **IEEE 1900.5 Working Group (WG) on Policy Language and Policy Architectures for Managing Cognitive Radio for Dynamic Spectrum Access Applications**: This standard defines a vendor-independent set of policy-based control architectures and corresponding policy language requirements for managing the functionality and behaviour of DSA networks;
- **IEEE 1900.6 Working Group on Spectrum Sensing Interfaces and Data Structures for Dynamic Spectrum Access and other Advanced Radio Communication Systems**: This standard defines the interfaces and data structures required to exchange sensing-related information in order to increase interoperability between sensors and their clients developed by different manufacturers; 1900.6a is an amendment providing procedures, protocols and message format specifications for the exchange of sensing related data, control data and configuration data between spectrum sensors and their clients. In addition, it adds specifications for the exchange of sensing related and other relevant data and specifies related interfaces between the data archive and other data sources;
- **IEEE 1900.7 White Space Radio Working Group**: This standard project develops solutions for Radio Interface for White Space DSA Radio Systems Supporting Fixed and Mobile Operation.

1.3.2 ETSI

ETSI focuses its work on CR solutions in the Reconfigurable Radio Systems Technical Committee (ETSI RRS). They are closely related to the current EC Mandate M/512 on Reconfigurable Radio System standardisation [36]. Current key activities relate to:

- **Licensed Shared Access (LSA)**: In alignment to the European Commission's Radio Spectrum Policy Group (RSPG) definitions on LSA [37], ETSI has issued a System Reference document (SRdoc) [38] and develops solutions for LSA

usage in the 2.3–2.4 GHz band in close cooperation with National Regulation Authorities in Europe in ETSI RRS;
- **Secondary usage of (TV) white spaces**: ETSI is in the process of finalizing a Harmonized Standard for enabling the introduction of TVWS systems into the European market [39]. Besides this rather regulation focused document, a number of technical documents are under development related to inter-database-exchange and coordinated/uncoordinated access to TVWS [40–43];
- **Mobile Device Architectures for enabling a Heterogeneous Networks approach**: This activity is performed in alignment to the current revision of the basic regulatory framework in Europe given by the Radio Equipment and Telecommunications Terminal Equipment Directive (R&TTE Directive) [44]. The new Directive is indeed expected to allow for Software reconfiguration of Mobile Devices. ETSI develops corresponding solutions for Mobile Device architectures and Interfaces [45–47] as well as novel Certification approaches [48] in support of simultaneous usage of multiple RATs.

1.3.3 ECMA

The European Computer Manufacturers Association (ECMA) has developed a standard for secondary usage of TVWS with a specific focus, among others, on home electronics equipment:

- **Secondary usage of TVWS**: ECMA has issued the standard 392 [49], which specifies a physical layer and a medium access sub-layer for wireless devices to operate in the TV frequency bands.

1.3.4 IETF

The Internet Engineering Task Force (IETF) is in the process of developing a protocol on how to access (TV) White Space databases:

- **Accessing (TV) White Space databases**: The Protocol to Access White Space database (PAWS) [50] is intended to enable a radio device to determine, in a specific location and at specific time, if any white space is available for secondary use.

The above had depicted the complex picture of global standardisation efforts in support of development of CR technologies.

1.4 Standardisation as Enabler of CR Developments

Vladislav V. Fomin[1], Gintare Sukareviciene[1] and Heejin Lee[2]

[1]Vytautas Magnus University, Kaunas, Lithuania
[2]Yonsei University, Seoul, Korea

1.4.1 Introduction

Standardisation plays important role in the technology innovation and the associated economic development of industries and nation-states [51–53]. However, predicting how standardisation process must unfold to maximize the chance of successful innovation is hardly possible due to the complexity of both standardisation and innovation processes, and their inter-dependencies [54–58].

The growing complexity of Information and Communication Technologies (ICT) in general [59] and ICT standards in particular [54], contribute to the development of "blurred ICT standardisation landscape" [60, 61]. With the "blurred landscape" scholars and politicians refer to the situation when there are several standards development organizations (SDOs) and/or consortia active in a particular field, and some areas of standardisation activities are overlapping [62]. Such situation makes it difficult to efficiently steer the overall innovation process (as it is critically dependent on the availability of technology standards). CR innovation process is not an exception here.

Recognizing the importance of the wireless communications in general, and CR-enabled technologies in particular, many (if not all) major SDOs and relevant regulatory organizations have embarked on developing standards or defining norms and regulation for one or another aspect of CR-related telecommunications, as may be seen in the previous Sect. 1.3. Those efforts, in one or another way, are aimed at developing a new generation of telecommunications services, a "more dynamic" one than the current telecommunications paradigm. "More dynamic" spectrum management here is understood as not being bound by rigid spectrum allocations with regard to particular technologies and services.

The switch from the current "command-and-control" spectrum management principle to the one based on dynamic spectrum allocation and spectrum sharing is not likely to happen overnight. As commented by Intel Corp.'s Markus Mueck [63], we are likely to see "islands of CR" within the existing telecommunications domain at first, with the presence of CR-enabled services growing gradually, as more stakeholders realize the advantages of the new paradigm and as the number of CR-related technology standards grows:

> I think that the more dynamic the system is—the more spectrum opportunities you will have. So actually it makes sense in the first step to look at something that is really just a

> little bit dynamic and you will get some more spectrum. Then once you run out of that, probably you will have to increase the level of dynamicity a little bit in order to free new resources. And so my expectations is that little by little we will move from LSA [Licensed Spectrum Access] to DSA [Dynamic Spectrum Access] depending, actually, on how fast we run out of spectrum availability [63].

This cumulative, versatile pattern of CR standardisation has sparked many different research threads. In this section we review some recent works on CR standardisation. In doing so, we aim to demonstrate the complexity of the CR development process, highlighting the major drivers and barriers in the development of this novel paradigm in the history of radio communications.

1.4.2 Assessing Standardisation Arena for the CR Infrastructure

Information and communication technologies have been among the fastest growing and innovating technologies in both production and use during the four past decades. The ever growing demand for wireless services, however, has one negative effect—the shortage of the "enabling resource"—the radio spectrum. In part, the deficit of the radio spectrum is due to lack of flexibility in the operational principles of the extant radio telecommunications paradigm. Until today, each communication system has always had an exclusively assigned frequency band, and each wireless device must have dedicated hardware for each wireless communication system being supported [64, p. 4].

This very principle of the telecommunications services/market has one important implication for the standardisation. Namely, development of novel technologies has usually been seen from the perspective of single standard development initiative. For example, the 2nd generation (2G) mobile telephony system GSM—likely the most often quoted success story of the European telecommunications policy and market developments—is a story of the (single) GSM standard development [65, 66]. Similarly, the 1G and the 3G telephony systems were studied and referred to as single standard initiatives [67–69]. With the development of the 3G mobile telecommunications services, however, we have seen a substantial diversification of standardisation efforts, with different technologies-as-standards being accepted under the umbrella of the Universal Mobile Telecommunications Services (UMTS) tag.

CR-related standardisation seems to have surpassed the previous generations of wireless communications in terms of the scope and complexity of the on-going efforts. Two main factors contributing to the complex setup are, first, the ambiguity with regard to what exactly consumers and markets will be served by CR standards, and, second, the fact that CR application requires horizontal interoperability between (and harmonization of) different technology standards, business practices, and regulatory policies [54]. The commentary by Markus Mueck [63] points sharply at this problem:

> ...The thing in standards is that it doesn't make sense to standardize, for example, a black box that one vendor sells. I mean one vendor can do whatever he wants within this black box. Standardisation is really about ...interfaces and building blocks that need to interoperate between different entities... So the problem with cognitive radio is to identify which is the black box that is done by the company on the custom basis, and which are the interfaces (or the components) to be defined [in a standard], because they are required to make different vendors interoperate [63].

The multi-threaded nature of CR standardisation implies there will be a much larger amount of coordination required, as compared to the previous generations of telecommunications systems [54, p. 2], [70, p. 108]. The need for (more) coordination is particularly true in the light of convergence of previously distinct markets with one another, such as e.g., telephony, Internet, and TV broadcasting as a result of or the precondition for the introduction of CR. This convergence is not just convergence of markets (business) or technologies. Delaere and Ballon [54, p. 3] identify four inter-related domains in the development of CR-enabled markets: institutional, technological, functional, and infrastructural. A comment by Rudi Bekkers [63] shows how these multi-domain specifics of CR may look from a standardisation perspective:

> I recently was in ITU meeting, global symposium for regulators in Poland, in Warsaw, and what I noticed there was that there were several sessions on [White Space] WS and in the end it was like a very big confusion among many of the participants: many people actually were talking about different things (all called them WS) but you could see that someone was just talking specifically about the area of bands; other people were talking about ... current digital dividend and ...how they can use the spectrum... and then in the end people's [discussions] started to [encompass] everything, because they actually were talking about totally different things. The worst is, actually, if a lawyer starts to talk about [WS]. I can tell you—then it is getting very tricky [63].

Given the variety of perspectives on what CR is (see Sect. 2.8), and multitude of opinions on what [parts of CR] must be standardized, identification of the main stakeholders and their interests in CR may shed some light on who will be driving the innovation process in general and standardisation process in particular. Fomin et al. [71] have mapped the main stakeholders in the CR development, identifying both drivers and barriers in the CR innovation, as well as associated variables in dependent (co-evolving) domains (see Tables 1.1, 1.2, 1.3).

Complimentary to the aforementioned research on co-evolutionary forces in CR development, Fomin et al. [73] looked at the variety of threads in the CR innovation process from a different angle. They collected and content-analysed 68 documents originated by COST Action IC0905 "TERRA". Within a two-year activity of this predominantly European think-tank, a diversity of directions, institutions, and technological concepts were covered (see Fig. 1.5), thus seconding experts' opinion on the need for substantial coordination efforts in the development and standardisation of CR [54, 70].

Some authors attempted to show the complexity of the CR standardisation by enlisting different workgroups (WGs) and SDOs in charge of one or another

Table 1.1 Inter-dependencies between drivers and barriers in the market domain of CR

Original domain: market		Co-evolutionary factors	
Drivers	Barriers	Technology	Policy
Consumer demand for lower price broadband wireless services	The strong position of 3G/4G mobile telephony services in case of eventual price wars	SW-based R&D drives prices down	Increasing competition lowers prices
Demand for broader supply and diversity of RATs	Uncertain business model for provision of innovative CR-based services	SW-base allows myriad variations and on-demand adaptability	Flexible/neutral regulations promote experimentation
Demand for license-exempt home devices	Interference concerns	Innovative solutions: sensing, geolocation, pilot channel	Regulations for license-exempt use with interference safeguards

Source [72]

Table 1.2 Inter-dependencies between drivers and barriers in the technology domain of CR

Original domain: technology		Co-evolutionary factors	
Drivers	Barriers	Market	Policy
Shift from HW to SW paradigm implies faster R&D cycles and time-to-market	Initial resistance to disruptive change of technological & manufacturing base	Strong demand for new/diverse RATs	Access to "new" spectrum implies wireless market growth
Cognitive features open up opportunistic spectrum access	Interference and CR type approval concerns	Formation of new market players to service CR users, i.e. GDB operators	Suitable "fail-proof" regulations to re-assure incumbents

Source [72]

CR-related standard [74], see also Sect. 1.3. Baldini et al. [62] developed an impressively long list of SDOs—their analysis re-mapped to better demonstrate the SDO involvement in a later work by Sukarevičienė and Fomin [75] (see Table 1.4).

Specifically, Sukarevičienė and Fomin [75] extended the earlier works of Baldini et al. [62] and Fomin et al. [71] by enlisting key stakeholders and naming specific CR services and standards associated with the stakeholders or the services in the context of Lithuanian DSA development (see Fig. 1.6).

Table 1.3 Inter-dependencies between drivers and barriers in the policy domain of CR

Original domain: policy		Co-evolutionary factors	
Drivers	Barriers	Market	Technology
Shift from administrative to market-based frequency allocation to ease growing pressure for new assignments	Institutional inertia and lack of confidence in as-yet unproven dynamic spectrum assignment processes	Strong demand for new/diverse RATs; state's obligation to aid consumers	Innovative solutions: sensing, geolocation, pilot channel
Addressing demand for flexible *ad hoc* emergency frequencies	Fragmentation and sovereignty concerns for control of emergency services	Demand for cost savings by using off-the-shelf equipment	CR easily re-configurable
Making use (and generating income) from swathes of seldom used spectrum such as in military bands	Reluctance of governmental spectrum owners to relinquish control of "their" spectrum	Meeting market demand for new spectrum while generating new income to the state/spectrum owners	Technological solutions that allow prompt vacating of bands by re-configurable CR

Source [73]

Fig. 1.5 Keywords by frequency of appearance (font size proportional to the frequency of its appearance): analysis of 68 documents originated by COST Action IC0905 "TERRA" [73]

1.4.3 Conclusions

The advent of CR-enabled telecommunications services, more than the previous generations of telecommunications, will integrate and embed significant technological innovations [55, p. 3]. Even when a single standard is considered, anticipatory technology standards are very challenging to formulate and enforce as they do not just record existing practices (e.g., quality standards), or establish some new practices that need just to be coordinated (e.g., measures) [55, p. 3].

Table 1.4 Main regulatory and standardization bodies for CR and DSA, which participate in spectrum regulation and related standardization processes in Europe. Adapted from [62]

CEPT: The European Conference of Postal and Telecommunications

ECO: The European Communication Office

Purpose and members	Relationships	Structure	
CEPT: harmonization of the regulatory requirements 48 countries across Europe	ECO, ECC, ITU, WRC-2012	Committees: • ECC • Com-ITU • CERP	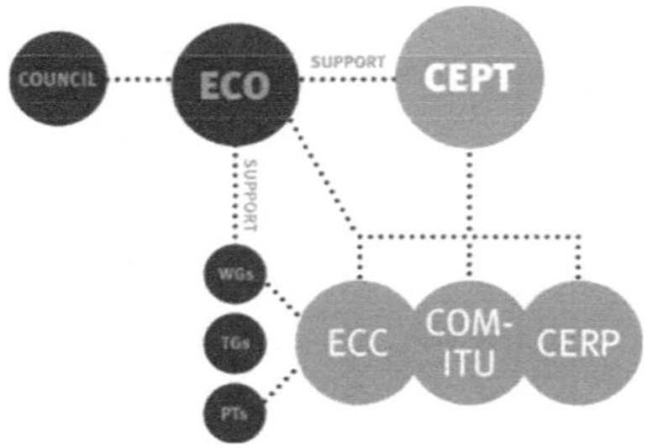
ECO: advice and support to the CEPT 32 CEPT countries	Secretariat of the CEPT	A team of 13 people	

Related standards, regulations, and projects:
Technical conditions for the use of the bands 821–832 MHz and 1785–1805 MHz for wireless radio microphones in the EU

Abbreviations used:
ECC: The Electronic Communications Committee
Com-ITU: The Committee for ITU Policy
CERP: The Committee for Postal Regulation

ETSI: The European Telecommunications Standards Institute

Purpose and Members	Relationships	Structural entities related to CR/DSA	
Responsible for most of the European telecommunication standardization activities 62 countries on 5 continents, large number of industry members	Together with CEN and CENELEC	ETSI TC: • ETSI-RRS • ETSI-ERM	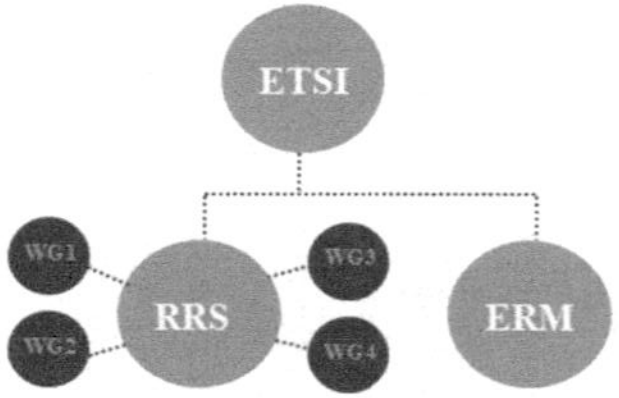

(continued)

Table 1.4 (continued)

Related standards, regulations, and projects: ETSI TC RRS, ETSI Specialist Task Force (STF) 386, e.g.: ETSI TR 102 799, ETSI TS 102 800	**Abbreviations used:** **CEN:** The European Committee for Standardization **CENELEC:** The European Committee for Electrotechnical Standardization **ETSI TC:** The European Telecommunications Standards Institute Technical Committee **ETSI-RRS:** The European Telecommunications Standards Institute Technical Committee on Reconfigurable Radio Systems **ETSI-ERM:** The European Telecommunications Standards Institute Technical Committee on Electromagnetic Radio Matters

EC: The European Commission
RSPG: The Radio Spectrum Policy Group
TCAM: The Telecommunications Conformity Assessment and market Surveillance

Purpose and members	Relationships	Structure	
EC: implementation of regulatory environment EU Member States	Mandates from CEPT and ETSI	32 DGs, e.g.: • INFSO • ENTR	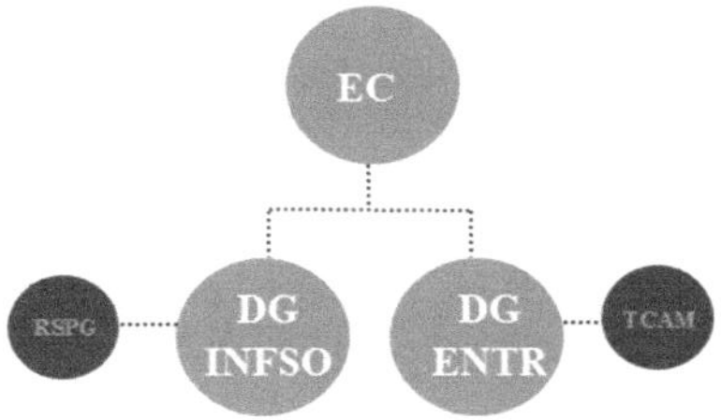
RSPG: development of radio spectrum policy EU Member States	Assists the EC (DG INFSO)		
TCAM: development of conformity assess and market surveillance policy EU Member States	Assists the EC (DG INFSO)		
	Assists the EC (DG ENTR). TCAM also collaborate with both CEPT and ETSI TCs		

(continued)

Table 1.4 (continued)

Related standards, regulations, and projects:
Examples of EU-funded research projects:
- E^3
- The FARAMIR project
- The CROWN project
- COGEU project

Abbreviations used:
DG: Directorate General
INFSO: Directorate General of the Information Society
ENTR: Directorate General of the Enterprise and Industry

IEEE: Institute of Electrical and Electronics Engineers

Purpose and members	Relationships	Structure
Foster technological innovation, excellence for the benefit of humanity Industry and institutions from more than 160 countries	IEC, ETSI, COM/SDB, DySPAN, and many others	6 subordinate boards, 15 committees (each board also consists of several committees)

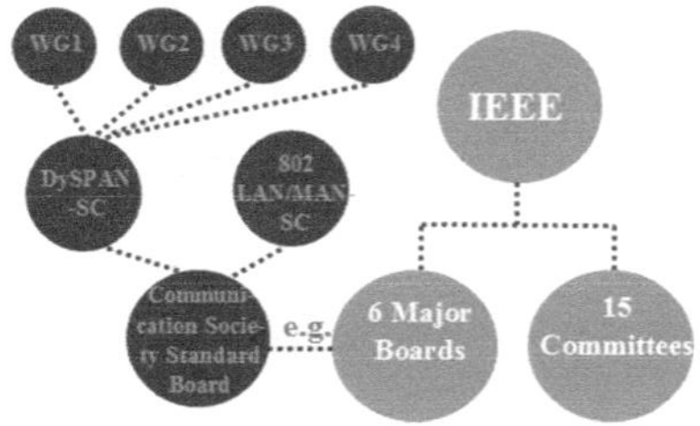

Related standards, regulations, and projects:
Series: 802.22
(e.g. 802.11af),
DySPAN-SC/P1900, 1900.1-1900.7, 802.16, 802.19, 802.18

Abbreviations used:
IEC: International Electrotechnical Commission
ETSI: The European Telecommunications Standards Institute
COM/SDB: The IEEE Sommunications Society (ComSoc) Standards Development Board
DySPAN: The IEEE Dynamic Spectrum Access Networks standards committee

(continued)

Table 1.4 (continued)

ECMA: The European association for standardizing information and communication systems

Purpose and members	Relationships	Structure	
Standardization of information and communication systems Company members from Asia, Australia, Europe, N. America	Liaisons: ETSI, ITU-T, CEN, JTC1, CENELEC, IEC, ISO	Two decision levels: managerial and technical	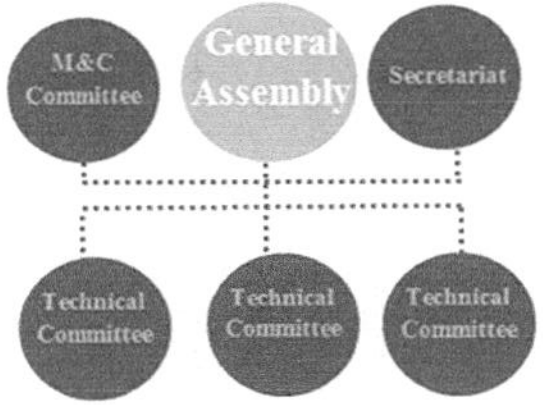

Related standards, regulations, and projects:
ECMA-392

Abbreviations used:
CEN: The European Committee for Standardization
CENELEC: The European Committee for Electrotechnical Standardization
ETSI: The European Telecommunications Standards Institute
IEC: International Electrotechnical Commission
ISO: International Organization for Standardization
ITU-T: Telecommunication Standardization Sector of the International Telecommunications Union
JTC1: ISO/IEC Joint Technical Committee 1-Information technology

IETF: The Internet Engineering Task Force

Purpose and members	Relationships	Structure	
Mission: to make the Internet work better Open to any interested individual	ITU-T SG and WP Liaisons	Working Groups (short-lived in nature)	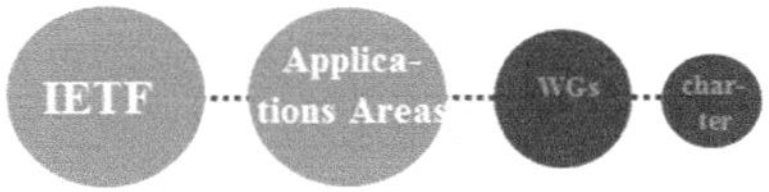

Related standards, regulations, and projects:
RFC 5378 (BCP 78), RFC 3979 (BCP 79)

Abbreviations used:
ITU-T: Telecommunication Standardization Sector of the International Telecommunications Union
ITU-T SG: ITU-T Study Group
ITU-T WP: ITU-T Work Programme

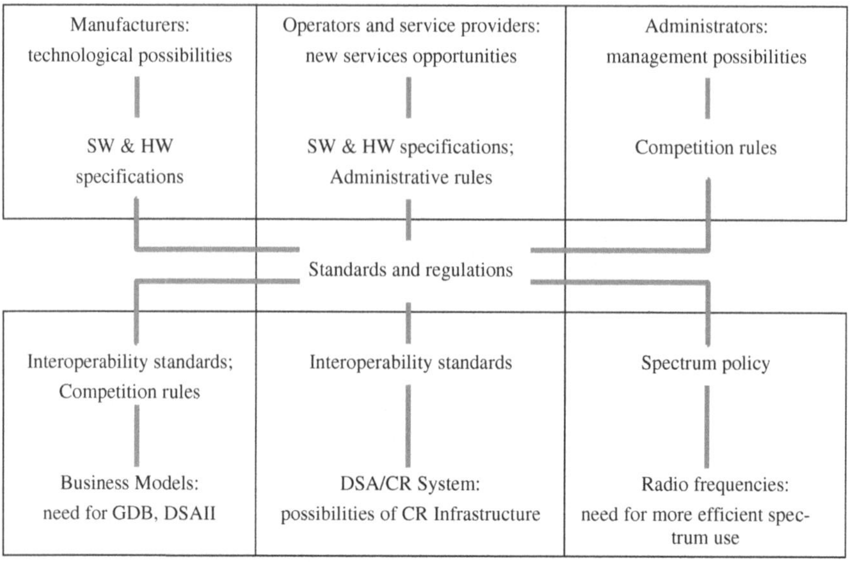

Fig. 1.6 Key elements of DSA infrastructure, adapted from [75]

Given the ambiguity surrounding the nature of future CR services, and potentially big stakes involved in developing CR-based markets, one may assume standardisation and regulatory efforts will be more and more directed towards interface standardisation. In this light, as commented by Intel Corp.'s Markus Mueck [63], "DSA standardisation is really about interface standardisation", and not about DSA-as-a-system. "The interesting thing here is that standardisation does not focus so much on DSA as such, but it focuses on the tools that are needed to implement DSA" [63].

However, standardisation focused (solely) on interfaces is not likely to give sufficient impetus to the CR innovation. Not only novel tools are needed for the paradigm-changing CR services. Orchestration of regulatory action (see Sects. 1.1 and 1.2), development of business models (see Chap. 4) and other similar measures are equally important to create basis for new CR-based communications solutions and new services. Ahead of time, very little is known about the possible market impacts of design choices made today, in the many threads of the CR standardisation efforts. In this regard, research reported in this section helps stakeholders participating in the standardisation process better estimate what choices would serve their interest best.

1.5 National Champions of CR Policy

1.5.1 US Regulatory Policy Developments and Visions

Jeff Schmidt

Spectrum Bridge, Orlando, FL, USA

FCC regulators first contemplated the idea of TVWS in 2004, in a Notice of Proposed Rule Making titled "Unlicensed Operation in the TV Broadcast Bands". Through much debate and contentious discussion within industry, the FCC published rules, that four years later made VHF and UHF spectrum available for "*new and innovative broadband products and services*" [76]. Since that time, further refinements and tangible progress have been made by industry (see Fig. 1.7). The FCC rules specify the use of a geo-location database that resides in the "cloud" which facilitates channel allocations without interference to incumbents. Incumbents are users of television broadcasting, land mobile radio, wireless microphones and radio astronomy. Establishing an ecosystem that would have a profound influence on future of CR technology may not have been the FCC's objective, but that was exactly the result. When CR was envisioned as a nascent technology, it was as an exotic technology that incorporated components of sensing, waveform synthesis and MAC layer adaptation and often referred to as Software Defined Radio (SDR). For various reasons, this form of SDR has not seen significant adoption, except in military and laboratory applications. This is partially due to complex and expensive radio requirements and a lack of regulatory policy to facilitate its use. However, the FCC's rules—elegant in their simplicity, changed the trend of CR. The confluence of this new regulatory landscape, widespread internet availability, cloud based computing resources and low cost radios have permanently shifted the CR paradigm from a self-intelligence to a shared-intelligence schema.

The basis of the FCC's TVWS rules is straightforward. They track geographic areas of incumbent operations, specify buffer requirements, determine channel adjacency constraints and utilize a geo-location database to allocate leftover or white space spectrum in terms of time, frequency and geography. White Space Devices (WSDs) report their location to the database, and spectrum is allocated in real-time.

Despite the appearance of simplicity, there are aspects that have caused debate, such as the path loss models to be used to determine protection and interference, how much noise margin should be allowed to ensure reliable operation and how portable devices will be managed. Fortunately, these parameters can be quickly adapted or modified, through the flexibility of a cloud based database. There is no need to re-program millions of individual devices with new operational parameters. The conservative nature of these rules has allowed industry to trudge forward without detriment to the incumbent ecosystems. As growth continues and the

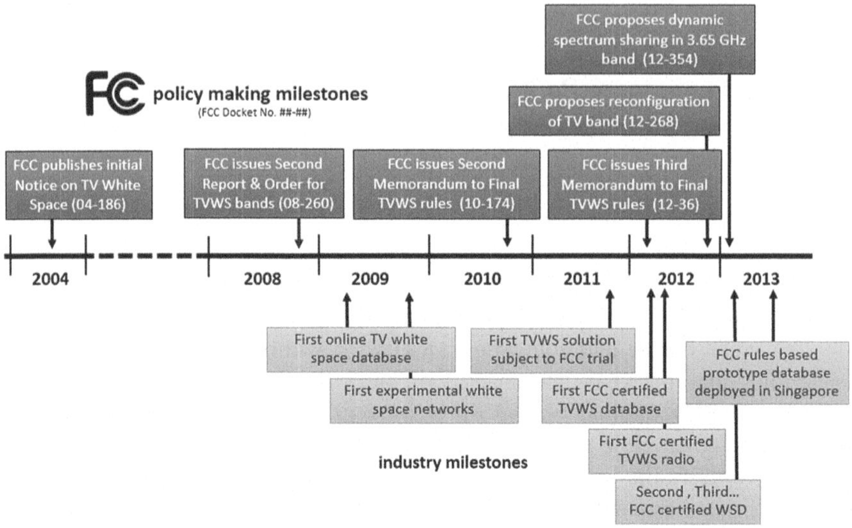

Fig. 1.7 TV white space: FCC policy making and industry progression timeline

effects of the incumbent and white space on one another is better quantified, it is expected that the FCC will consider changes to these rules.

The merits and acceptance of CR when coupled with geo-location database technology have been recognized as a positive development and served as an exemplary model for spectrum sharing. This has been affirmed through the continued development of TVWS radios and geo-location database technology in the US and unabated discussions on how this technology can be applied in other bands. In fact, the US rules have been adopted as a baseline for discussion in countries such as Canada and Singapore and several additional proceedings have been opened in the US to consider this model in other bands.

Unfortunately, widespread deployment of TVWS in the US has been lethargic and limited to a few applications in which it is currently well suited. Most applications have been targeted at smart agriculture, industrial telemetry and some rural broadband applications. Suitability of TVWS spectrum for these applications is driven by the unique combination of availability (in rural areas), excellent propagation characteristics (in harsh environments) and marginal broadband data rates.

Narrow adoption is ultimately the consequence of several policy based impediments: inflexible and particularly harsh out of band emissions requirements and the uncertainty surrounding re-allocation of UHF frequencies to accommodate changing television technology and growing demand for broadband spectrum. The out of band emissions requirements specified by the FCC for WSDs do not allow emissions to exceed -55 dB at the band edge, see Fig. 1.8. Unfortunately this is not congruent with standards based technology and every radio certified to date has required extensive filtering mechanisms.

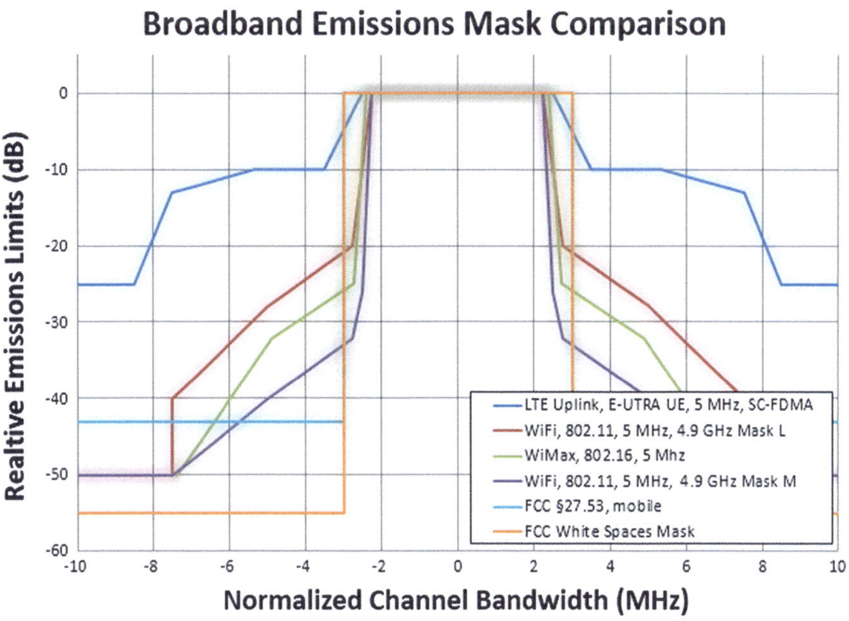

Fig. 1.8 Emissions mask comparisons: FCC TVWS and common broad band technologies

Specifically, the consequences of this limit are lower throughput (bits/Hz) and less transmit power resulting in a diminished link budget. This singular emissions limit is in contrast to the more flexible ETSI emissions device class model described in EN 301 598. Although the ETSI model is more complicated by virtue of variable transmit power limits and additional calculation requirements, it can be easily implemented within a geo-location database. The ETSI limits also permit the use of standards based technologies such as 802.11 Wi-Fi, with a bit of filtering. The only negative (albeit necessary) is the trade-off between undesired emissions and available TVWS, but it is better to have options! The second major inhibitor to TVWS adoption in the US is the uncertainty surrounding the potential repurposing of TV broadcast spectrum for licensed broadband use. In some locations within the US, there exists miniscule amounts of TVWS, especially near metropolitan areas. The risk to network operators is that when TVWS is reduced to a few channels, a real possibility exists that it may disappear altogether—after considerable capital investment and commitments to customers are made. Although the FCC has stated, "*In the white space … we propose measures that…would make a substantial amount of spectrum available for unlicensed uses, including a significant portion that would be available on a uniform nationwide basis…Television white spaces will continue to be available for unlicensed use in the repacked television band*" [77] it does not appear that this issue will be resolved soon, as illustrated in Fig. 1.9.

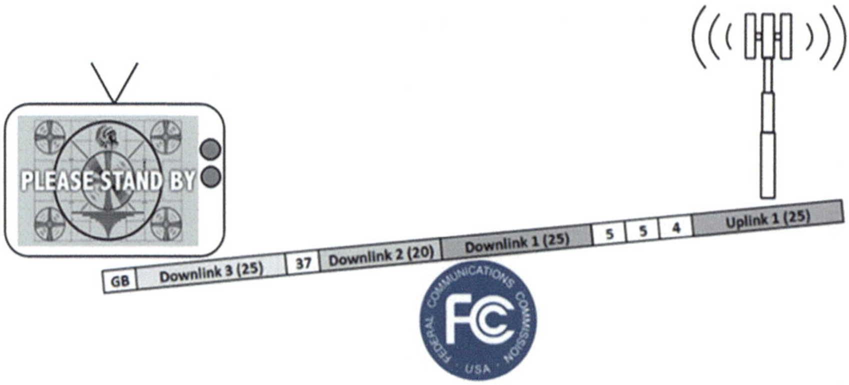

Fig. 1.9 The delicate balancing act the Federal Communication Commission facilitates between large wireless industry contingents

Fig. 1.10 The distinction between white and grey space

One advantage of this proposal is that a critical mass of TVWS might be dedicated to permanent unlicensed use through the designation of guard bands which would separate TV broadcast spectrum from licensed broadband spectrum. What remains unclear is what new constraints might appear. For example, what will the maximum allowable transmit power be in the guard band? And will portable devices be permitted to operate in a guard band? It is also unclear how long this protracted process will take.

There are other issues impeding the adoption of TVWS, but they are more technical than regulatory. A prime example is the unintended consequence of mixing high power television technology with low power broadband technology. The effects of this problem are illustrated in Fig. 1.10.

Because the FCC's primary goal in defining rules for TVWS operation is to protect incumbents, it has been left to industry to develop innovative solutions to deal with this challenge. Fortunately a geo-location database can be very effective

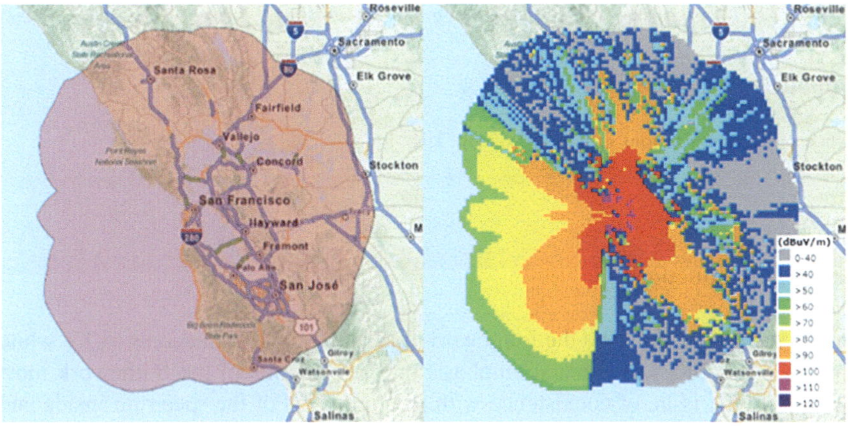

Fig. 1.11 Comparison between F-curve (R6602) and Longley-Rice contours near San Francisco, CA

in predicting and mitigating the interference between dissimilar ecosystems. This can be improved even more through crowdsourcing and feedback of interference data, in a way similar to how our smartphones forewarn us of traffic congestion on the roads we travel.

Another area in which the US model can be optimized is through the use of more accurate and comprehensive path loss and coverage models. The FCC currently utilizes the R 6602 [78] model for TV coverage, commonly referred to as the F-curve model. Although this model is fairly simple to implement and provides a more than adequate service contour prediction for television, it does not yield the resolution provided by other models such as Longley-Rice [79] or ITU-R P.1546. This is not to say that the R 6602 model is inadequate for its intended purpose, which is to define incumbent protection, but it is not well suited for optimizing the amount of TVWS available. This is illustrated in Fig. 1.11, where large areas are blocked by mountains and protected, but have no real opportunity for television coverage. One reason for the difference is that the Longley-Rice model makes significantly better use of terrain data. The Longley-Rice model also yields high resolution signal and coverage data which is compatible with the EN 301 598 pixel based methodology for defining incumbent protection. The pixel based approach is more complex in terms of computing available TVWS and data storage, but presents a trade-off that is feasible when considering the low cost of cloud based computational availability.

Although the growth of TVWS adoption and evolution of policy remains slow in the US, the promise of geo-location database technology remains positive. In fact the FCC should be affirmatively recognized as creating the first working set of TVWS rules and certifying the first TVWS radios and database platforms. Other implications of these rule making efforts have given rise to a new trend in CR. Nevertheless, it is expected that practical experience and regulatory efforts worldwide will have the biggest influence on US policy evolution.

1.5.2 UK Framework for Access to TV White Spaces

Hamid Reza Karimi

Office of Communications (Ofcom), London, UK

1.5.2.1 Introduction

In this section we describe the framework for database-assisted access to TV white spaces in the UK, with special emphasis on the elements of the framework most relevant to the issue of coexistence with existing users of the spectrum inside and outside the UHF TV band (470–790 MHz). Note that at the time of writing, the final regulatory rules for access to TV white spaces in the UK are subject to public consultation.

1.5.2.2 Database-Assisted Access to TV White Spaces

White space devices (WSDs) operating in the UHF TV band in the UK will be licence exempt equipment that share the spectrum with the Digital Terrestrial TV (DTT) and Programme Making and Special Events (PMSE) services. These two licensed services are the primary users of the band, and as such, Ofcom must ensure a low probability of harmful interference to these services.

The requirement for a low probability of harmful interference also extends to services outside the UHF TV band. These include mobile networks above the band (791–862 MHz), and a range of uses such as emergency services, PMSE, scanning telemetry, short range devices, business radio, and maritime radio below the band (450–470 MHz).

The frequency allocations for the above services are illustrated below (see Fig. 1.12). Note that channels 31 to 37 are currently cleared of DTT transmissions, but are in use by PMSE, with plans for use by high definition DTT broadcasts in the near future.

By itself, a WSD does not have access to the requisite information about DTT and PMSE usage of the band to be able to transmit without there being a substantial risk of causing harmful interference to existing users. Therefore, a WSD must contact an appropriate repository—a white space database (WSDB)—and communicate information about itself and its geographic location. The WSDB will respond to the WSD with a set of Operational Parameters including the frequencies and maximum powers at which the WSD can transmit in order to ensure a low probability of harmful interference to the primary users.

The following are some of the key elements of the UK's proposed regulatory framework (see Fig. 1.13):

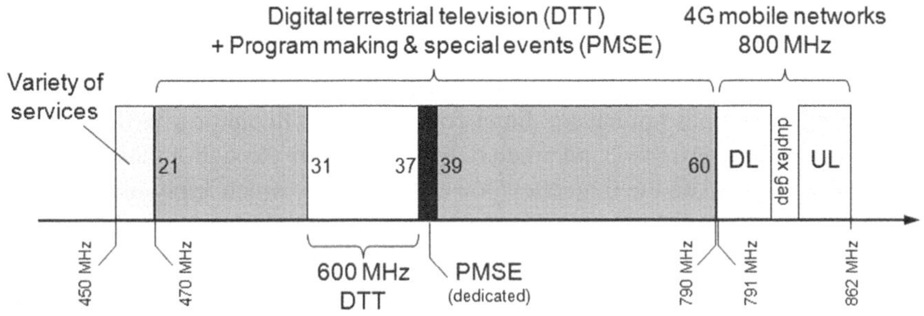

Fig. 1.12 The UHF TV band (470–790 MHz) and its users

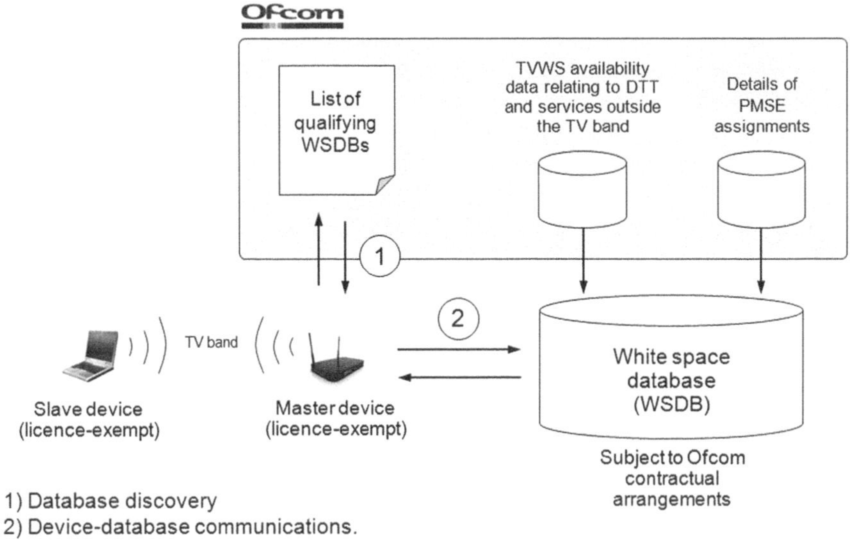

Fig. 1.13 Proposed framework for authorising the use of TV white spaces

- WSDs will be permitted to transmit in the UHF TV band provided that there is a low probability that they will cause harmful interference to existing licensed users within the band (DTT and PMSE) as well as users outside the band;
- Compliance with the licence exemption regulations will require that WSDs operate according to the frequency/power parameters (restrictions) that they receive from a WSDB. They will be required to obtain such parameters from a qualifying WSDB. The qualifying WSDB will generate the frequency/power parameters for WSDs on the basis of information relating to the existing users that Ofcom will regularly make available;
- WSDs will be able to identify qualifying WSDBs by consulting a list on a website maintained by Ofcom, and selecting a preferred WSDB from that list. This is the so-called "database discovery". The choice of preferred WSDB will be for the master WSD to determine itself;

- WSDs are categorised as masters or slaves. A master WSD is required to have a communications link to access Ofcom's list of qualifying WSDBs, and a communications link to query one of the qualifying WSDBs. A slave WSD, on the other hand, does not have a direct connection to Ofcom or a WSDB; it will obtain its frequency/power parameters from a WSDB through a master WSD;
- Ofcom will calculate the frequency/power restrictions which apply in relation to interference from WSDs to DTT (both in the UK and across borders). The results of these calculations will be communicated to the WSDBs. These will also include any additional *location agnostic* frequency/power restrictions that may apply in relation to interference to services inside or outside the UHF TV band. Ofcom will provide scheduled updates to the above data whenever there is a relevant change to the planning of DTT or other services. It is expected that these updates will occur once or twice a year. On certain occasions, there may be unscheduled updates to the above data. These may be triggered by an interference management process or by the fine-tuning of Ofcom's coexistence modelling parameters;
- Ofcom will also provide to WSDBs information on geolocated PMSE assignments throughout the UK. This information will be updated on a scheduled three-hourly basis. WSDBs will use this information to calculate frequency/power restrictions in relation to interference from WSDs to PMSE. On certain occasions, there may be unscheduled updates to the above information. These may be triggered by an interference management process;
- The WSDBs will combine the frequency/power restrictions calculated by Ofcom with those they calculate themselves in relation to PMSE, and convey these to the relevant WSDs.

For the purposes of this section, we use the terms frequency/power restrictions, WSD emission limits, and TVWS availability data interchangeably.

1.5.2.3 Interactions Between Databases and Devices

In November 2012 Ofcom published "A consultation on white space device requirements" where it outlined proposals for the operation of WSDs and the nature of the data exchanged between WSDs and WSDBs. These proposals (among others) were subsequently incorporated into the draft European harmonised standard EN 301 598 which is currently subject to public consultation.[5] Details of EN 301 598 are presented in Sect. 2.7 of this book. Here we summarise some of the key elements of the WSDB-WSD interactions implied by EN 301 598.

As noted earlier, the first operation of a master WSD is database discovery. This is where the device consults a web listing of qualifying WSDBs. Ofcom may

[5] Draft ETSI EN 301 598 V1.0.0 (2013-07), "White space devices (WSD); Wireless access systems operating in the 470 MHz to 790 MHz frequency band; Harmonized EN covering the essential requirements of article 3.2 of the R&TTE Directive".

occasionally update this list. For this reason, master WSDs must repeat database discovery with a minimum regularity as specified by Ofcom.

Having selected a WSDB from the web list, the master WSD may then initiate communications with that WSDB. WSDBs and WSDs are required to exchange the following parameters:

Device Parameters—These are communicated from a WSD to a WSDB, and identify specific characteristics of the WSD (including its location).

Operational Parameters—These are generated by a WSDB and communicated to WSDs. They specify the frequency/power restrictions (and other instructions) which WSDs must comply with when transmitting in the UHF TV band. There are two types of Operational Parameters:

(a) Specific Operational Parameters account for the Device Parameters of a specific WSD. In this way, for example, a WSD with a lower antenna height, a more stringent emission mask, and a more benign signal structure, would benefit from greater TVWS availability.
(b) Generic Operational Parameters are intended for slave WSDs whose Device Parameters are not known. A WSDB will communicate Generic Operational Parameters to a master WSD, which in turn will broadcast these to all slave WSDs in its coverage area. Generic Operational Parameters account for certain characteristics of the serving master WSD (e.g., location, power, and hence coverage area), but are based on assumed default values for the Device Parameters of the slave WSDs.

Channel Usage Parameters—These are reported by a WSD to inform a WSDB of the *actual* radio resources (channels and powers) that will be used by the WSDs.

The interactions between master WSDs, slave WSDs and WSDBs are described in more detail in Sect. 2.7 of this book in the context of the ETSI EN 301 598 harmonized standard.

1.5.2.4 Emission Limits for Coexistence with Existing Services

Here we describe the UK's proposed approach for the calculation of WSD emission limits to ensure a low probability of harmful interference to existing users of the radio spectrum.

Emission limits in relation to DTT use in the UK. In relation to DTT, the derivation of location-specific TVWS availability is formulated as the following problem: Calculate the maximum permitted WSD in-block EIRP, $P_{\text{WSD-DTT}}(i, F_{\text{WSD}})$ in dBm/(8 MHz), for a WSD located in a geographic pixel indexed as i, and radiating in channel F_{WSD}, subject to a target reduction in DTT signal-to-interference-plus-noise ratio in any channel $F_{\text{DTT}} = 21$ to 60. For a UK-wide picture, the above would need to be performed for each pixel in the UK and for each WSD channel ($F_{\text{WSD}} = 21$ to 60), accounting for the nationwide quality of DTT. The result can be interpreted as 40 maps of the UK with the maximum permitted WSD

EIRP depicted in each pixel. Ofcom will be responsible for generating UK-wide TVWS availability datasets in relation to DTT.

In the approach adopted by the FCC, WSDs are permitted to radiate at up to a fixed maximum power so long as they are located outside pre-defined geographic exclusion zones surrounding TV transmitters. The exclusion zones correspond to areas where the received DTT field strength exceeds FCC-defined thresholds based on FCC-defined propagation models.

In the approach proposed by Ofcom, there are no explicit exclusion zones. Here, it is the in-block EIRP of the WSDs (rather than their geographic location) that is explicitly restricted. The approach permits WSDs to communicate at greater EIRPs in areas where DTT field strength is greater; i.e., where DTT is more robust to interference. Furthermore, it is proposed to cap the maximum in-block EIRP of all WSDs at 36 dBm/(8 MHz). It is considered that such a cap on the maximum permitted power is important in avoiding the overloading of nearby DTT receivers. This value is also in line with the FCC limit for fixed devices, and is a sensible value which caters for most of the envisaged TVWS use cases.

Emission limits in relation to PMSE. In relation to PMSE, the derivation of location-specific TVWS availability is formulated as the following problem: Calculate the maximum permitted WSD in-block EIRP, $P_{\text{WSD-PMSE}}(j, F_{\text{WSD}})$ in dBm/(100 kHz), for a WSD located in a geographic location indexed as j, and radiating in channel F_{WSD}, subject to a given PMSE wanted-to-unwanted power ratio in any channel $F_{\text{DTT}} = 21$ to 60. For a UK-wide picture, the above would need to be repeated for each WSD location in the UK and for each WSD channel ($F_{\text{WSD}} = 21$ to 60), accounting for each licensed PMSE assignment.

WSDBs will be responsible for performing the above calculations. The WSDBs will need to account for WSD spectrum emission class, reported WSD antenna height, and WSD type (fixed or portable/mobile) in performing the calculations. In practice, WSDBs do not need to develop a UK-wide picture, as the calculations can be performed in real time in response to queries by individual WSDs.

Emission limits in relation to cross border DTT. In relation to cross border DTT, the derivation of location-specific TVWS availability is formulated as the following problem: Calculate the maximum permitted WSD in-block EIRP, $P_{\text{WSD-XB}}(i, F_{\text{WSD}})$ in dBm/(8 MHz), for a WSD located in a geographic pixel indexed as i, and radiating in channel F_{WSD}, subject to the received field strength in neighbouring countries not exceeding relevant international coordination trigger thresholds in channel F_{WSD}. For a UK-wide picture, the above would need to be performed for each WSD pixel in the UK and for each WSD channel ($F_{\text{WSD}} = 21$ to 60). In practice, only WSD pixels near the UK coastlines or land borders need to be examined since pixels in-land are unlikely to be subject to any cross-border restrictions. Ofcom will be responsible for generating UK-wide TVWS availability datasets in relation to cross border DTT.

Calculation of location-agnostic emission limits. Location-agnostic WSD emission limits will apply in the context of seeking to ensure a low probability of harmful interference to uses above and below the UHF TV band, as well as PMSE usage in channel 38 (the latter use is UK-wide and non-geolocated). These limits are not location-specific because information on the locations of the above uses is not readily available and therefore cannot be exploited in the database-assisted framework for access to TV white spaces. As a result, the WSD emission limits are simply specified by Ofcom as location-agnostic limits, $P_{LA}(F_{WSD})$ in dBm/(8 MHz), in each channel $F_{WSD} = 21\ldots60$.

Combining of emission limits by Ofcom. As explained above, Ofcom will be responsible for calculating the individual limits $P_{WSD\text{-}DTT}(i, F_{WSD})$, $P_{WSD\text{-}XB}(i, F_{WSD})$, and $P_{WSD\text{-}LA}(F_{WSD})$ across the UK. For a WSD located in geographic pixel i, and radiating in channel F_{WSD}, Ofcom will then calculate the overall EIRP limit, $P_1(i, F_{WSD})$ in dBm/(8 MHz), as the minimum of the above individual limits. Ofcom will then communicate the UK-wide values of $P_1(i, F_{WSD})$ to the WSDBs. Ofcom proposes to generate a unique set of individual limits for each combination of five WSD spectrum emission classes, three WSD technology (protection ratio) categories, and a number of representative WSD antenna heights, all for fixed WSDs. Limits for portable/mobile WSDs will be inferred by WSDBs from the limits provided for fixed WSDs.

Combining of emission limits by databases. As well as receiving the limits $P_1(i, F_{WSD})$ in dBm/(8 MHz) from Ofcom, WSDBs will calculate the limits $P_{WSD\text{-}PMSE}(j, F_{WSD})$ in dBm/(100 kHz) in relation to PMSE. Then, for a WSD at geographic location j (which falls within pixel i), and radiating in channel F_{WSD}, a WSDB will calculate the overall EIRP spectral density limit, $P_0(j, F_{WSD})$ dBm/(100 MHz), as the minimum of $P_1(i, F_{WSD}) - 10\log_{10}(80)$ and $P_{WSD\text{-}PMSE}(j, F_{WSD})$. The values $P_0(j, F_{WSD})$ dBm/(100 kHz) and $P_1(i, F_{WSD})$ dBm/(8 MHz) form the basis of the Operational Parameters which WSDBs communicate to WSDs.

Power adjustments by Ofcom (volume dial). It may be necessary for Ofcom to adjust the emission limits $P_1(i, F_{WSD})$ and $P_0(j, F_{WSD})$. Such adjustments, $\Delta(i, F_{WSD})$, will be communicated by Ofcom to the WSDBs. These might be on a location-specific and/or channel-specific basis, and may be triggered by an interference management process or by a fine tuning of Ofcom's coexistence modelling parameters.

Multiple devices and interference aggregation. In the framework presented above, it is implicitly assumed that at any one time only one WSD radiates per pixel/location and per DTT channel. In practice, one or more WSDBs may serve multiple WSDs in the same geographic area, and this may result in an aggregation of interferer signal powers.

It is believed that such aggregation of interference is unlikely to be problematic in the short term, since (a) the calculated WSD emission limits are cautious, (b) interference tends to be dominated by the nearest interferer, (c) many WSDs will implement polite protocols and are not likely to transmit simultaneously and/or congregate in the same DTT channels when in close geographic proximity, (d) even if WSDs did transmit simultaneously and in the same DTT channels, the composite signal would increasingly appear noise-like and would render the time–frequency structure of the aggregate signal more benign in the context of interference to existing services.

In the longer term, there are three high-level mitigation options in the event that interference aggregation were to become a problem:

(1) Direct reductions in WSD emission limits to incorporate fixed added margins for interference aggregation.
(2) Rule-based reductions in WSD emission limits as a function of the number of WSDs which a WSDB already serves at any given location.
(3) Rule-based reductions in WSD emission limits informed by inter-WSDB communications.

1.5.2.5 Conclusions

We have described the UK framework for database-assisted access to TV white spaces, and presented some of the key components of the data exchanged between WSDBs and WSDs, as specified in the ETSI harmonised standard EN 301 598.

We have also summarised at a high level the approach proposed by Ofcom for calculating the WSD emission limits in relation to the various existing uses of the spectrum inside and outside the UHF TV band. We have explained how the various emission limits must be combined to derive location-specific and frequency-specific limits P_0 dBm/(100 kHz) and P_1 dBm/(8 MHz) which form the basis of Operational Parameters that WSDBs communicate to WSDs.

References

1. Radio Regulations: Articles, vol 1. Geneva: International Telecomunication Union (2012)
2. Radio spectrum management for a converging world. In: Workshop on Radio Spectrum Management for a Converging World. ITU, Geneva, 16–18 Feb (2004)
3. Directive 1999/5/EC of the European Parliament and of the Council of 9 March 1999 on radio equipment and telecommunications terminal equipment and the mutual recognition of their conformity. Off. J. **L 091**, 0010–0028 (1999)
4. Directive 2002/21/EC of the European Parliament and of the Council of 7 March 2002 on a common regulatory framework for electronic communications networks and services (Framework Directive). Off. J. **L 108**, 0033–0050 (2002)
5. Directive 2002/20/EC of the European Parliament and of the Council of 7 March 2002 on the authorisation of electronic communications networks and services (Authorisation Directive). Off. J. **L 108**, 0021–0032 (2002)
6. Decision No 676/2002/EC of the European Parliament and of the Council of 7 March 2002 on a regulatory framework for radio spectrum policy in the European Community (Radio Spectrum Decision)
7. Directive 2009/140/EC of the European Parliament and of the Council of 25 November 2009 amending Directives 2002/21/EC on a common regulatory framework for electronic communications networks and services, 2002/19/EC on access to, and interconnection of, electronic communications networks and associated facilities, and 2002/20/EC on the authorisation of electronic communications networks and services. Off. J. **L 337**, 0037–0069 (2009)

8. Anker, P.: Does cognitive radio need policy innovation? Competition Regul. Netw. Ind. **11**(1), 2–26 (2010)
9. Nekovee, M.: Dynamic spectrum access—concepts and future architectures. BT Technol. J. **24**(2), 111–116 (2006)
10. From Semaphore to Satellite. International Telecommunications Union, Geneva (1965)
11. Kirby, R.C.: History and trends in international radio regulations. In: IEE International Conference on 100 Years of Radio (1995)
12. Robinson, J.O.: Spectrum Management Policy in the United States: An Historical Account, OSP working paper 15 (1985)
13. Codding, G.A.: The International Telecommunication Union. An Experiment in International Cooperation. Brill, Leiden (1952)
14. Coase, R.: The federal communications commission. J. Law Econ. **2,** 1–40 (1959) (October)
15. Coase, R.: The problem of social cost. J. Law Econ. **3**(1), 1–44 (1960)
16. Hazlett, T.: The Wireless Craze, The Unlimited Bandwidth Myth, The Spectrum Auction Faux Pas, and the Punchline to Ronald Coase's 'Big Joke': An Essay on Airwave Allocation Policy, AEI-Brookings Joint Center Working Paper No. 01-2. SSRN: http://ssrn.com/abstract=286932 (2001)
17. Baumol, W.J., Robyn, D.: Toward an Evolutionary Regime for Spectrum Governance—Licensing or Unrestricted Policy. AEI-Brookings Joint Center for Regulatory Studies, washington (2006)
18. Cave, M., Doyle, L., et al.: Essentials of Modern Spectrum Management. Cambridge University Press, Cambridge (2007)
19. Mueller, M.: New Zealand's revolution in spectrum management. Inf. Econ. Policy **5**(2), 159–177 (1993)
20. Lemstra, W., Anker, P., et al.: Cognitive Radio: enabling technology in need of coordination. Competition Regul. Netw. Ind. **12**(3), 210–235 (2011)
21. Faulhaber, G.R., Farber, D.J.: Spectrum management: property rights, markets and the commons. In: Craven, F., Wildeman, S. (eds.) Rethinking Rights and Regulations: Institutional Response to New Communications Technologies. MIT Press, Cambridge (2003)
22. Faulhaber, G.R.: The Question of Spectrum: Technology, Management, and Regime Change, pp. 05–15. AEI-Brookings Joint Center for Regulatory Studies, Washington (2005)
23. Olafsson, S., Glover, B., et al.: Future management of spectrum. BT Technol. J. **25**(2), 52–63 (2007)
24. Anker, P.: Cognitive radio, the market and the regulator. In: 2010 IEEE Symposium on New Frontiers in Dynamic Spectrum Access Networks, IEEE (2010)
25. Report ITU-R SM.2152 Definitions of Software Defined Radio (SDR) and Cognitive Radio System (CRS). ITU, Geneva (2009)
26. CEPT Report 24. A preliminary assessment of the feasibility of fitting new/future applications/services into non-harmonised spectrum of the digital dividend (namely the so-called "white spaces" between allotments). http://www.erodocdb.dk/ (2008)
27. ECC Report 159. Technical and operational requirements for the possible operation of cognitive radio systems in the 'white spaces' of the frequency band 470–790 MHz. http://www.erodocdb.dk/Docs/doc98/official/pdf/ECCREP159.PDF (2011)
28. ECC Report 185. Complementary Report to ECC Report 159: Further definition of technical and operational requirements for the operation of white space devices in the band 470–790 MHz. http://www.erodocdb.dk/Docs/doc98/official/pdf/ECCREP185.PDF (2013)
29. ECC Report 186. Technical and operational requirements for the operation of white space devices under geo-location approach. http://www.erodocdb.dk/Docs/doc98/official/pdf/ECCREP186.PDF (2013)
30. Decision No 243/2012/EU of the European Parliament and of the Council of 14 March 2012 establishing a multiannual radio spectrum policy programme. http://ec.europa.eu/digital-agenda/en/rspp-roadmap-wireless-europe
31. COM(2012)478. Communication from The Commission To The European Parliament, The Council, The European Economic And Social Committee And The Committee Of The

Regions on Promoting the shared use of radio spectrum resources in the internal market. https://ec.europa.eu/digital-agenda/sites/digital-agenda/files/com-ssa.pdf (2012)
32. IEEE 802.22 Working Group on Wireless Regional Area Networks. http://www.ieee802.org/22/
33. IEEE 802.11 TASK GROUP AF, Status of project P802.11af. http://www.ieee802.org/11/Reports/tgaf_update.htm
34. IEEE 802.19 Wireless Coexistence Working Group (WG). http://www.ieee802.org/19/
35. IEEE DySPAN Standards Committee (DySPAN-SC). http://grouper.ieee.org/groups/dyspan/
36. Standardisation Mandate to CEN, Cenelec and ETSI for Reconfigurable Radio Systems. http://www.etsi.org/images/files/ECMandates/m512.pdf
37. Radio Spectrum Policy Group 2011. Report on CUS and other spectrum sharing approaches: Collective Use of Spectrum. http://rspg-spectrum.eu/_documents/documents/meeting/rspg26/rspg11_392_report_CUS_other_approaches_final.pdf (2011)
38. Electromagnetic compatibility and Radio spectrum Matters (ERM); Mobile broadband services in the 2300 MHz–2400 MHz frequency band under Licensed Shared Access regime—System Reference Document, ETSI (2012)
39. ETSI EN 301 598: White Space Devices (WSD); Wireless Access Systems operating in the 470 MHz to 790 MHz frequency band; Harmonized EN covering the essential requirements of article 3.2 of the R&TTE Directive, to appear
40. ETSI TS 102 946: Reconfigurable Radio Systems (RRS); System requirements for Operation in UHF TV Band White Spaces, to appear
41. ETSI TS 103 143: Reconfigurable Radio Systems (RRS); System Architecture for WSD GLDBs, to appear
42. ETSI EN 303 144: Reconfigurable Radio Systems (RRS); Cognitive Radio Systems and geo-location databases, to appear
43. ETSI TS 103 145: Reconfigurable Radio Systems (RRS); System Architecture for TVWS use, to appear
44. Review of the Directive 1999/5/EC, Commission proposal for the revision of the R&TTE Directive. http://ec.europa.eu/enterprise/sectors/rtte/documents/legislation/review/
45. ETSI EN 102 969: Reconfigurable Radio Systems (RRS); Radio Reconfiguration related Requirements for Mobile Devices, to appear
46. ETSI EN 103 095: Reconfigurable Radio Systems (RRS); Radio Reconfiguration related Architecture for Mobile Devices, to appear
47. ETSI EN 103 146-1: Reconfigurable Radio Systems (RRS); Mobile Device Information Models and Protocols: MURI, to appear
48. ETSI TR 102 967: Reconfigurable Radio Systems (RRS); Dynamic Declaration of Conformity, to appear
49. ECMA-392, MAC and PHY for Operation in TV White Space, Dec 2009
50. IETF PAWS status page. http://tools.ietf.org/wg/paws/
51. Brykman, L.: Better standards to enhance European competitiveness. In: Proceedings of the 7th Biennial Standardisation and Innovation in Information Technology conference (SIIT 2011), pp. 193–194, Berlin (2011)
52. DTI, The Empirical Economics of Standards. Department of Trade and Industry (DTI), London (2005)
53. Blind, K.: The Economics of Standards. Theory, Evidence, Policy. Edward Elgar, Cheltenham, UK (2004)
54. Delaere, S., Ballon, P.: Multi-level standardisation and business models for cognitive radio: the case of the cognitive pilot channel. In: The 3rd IEEE Symposium on New Frontiers in Dynamic Spectrum Access Networks (DySPAN'08), Chicago, IL, (2008)
55. Lyytinen, K.J., Keil, T., Fomin, V.V.: A framework to build process theories of anticipatory information and communication technology (ICT) standardizing. Int. J. IT Stand. Stand. Res. (JITSR) **6**, 1–38 (2008)
56. Isaak, J.: The role of individuals and social capital in POSIX standardisation. Int. J. IT Stand. Stand. Res. **4**, 1–23 (2006)

57. Bousquet, F., Fomin, V.V., Drillon, D.: Standardisation: a major tool for competitive intelligence. In: 14th EURAS Annual Standardisation Conference Standardisation and Corporate Intelligence, pp. 27–36. Cergy, France (2009)
58. Jakobs, K.: Information Technology Standards, Standards Setting and Standards Research: Mapping the Universe (2003). http://www-i4.informatik.rwth-aachen.de/~jakobs/Papers/stanhope.pdf
59. Schneberger, S.L., McLean, E.R.: The complexity cross—implications for practice. Commun. ACM **46**, 216–225 (2003)
60. European Commission, Final report of the study on the specific policy needs for ICT standardisation, European Commission (2007)
61. Jakobs, K., Blind, K.: ICT standardisation policy in Europe—recent past, presence, and future(?). In: Proceedings of the 7th Biennial Standardisation and Innovation in Information Technology conference (SIIT 2011), pp. 101–114. Berlin (2011)
62. Baldini, G., Holland, O., Stavroulaki, V., Tsagkaris, K., Demestichas, P., Polydoros, A., Karanasios, S., Allen, D.: The evolution of cognitive radio technology in Europe: regulatory and standardisation aspects. Telecommun. Policy **37**(2–3), 96–107 (2013)
63. Mueck, M., Bekkers, R.: Interveiw with Markus Dominik Mueck, INTEL Mobile Communications Group and ETSI TC RRS Chairman, and Rudi Bekkers, TU Eindhoven, on important trends in Cognitive Radio Standardisation, 25 Sept (2013)
64. Kloch, C., Bilstrup, B., Vesterholt, C.K., Pedersen, T.P.: Roadmap for Software Defined Radio—A Technology, Market and Regulation Perspective. Danish Technological Institute (2009)
65. Manninen, A.: Elaboration of NMT and GSM Standards—From Idea to Market. University of Jyväskylä, Jyväskylä (2002)
66. Bekkers, R.: The Development of European Mobile Telecommunications Standards. An assessment of the success of GSM, TETRA, ERMES and UMTS. Technische Universiteit Eindhoven, Eindhoven (2001)
67. West, J., Fomin, V.V.: When government inherently matters: national innovation systems in the mobile telephony industry, 1946–2000. In: Academy of Management Conference, Washington, DC (2001)
68. Stewart, J., Shen, X., Wang, C., Graham, I.: From 3G to 4G: standards and the development of mobile broadband in China. Technol. Anal. Strateg Manag. **23**, 773–788 (2011)
69. Fomin, V.V.: The Process of Standard Making. The Case of Cellular Mobile Telephony. Jyväskylä University, Jyväskylä (2001)
70. Schmidt, S.K., Werle, R.: Coordinating Technology. Studies in the International Standardisation of Tellecommunications. The MIT Press, Cambridge (1998)
71. Fomin, V.V., Medeišis, A., Vitkutė-Adžgauskienė, D.: Pre-Standardisation of Cognitive Radio Systems. Int. J. IT Stand. Stand. Res. (JITSR) **10**, 1–16 (2012)
72. Fomin, V.V., Medeišis, A., Vitkutė-Adžgauskienė, D.: The Role of Policy in the Development of Cognitive Radio Systems: Co-Evolutionary Perspective. In: The 7th International Conference on Standardisation and Innovation in Information Technology (SIIT 2011), pp. 79–90. Berlin (2011)
73. Fomin, V.V., Zoo, H.J., Lee, H.: Cognitive Radio technology development at early stage: content analysis of COST Action documentation. Int. J. IT Stand. Stand. Res, forthcoming
74. Mueck, M., Piipponen, A., Dimitrakopolous, G., Tsagkaris, K., Casadevall, F., Demestichas, P., Pérez-Romero, J., Sallent, O., Baldini, G., Filin, S., Harada, H., Debbah, M., Haustein, T., Gebert, J., Deschamps, B., Bender, P., Street, M., Kalliojärvi, K., Kandeepan, S., Lota, J., Hayar, A.: ETSI reconfigurable radio systems—status and future directions on software defined radio and cognitive radio standards. IEEE Commun. Mag. **48**, pp. 78–86 (2010)
75. Sukarevičienė, G., Fomin, V.V., Modeling Dynamic Spectrum Access Infrastructure Development As a Techno-Economic System. Research in progress. In: EURAS 2013, Brussels (2013)

76. Unlicensed Operation in the TV Broadcast Bands, Additional Spectrum for Devices Below 900 MHz and in the 3 GHz Band, Dockets No. 04-186, 02-380, FCC Second Report and Order and Memorandum Opinion and Order (2008)
77. Expanding the Economic and Innovation Opportunities of Spectrum Through Incentive Auctions, Docket No. 12-268, FCC Notice of Proposed Rulemaking, para. 9 (2012)
78. Damelin, J., Daniel, W.A., Fine, H., Waldo, G.V.: Development of VHF and UHF Propagation Curves for TV and FM Broadcasting. Technical no. R-6602. FCC, Washington (1966)
79. Longley, A.G., Rice, P.L.: Prediction of tropospheric radio transmission loss over irregular terrain. A computer method-1968, Technical Report ERL 79-ITS 67, U.S. Government Printing Office, Washington, July 1968

Chapter 2
Deployment Scenarios for Cognitive Radio

Luca De Nardis and Oliver Holland

Abstract This chapter presents a selection of potential deployment and application scenarios for Cognitive Radio (CR). The chapter goes beyond a simple review of scenarios by considering the viewpoints of several key players in wireless communication research and applications: regulators, standardisation bodies, researchers from the engineering and economic/business communities, industrial partners and companies. In this framework, two key issues related to scenario definition are addressed: (1) An analysis of players that determine the evolution of scenarios, including both technical and economic/business aspects; (2) Study of approaches for classification of CR deployment scenarios, with the aim of identifying a set of elements that allow creating taxonomy capable of fitting existing and new scenarios relevant to CR and SDR. The chapter opens with an overview of CR scenarios proposed by ITU-R in Sect. 2.1. It is followed by Sect. 2.2 that describes the CR use cases envisaged by ETSI. Section 2.3 offers examples of CR scenarios developed in several research projects. The impact of different regulatory and environmental conditions on application scenarios is addressed in Sect. 2.4, which provides a comparison between feasible scenarios in India and Finland. Section 2.5 highlights the issue of growing spectrum demand for mobile services and suggests how the CR may be positioned to help meeting that demand. This is followed up in Sect. 2.6 which provides analysis of upcoming mobile scenarios in Europe focusing on the concept of Licensed Shared Access as defined based on activities carried out in ETSI and CEPT. Next, the chapter moves on to aspects related to planning and classification of scenarios. Section 2.7 proposes a scenario planning methodology aiming to support planning and classification of scenarios for CR and to help identify relevant business models. Section 2.8

L. De Nardis (✉)
Sapienza University of Rome, Rome, Italy
e-mail: lucadn@newyork.ing.uniroma1.it

O. Holland
Centre for Telecommunications Research, King's College London, London, UK
e-mail: oliver.holland@kcl.ac.uk

A. Medeisis and O. Holland (eds.), *Cognitive Radio Policy and Regulation*, Signals and Communication Technology, DOI: 10.1007/978-3-319-04022-6_2,

proposes an approach to the definition of a taxonomy of CR application scenarios, aiming at fitting present and future applications in a coherent framework. Finally, Sect. 2.9 offers an example of very practical application scenario for deployment of White Space Devices in TV Bands as established by a harmonised European standard EN 301 598.

2.1 ITU-R Scenarios

Miia Mustonen and Marja Matinmikko

VTT Technical Research Centre of Finland, Oulu, Finland

This section follows on the previous introduction of ITU-R activities in Sect. 1.1. It may be noted that ITU-R has published two reports on scenarios for Cognitive Radio Systems (CRS). One is about land mobile radio service [1], whereas the other is more specifically about International Mobile Telecommunications (IMT) systems [2]. In the case of land mobile service, a CRS may be implemented in frequency bands allocated exclusively to the mobile service or frequency bands that have multiple radiocommunication service allocations.

2.1.1 Land Mobile Radio Service

The following scenarios for a CRS in the land mobile service have been identified in [1]:

- **Scenario 1: Reconfiguration of connections between terminals and radio systems** In this scenario, multiple radio systems employing different radio access technologies (RATs) are deployed on different frequency bands to provide wireless access. Reconfigurable terminals are able to dynamically adjust their operational parameters and protocols to use different RATs and to obtain knowledge required for these decisions. Alternatively, radio systems may assist terminals (e.g. using control channel). If terminals are able to communicate with other radio systems but not to reconfigure their parameters, additional multi-RAT nodes can be deployed to bridge these terminals to multiple radio systems;
- **Scenario 2: An operator improving the management of its spectrum** Exploiting CRS capabilities, a network operator managing two or more RATs can dynamically and jointly manage the resources of the deployed RATs, in order to adapt the network to the dynamic behaviour of the traffic and to globally maximize the capacity;
- **Scenario 3: An enabler of cooperative spectrum access** Using CRS technology, parts of the spectrum remaining unused due to variations in the

spectrum occupancy can be utilized. The capability to predict these variations or to exchange information on the spectrum usage among systems allows operators to agree on spectrum sharing and avoid mutual interference. Also cooperation between public land mobile networks and private networks can be considered;

- **Scenario 4: An enabler for opportunistic spectrum access** In this scenario, the CRS can access parts of unused spectrum in bands shared with other radio systems and services without causing harmful interference. This can be done by identifying the unused spectrum resulting from variations over time or geographic area e.g. based on a real-time radio environment analysis. The identified spectrum could then be used for different types of communication e.g. device-to-device communications. When the spectrum availability is more static e.g. TV white spaces, the CRS can locate and utilize the unused parts of the spectrum through its capabilities. Alternatively, CRS technology can enable multiple radio systems to share a band with equal access to it.

From the above scenarios, the first two scenarios are intra-system scenarios, allowing an operator to obtain more efficient use of its resources within the networks that it is managing. The latter two are inter-system scenarios, where CRS technology is used for providing more flexible use of spectrum between different operators or systems.

2.1.2 IMT Systems

IMT systems are operating in a globally harmonized and regulated spectrum bands as will be described in more detail in Sect. 2.5. The introduction of CRS capabilities and their applicability to IMT systems should be carefully evaluated. An IMT system employing CRS technology should still meet the minimum requirements for IMT systems [3] and it should not cause harmful interference or quality-of-service (QoS) degradation to the existing IMT systems. In the case of IMT, only intra-operator scenarios were identified in ITU-R [2], see also Fig. 2.1:

- **Scenario 1: Updating a network for optimized radio resource usage** A CRS management entity enables a mobile network operator (MNO) to manage radio resources more efficiently and optimize the network performance in terms of QoS. Load balancing can be done between different applications on a specified RAT or among different RATs when traffic varies from one area to the other depending on the time of day. This can be done autonomously e.g. based on measurements. Additionally, combined knowledge of the radio environment and geo-location information can be used for making optimal handover decisions for intra- or inter-RAT and intra- or inter-frequency band handovers;
- **Scenario 2: Upgrading an existing radio interface or a network with a new radio interface** An MNO of a RAT can decide to deploy a new RAT in the same frequency band to eventually replace the first one and provide all mobile services. During a transition period the legacy mobile devices, that only have

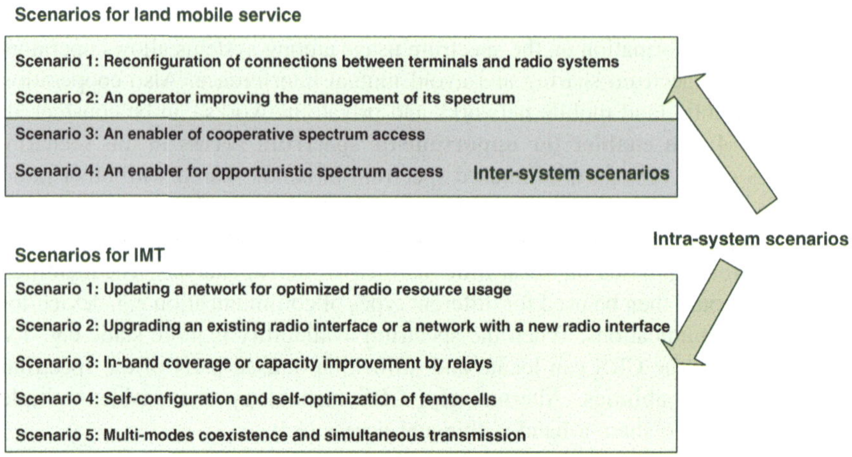

Fig. 2.1 CRS scenarios from ITU-R in [1] and [2]

access to the first technology, coexist with multi-mode mobile devices accessing both technologies. To guarantee QoS during this period, reconfigurable base stations (RBSs) together with appropriate mechanisms are needed to manage the radio resources. Using CRS technology, either a radio equipment can perform the communication by aggregating the bandwidth allocated to each RAT or the network can determine the combination of RAT and available spectrum that provides the best connectivity and adapt the radio accordingly;

- **Scenario 3: In-band coverage or capacity improvement by relays** The deployment of relays is one IMT-Advanced technology feature to alleviate problems such as high channel impairments (e.g. high shadowing) and high traffic demand. The CRS may detect and locate the need for coverage and capacity improvement and identify the available resources. Relay parameters could be optimized using radio environment map (REM) to supply geo-location information on the coverage or capacity indicators;
- **Scenario 4: Self-configuration and self-optimization of femtocells** Initial dimensioning and planning of a femtocell network is challenging since it is difficult to evaluate how many femtocells will be deployed and where back-hauling is provided. Radio access is achieved by the RAT that defines the femtocell (IMT or IMT Advanced). Femtocells, being plug-and-play type devices, are assumed to be autonomous in specific operations as long as they are used in the operators licensed frequency bands. Furthermore, interference mitigation with neighbouring femtocells and the macro cells of the same RAT is needed to prevent QoS degradation. Systems involving femtocells are expected to be highly dynamic, therefore issues like self-configuration, self-optimization are of primary importance;
- **Scenario 5: Multi-modes coexistence and simultaneous transmission** In order to achieve multi-modes coexistence and avoid harmful interference among different RATs, it could be possible to use different frequency bands or reuse the

same frequency band with an optimum transmission power and acceptable separation distance decided by the CRS. In multi-modes simultaneous transmission, both base stations and terminals should be reconfigurable, supporting operation in different modes among multiple radio interfaces and transmitting data by using multiple radio interfaces.

2.1.3 Discussion

The CRS scenarios identified by the ITU-R are valuable input to the research community in the development of the CRS technology. The CRS technology is a toolbox of techniques that could be applied to different scenarios to target different goals. The scenarios identified by the ITU-R indicate that the CRS technology could be applied in both intra-system scenarios as well as between mobile communication system and other radio systems.

2.2 ETSI Use Cases for CR Deployment

Markus Mueck

ETSI Reconfigurable Radio Systems Technical Committee, Sophia Antipolis, France

The ETSI Reconfigurable Radio Systems (RRS) Technical Committee (TC) is the centre of competence for CR technologies in the European Telecommunications Standards Institute (ETSI). This section describes key scenarios for CR deployment as being developed by ETSI RRS.

2.2.1 (TV) White Space Communication

A CR vision has been published in [4] and a number of active Working Items are currently dealing with TVWS related standardization. In the sequel, key Use Cases and Scenarios are presented as they have been derived by ETSI RRS [5].

2.2.1.1 Use Case: Mid-/Long Range Wireless Access Over White Space Frequency Bands

Internet access is provided from a base station to the end users by utilizing White Space frequency bands over ranges similar to today's cellular systems, e.g. in the range of 0–10 km.

Fig. 2.2 Mid-/long range wireless access, no mobility

This Use Case is considered in further detail in the following Scenarios:

- **Mid-/long range, no mobility**
  In this scenario, wireless access is provided from a base station towards fixed devices, e.g. a fixed mounted home base station/access point. The geo-location from both the base station as well as from the fixed device is well-known (Fig. 2.2).
- **Mid-/long range, low mobility**
  In this scenario, wireless access is provided from a base station towards mobile devices where the users have low mobility, e.g. they are staying at their location or walking. In that respect, sensing results for primary users retrieved for the current location are not getting invalid due to the mobility of the user. The geo-location from the base station is well-known. The geo-location from the mobile device must be determined during operation, e.g. via GPS or cellular positioning systems.
- **Mid-/long range, high mobility**
  In this scenario, wireless access is provided from a base station towards mobile devices and the mobile devices may move fast, e.g. because a user is in a car or a train. In that respect, sensing results for primary users retrieved for the current location may get invalid quickly due to the mobility of the user. Thus, this use case sets high constraints for the detection of primary users and it can be questioned if high mobility will be supported in TV White Spaces at all.

2.2.1.2 Use Case: Short Range Wireless Access Over White Space Frequency Bands

Internet access is provided via short range wireless communication (e.g. in the range of 0–50 m) from an access point or base station to the end users by utilizing White Space frequency bands.

This Use Case is considered in further detail in the following Scenarios:

- **Networks without coexistence management**
  In this scenario one or more independent networks access white space frequency bands. The access points has to have knowledge on the incumbent users of the spectrum (e.g. via white space incumbent geo-location database). However, in

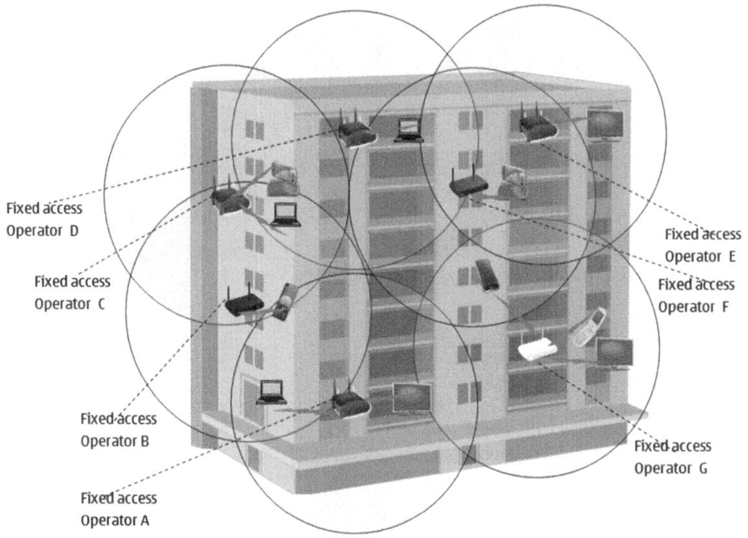

Fig. 2.3 Use case scenario of Internet access by networks with distributed coexistence management in white spaces frequency bands

this first scenario, the different networks are uncoordinated and thus they have no knowledge on other secondary networks and users operating in the white space bands.

- **Networks with distributed coexistence management**
  In this scenario multiple networks access white space frequency bands. The different networks are independent and the backbone connectivity is provided by different network operators. This kind of scenario can happen e.g. in an apartment house, where residents independently acquire their own local area access points operating in white space frequency band. These access points can be operated and maintained e.g. by the residents themselves or the Internet Service Providers (Fig. 2.3).
  In order to work properly, this scenario requires effective coexistence mechanisms for white space frequency access.
- **Networks with centralized coexistence management**
  In this case the White Space networks in the proximity are operated in coordinated manner by a White Space operator. Examples of this kind of usage can be small scale corporate networks, networks for academic institutions etc (Fig. 2.4).
- **Hybrid of networks with distributed and centralized coexistence management**
  This scenario combines the above two scenarios, i.e. in the same neighbourhood there are both networks leveraging centralized coexistence management and networks leveraging distributed coexistence management. Examples of where this kind of scenario can happen are combinations of public and private places,

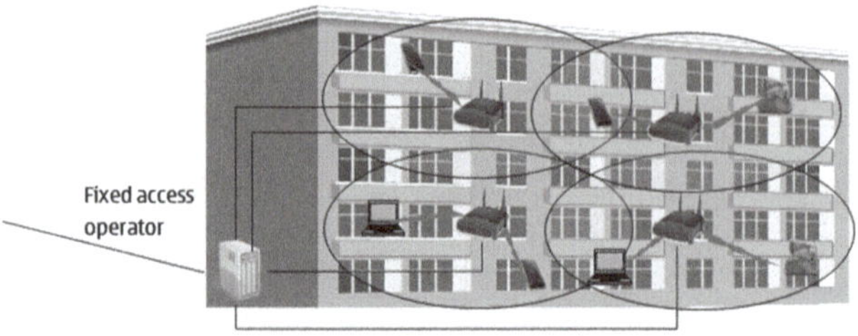

Fig. 2.4 Short range wireless access, coordinated networks

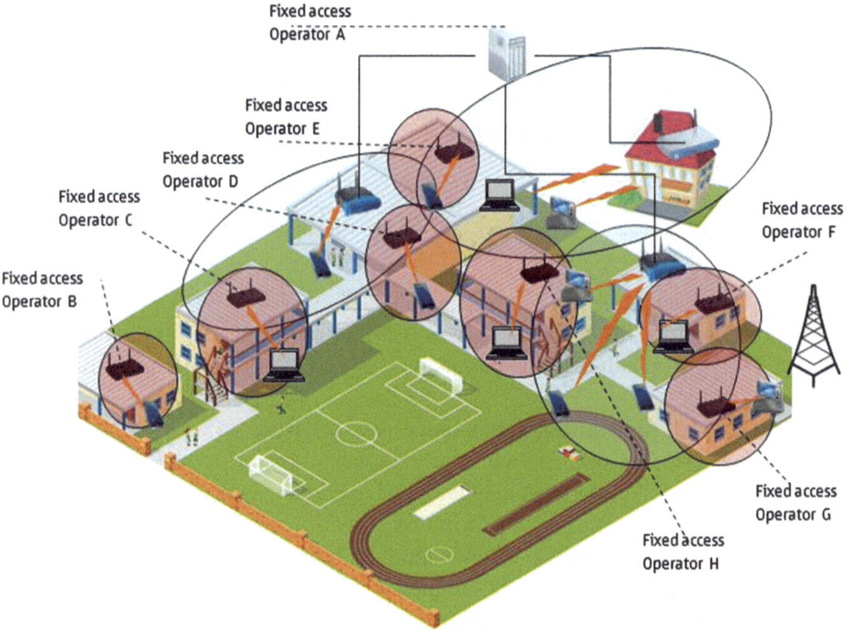

Fig. 2.5 Use case scenario of internet access by hybrid of networks with distributed and centralized coexistence managements on white space frequency bands (*NOTE*: The coverage area of networks with centralized coexistence management is shown in blue, the coverage area of networks with distributed coexistence management is shown in red.)

like campus areas and shopping malls, where e.g. the "official" local area networks, operating under centralized coexistence management, are complemented by independent access points set up independently by some individuals (Fig. 2.5).

The overall coexistence management in this scenario is distributed, due to the existence of the independent networks.

2.2.1.3 Use Case: Ad-hoc Networking Over White Space Frequency Bands

In this use case the devices (user devices and other devices like access points) communicate with each other to share information, to run joint applications or services, or to execute other similar tasks. The communication happens by forming an ad hoc network operating on White Space frequency band. There can be two or more devices in the ad hoc network formed.

- **Device-to-device connectivity**
  In this case two devices (which can be similar or different) connect in peer-to-peer manner to exchange information between each other. The information can be e.g. multimedia content, or control information like measurement results shared between the devices. The devices can be similar from their capabilities (like two mobile devices), or then completely different (like a mobile device and external printer or display).
- **Ad-hoc networking**
  In this case the devices form an ad hoc network to communicate and collaborate with devices in the neighborhood. As an example the devices can be operating a localized social networking service (possibly maintained by a service provider).
- **Infrastructure supported ad-hoc networking**
  In this scenario, the infrastructure supports the creation of ad-hoc networks by providing information and knowledge about policies, available resources, context and profiles. The users receive information about e.g. the proximity of other users and the available resources (including available white space frequencies) in the neighborhood from the infrastructure via the base station and thus an ad-hoc network can be created using white space frequency bands. The link between the base station and the terminals can be realized over licensed, unlicensed or white space frequency bands.

2.2.1.4 Use Case: Combined Ad-hoc Networking and Wireless Access Over White Space Frequency Bands

This use case presents a combination of the ad-hoc networking use case with the short range and/or mid-long range wireless access use cases as described above.

- **Expanding the coverage of the infrastructure**
  In this scenario, a device is out of coverage of the infrastructure. An ad-hoc network is created with other devices where at least one of the other devices has access to the infrastructure. The other devices have relaying/forwarding functions in order to route the traffic from the first device towards the internet and vice versa. In such a scenario, for example, a first user is out of coverage of the infrastructure. However, this user can create an opportunistic ad-hoc network with a second user where the second user relays the traffic from the first user to the network.

- **Resolving cases of congested access to the infrastructure**
  In this scenario, one part of the network is congested. As an example, a first access point is congested and thus a user will have a very bad QoS when being connected with the first access point. While this user is not able to connect directly to another access point, the situation can however be improved when an opportunistic ad-hoc network is created with a second user where the second user relays the traffic to the network.
- **Direct device-to-device links in TVWS managed by access points or femto cells**
  In this scenario, several wireless devices are connected to the internet through an infrastructure device such as an access point or a femtocell. The infrastructure device may have wireless or fixed access to the internet. The wireless devices can, for example, be communicating to an access point over the unlicensed ISM bands, thus representing a standard home or office network setup. They could also be operating in the TVWS bands. Under the management of the access point or femtocell, certain devices may start a direct device-to-device link or point-to-point communication over TVWS frequencies.

2.2.1.5 Use Case: Sporadic Use of TV White Space Frequency Bands

In this use case, TVWS slots are only available sporadically for secondary users, such as for example multi-mode user terminals being able to operate, among other systems, cellular systems in licensed and unlicensed spectrum. The supported unlicensed spectrum is assumed to include TVWS, i.e. the 470–790 MHz range in Europe/Region 1 (Fig. 2.6).

The time-limited switch of a Base Station to operate in unlicensed spectrum (in particular TVWS) is leading to a number of advantages which are further detailed in the following Scenario descriptions:

- **Lighter infrastructure deployment through larger cell sizes**
  Due to the improved propagation characteristics in the TVWS bands compared to typical licensed bands, a large cell size is chosen which will lead to a lighter infrastructure deployment and thus to an overall reduced CAPEX/OPEX.
- **Increased spectral efficiency through reduced propagation loss**
  Due to the improved propagation characteristics in the TVWS bands compared to typical other licensed bands, a possible deployment choice is to keep a cell size as it is the case for the licensed band deployment. The following advantages are observed:

(i) Due to the improved propagation characteristics in the TVWS bands, a higher QoS is achieved within the given cell. However, those propagation characteristics may also increase interference issues which require an adequate handling (e.g. suitable frequency reuse-factor for TVWS, power management, etc.);

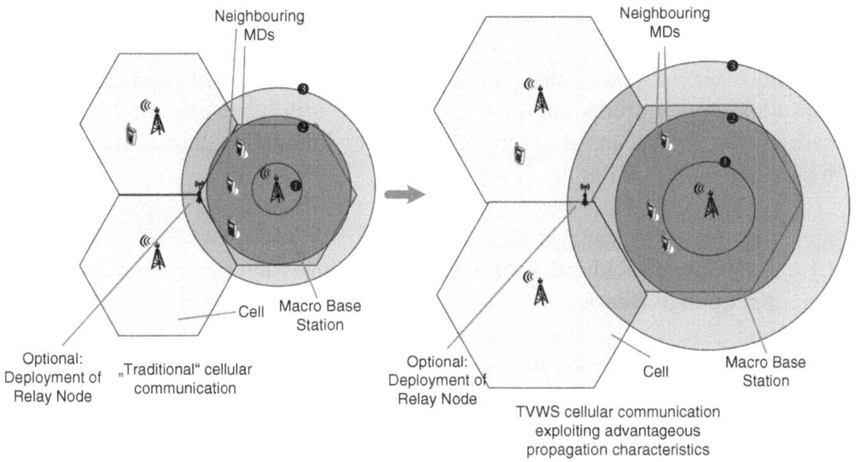

Fig. 2.6 Reduced propagation loss in TVWS and thus improved coverage (the symbols ❶, ❷, ❸ indicate decreasing throughput levels, QoS, etc.)

(ii) Due to the improved propagation characteristics in the TVWS bands, an identical QoS is achieved within the given cell at a lower RBS/MD output power level. The inherent power consumption can be reduced;

(iii) A hybrid solution of item (i) and (ii) is possible, i.e. a moderate reduction of the RBS output power levels combined with a moderate improvement of the QoS.

- **Increased spectral efficiency through extended macro diversity**
  Due to the improved propagation characteristics in the TVWS bands compared to typical other licensed bands, a possible deployment choice is to keep a cell size (or to increase it only slightly) as it is the case for the licensed band deployment. Then, joint operation of neighboring RBS can be exploited in order to achieve a higher Macro-Diversity gain in the UL (multiple RBS are decoding jointly the received signals) or in the DL (multiple RBS are contributing to jointly optimized transmission).
- **TVWS Band-Switch in case that incumbent user re-enters**
  The TVWS usage rules vary over the geographical regions. Typically, when an incumbent system arrives, the corresponding band is no longer available for opportunistic spectrum access by secondary systems. In this scenario, it is suggested that the secondary user switches to another TVWS channel that is still available for opportunistic access—if such a channel is available. Otherwise, the secondary user is assumed to switch to operate in a suitable licensed band.
- **Carrier Aggregation between IMT bands and TV WS band**
  When performing carrier aggregation of TVWS bands with IMT bands, the system may potentially employ either TDD or FDD on TVWS bands. Since operators are adopting either TDD or FDD (depending on their licensed

spectrum) and chip vendors are supporting both TDD and FDD versions of LTE, both techniques are important to consider. The TVWS bands can be used for downlink only, uplink only, or both uplink and downlink transmission. In downlink only or uplink only, the TVWS bands will inherently use FDD. If the carrier(s) in TVWS are used for both uplink and downlink, then FDD and TDD modes are both possible for TVWS.

2.2.1.6 Use Case: Backhaul Link Using TV White Space Frequency Bands

In current mobile networks, there are many access points which have different capabilities, e.g. they can support different coverage. These access points can be connected with wireless backhaul link. In 3GPP LTE for example, the backhaul link between the base station and relay is wireless. The wireless backhaul link operated on the TVWS can obtain the following advantages: improve the access link capacity; supporting the existing commercial terminals; providing a simple wireless environment; providing a better channel quality because of the good propagation performance of the TVWS bands; improve the capacity of the backhaul link; supporting the sensing capability and scalable spectrum bands with lower cost.

- **Relay node backhaul link**
  In this scenario, a city or a rural area is composed by many macro cells. In a macro cell there are some hotspots or blind areas in which relays can provide the coverage. The relay which has a fixed location e.g. on the roof, could be connected to the macro cell BS with wireless backhaul link. According to the time and area in which wireless backhaul link is operated, central control point can select a TVWS spectrum for this backhaul link so that the TV service and other macro cells do not suffer harmful interference (Fig. 2.7).

2.2.1.7 Use Case: MBMS Operating in TV White Space Frequency Bands

Broadcasting services are widely used when transmitting the same content to a group of users at the same time in many telecommunication systems (e.g. UMTS/ LTE). When performing broadcasting services, specific radio resources need to be reserved. Moreover, with the increased categories of the applications, especially some broadband applications, e.g. live TV program broadcasting, the radio resources will become quite insufficient. Therefore, the utilization of TVWS to transmit broadcasting services could meet the requirements for more radio resources. To be simplified, this use case takes the 3GPP LTE Multimedia Broadcast Multicast Service (MBMS) for example. However, other broadcasting services (e.g. UMTS) are not excluded.

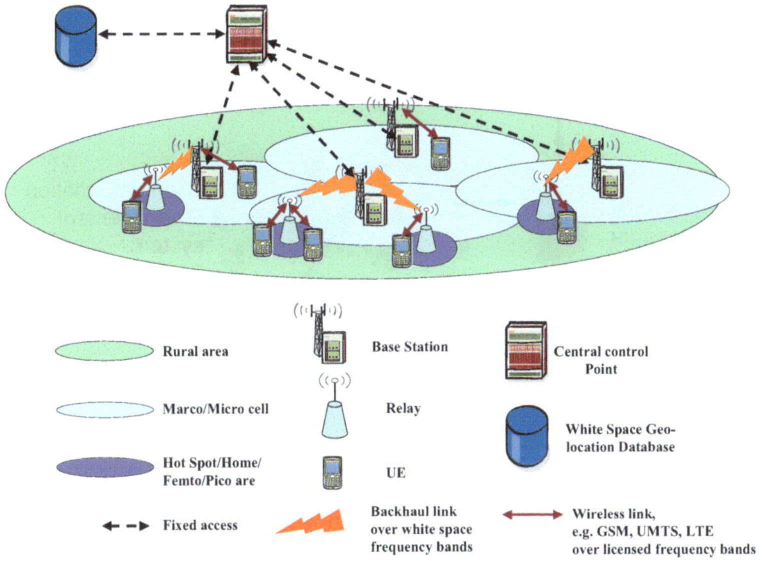

Fig. 2.7 Wireless backhaul link operated in TVWS

- **LTE MBMS in TV white space frequency bands**
  In LTE, the MBSFN (Multimedia Broadcast multicast service Single Frequency Network) mode is realised in the manner that a group cells transmit identical waveforms with the same frequency at the same time in order to promote the frequency efficiency. Those group cells constitute an MBSFN area. A central control entity called MCE (Multi-cell/multicast Coordination Entity) is used for admission control and allocation of the radio resource used by all the eNBs in the same MBSFN area to ensure the MBSFN transmission.

2.2.1.8 Use Case: Machine to Machine Communications Systems Operating in White Space TV Bands

There is a strong need for communication from a wide range of machines including smart meters, smart city infrastructure (traffic lights, temperature sensors, etc.), healthcare monitoring, asset tracking and much more. These typically require long battery life (more than 5 years) and facilities such as broadcast and emergency alert. Machine communications or M2M within white space is perceived as an application of high value with a high probability of emerging.

- **Scenario for M2M systems operating in white space TV bands**
  A network of base stations is deployed, typically providing contiguous coverage across a wide area which may be much greater than the area covered by a single TV transmitter (and so white space availability is likely to vary throughout the

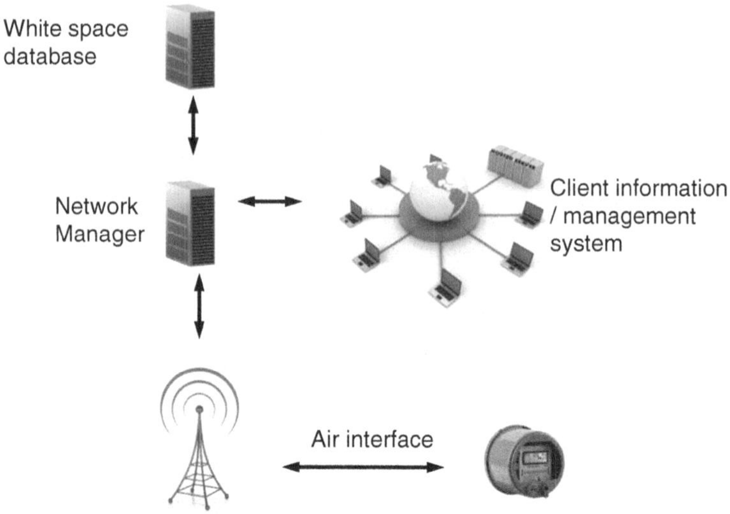

Fig. 2.8 Example schematic of an M2M communications network

coverage area). The base stations link to a network manager which provides a range of functions. One of these is the frequency assignment to the base stations. To do this, the network manager consults a white space database for each of the base stations and receives information on available frequencies which may differ from base station to base station. The network manager uses such information to optimally assign the frequencies to the base stations, for example, minimising the number of adjacent cells that use the same frequency (Fig. 2.8).

2.2.2 *Reconfigurable Radio Platforms as Enabler for CR*

In order to enable the market introduction of reconfigurable radio platforms as enabler for advanced Cognitive Radio technology, a revision of the basic regulation framework for wireless communication (the R&TTE Directive) is currently in preparation in Europe [6]. In particular, this future regulation framework will allow for 3rd party Software Components being acquired on a per-device, per-user basis—so called *RasioApps* [7, 8]. Each Mobile Device (MD) may thus have a unique (radio or other) configuration which is specifically tailored to its owner(s) and user(s). Under the current regulatory procedures, the type approval or self-declaration based approach that is valid for a multitude of MDs (up a millions of devices of a given type) is designed to be applied for the initial version of MD delivered by the Original Equipment Manufacturer (OEM). Extended procedures are required to support subsequent reconfiguration by third parties and ensure continuing compliance of the MD.

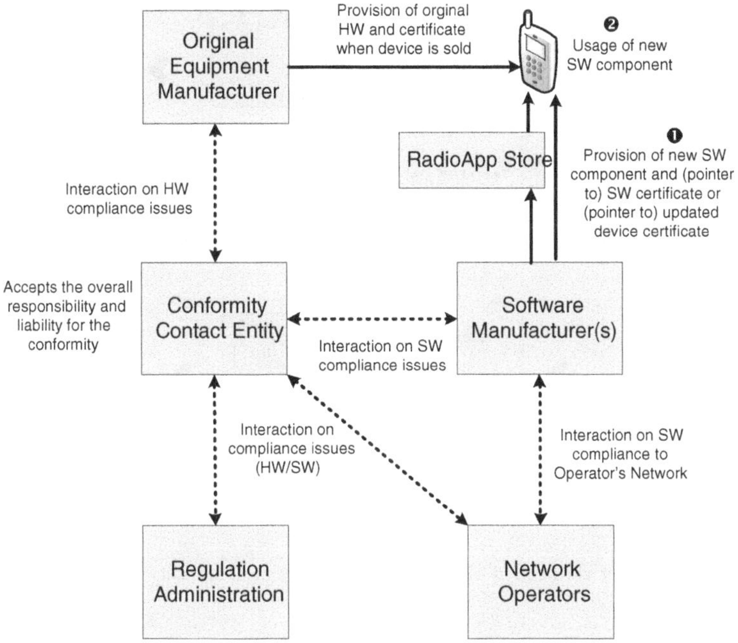

Fig. 2.9 Responsibility for conformity of the MD is transferred to the Conformity Contact Entity after reconfiguration although the Original Equipment Manufacturer may be involved in the overall reconfiguration process [7]

As further detailed in [7], a combination of Horizontal and Vertical market models is expected to develop. It both cases the regulator expects that there be one single legal entity that takes responsibility for the conformity of the MD to all regulations. In one typical example of MD reconfiguration, the resulting responsibility structure is outlined in Fig. 2.9. In this case, a i *Conformity Contact Entity* is introduced that takes the overall responsibility for conformity of the device reconfiguration, for example as a consequence of the installation of *RadioApps*.

In this context, ETSI standard [7] details seven key use cases which illustrate the functioning of the reconfiguration ecosystem. These are summarized in the sequel.

2.2.2.1 OEM Establishing Initial Conformity of MD Platform

The establishment of the initial conformity and certification of the reconfigurable equipment platform by the OEM is illustrated in Fig. 2.10. This use case is very similar to the conformity testing and declaration of conformity certification that is currently used by manufacturers (OEMs) for non-reconfigurable equipment. As shown in Fig. 2.10, the following basic steps can be identified:

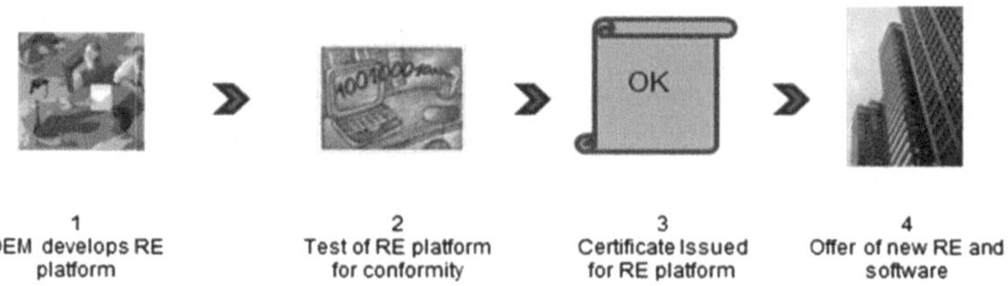

Fig. 2.10 Use case in which OEM develops, certifies conformity and deploys an MD

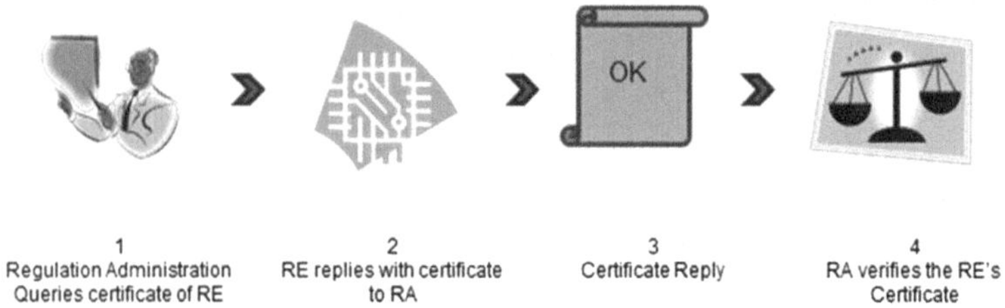

Fig. 2.11 Use case of verification of the MD's certificate of conformity

(1) The OEM designs and develops the MD.
(2) The MD is then tested, for example, by the Notified Body and the Service provider for conformity to applicable legislation and standards and a declaration of conformity is made.
(3) From the Declaration of Conformity (DoC), the OEM creates a Certificate of Conformity (CoC) for its conforming MD.
(4) The certificated MD may then be placed on the market.

2.2.2.2 Certificate Verification of Reconfigurable MD

The Use Case for the verification of a Certificate of Conformity for equipment MD is illustrated in Fig. 2.11. The case may be used, for example, by a national regulator to obtain the identity of the OEM or the conformity contact entity or by a reconfigurable software vendor to determine the current configuration of the MD. The following basic steps can be identified:

(1) To verify the conformity of the equipment the inquiring party, in this illustration shown as the Regulation Administration (RA) (but it may be any other party), queries the MD to request its certificate of conformity.
(2) The MD replies to the RA (or other inquiring party) with its current certificate (or certificates).
(3) The RA (or other inquiring party) receives the certificate (or certificates).

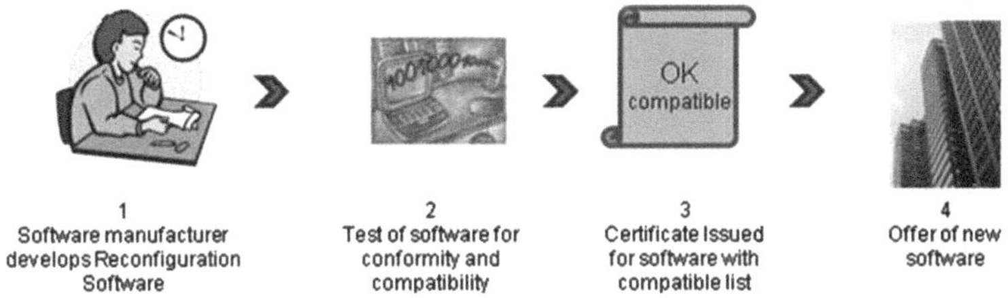

Fig. 2.12 Use case in which software manufacturer develops and certifies reconfiguration software

(4) The RA (or other inquiring party) verifies the authenticity of the MD's certificate to assure the conformity of the MD (including its conformity for operation in the current Member State or current service provider's network).

2.2.2.3 Establishing Conformity of Reconfiguration Software

The Use Case for establishing the conformity of newly developed reconfiguration software or software components for MD is illustrated in Fig. 2.12. The following steps can be identified:

(1) The Software Manufacturer develops reconfiguration software for applicable reconfigurable equipment together with a list of compatible MD.
(2) The reconfiguration software is tested by, for example, the Notified Body and the Service Provider for compatibility and conformity/compliance to regulations and standards for operation on the stated reconfigurable equipment and networks. With successful testing, a declaration of conformity is made.
(3) From the DoC, the software manufacturer creates a Certificate of Conformity and a compatibility list for the conforming software product.
(4) The certificated reconfiguration software may then be placed on the market (putting into service in a Member State is subject to the verification procedure described in use case 2)

This Use Case is very similar to that of the initial OEM conformity (first use case described in 2.2.2.1) with the addition of the compatibility information list associated with the certificate.

2.2.2.4 OEM Upgrade (Individual or en-masse)

In some Use Cases, the reconfiguration software components may be developed by a team that includes the original equipment manufacturer and the holder of the initial certificate of conformity under which the equipment was initially marketed. This Use Case may include, for example "bug fixing", SW Upgrades, new features

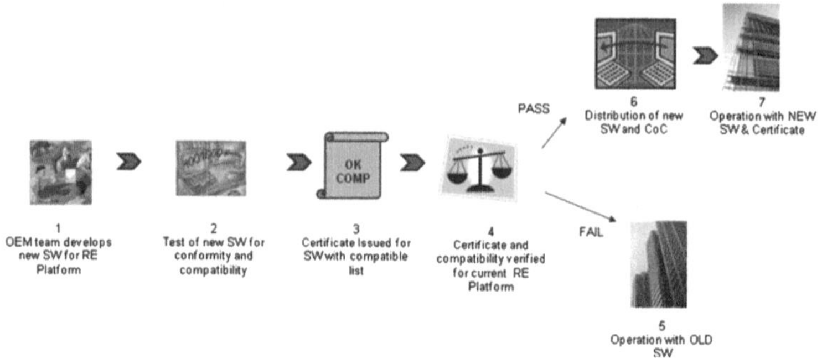

Fig. 2.13 Use case in which OEM team upgrades applicable MD en masse

and enablement of new technology such as new radio access technologies or new bands of operation.

This use case is illustrated in Fig. 2.13. This illustration considers a case where multiple MD are to be reloaded with new software by the OEM. The following steps can be identified:

(1) The OEM team develops new SW for the MD.
(2) The new SW is tested for conformity and for compatibility with the intended MD and software configuration and a declaration of conformity is made.
(3) From the DoC, the OEM team creates a Certificate of Conformity together with the compatibility list.
(4) As part of the distribution of the new configuration to the individual MD, the new components are verified for compatibility with the individual MD's current configuration.
(5) If the new reconfiguration is not compatible with the MD's current configuration, the new configuration is not loaded by the MD and the MD may continue using its previous SW and certificate of conformity.
(6) If the new reconfiguration is compatible with the MD's current configuration, the software components and associated certificate of conformity may be loaded by the MD for its use and the MD operates with the new SW and certificate of conformity (putting into service in a Member State is subject to the verification procedure described in use case (2).

2.2.2.5 Third Party Reconfiguration (Individual or en-masse)

In some Use Cases the reconfiguration components (e.g. software or database updates) may be provided by a team that is not associated with the original equipment manufacturing team that was responsible for the original declaration of conformity. In these Use Cases it is necessary to establish and verify the compatibility of the new configuration with the MD's current configuration.

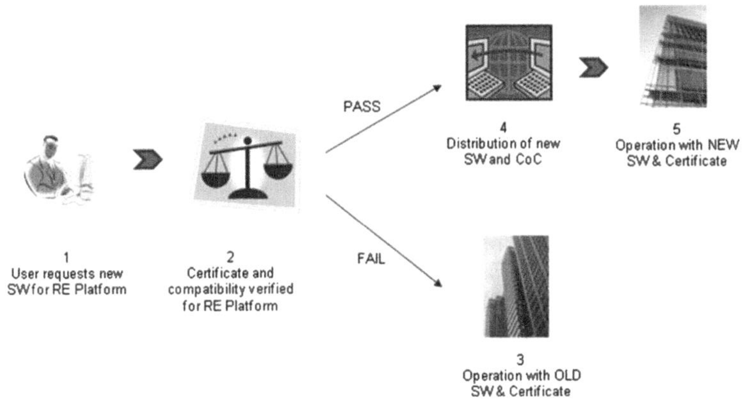

Fig. 2.14 Use case in which user upgrades SW MD platform

This use case may include, for example, new features, SW upgrades and enablement of new technology such as new radio access techniques or new bands of operation. In this case, the reconfiguration components may be directed to all the Reconfigurable Equipment, or to individual MD, but come from a source or sources outside the teams responsible for the MD's previous configuration.

This use case is also applicable when operational/reconfiguration database(s) that may be used by MD are updated. If the database information or protocols that affect the reconfigurable radios are updated, then a new certificate indicating transfer of responsibility to the new team for the database information or protocol updates is obtained by the reconfigured database.

This use case is illustrated in Fig. 2.14. This figure illustrates the individual MD reconfiguration use case. The following steps can be identified:

(1) The user requests new SW for the MD.
(2) The new SW certificate is verified for compatibility with the MD's current configuration.
(3) If the new reconfiguration is not compatible with the MD's current configuration, the new configuration is not loaded by the MD and it may continue using the old SW and certificate.
(4) If the new reconfiguration is compatible with the MD's current configuration, the software components and associated certificate of conformity may be loaded by the MD for its use and the MD operates with the new SW and certificate.

2.2.2.6 Conformity Enforcement of Reconfigurable Equipment

The Use Case for conformity enforcement of reconfigurable equipment, for example to halt improper operations, is illustrated in Fig. 2.15. The following steps can be identified:

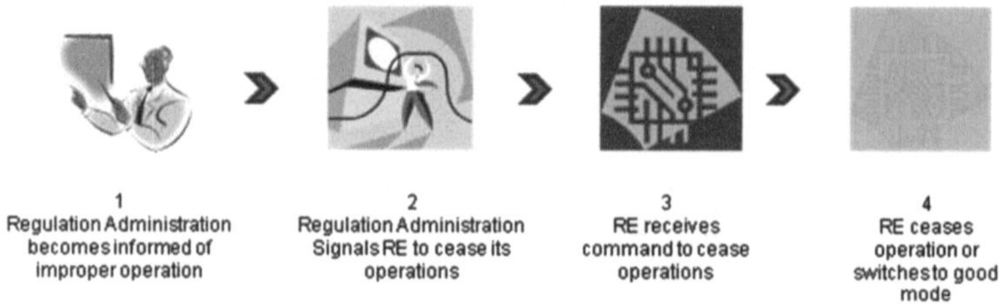

Fig. 2.15 Use Case for enforcement in the event of improper operation of reconfigurable equipment

(1) The Regulation Administration or another appropriate body becomes aware of improper operation of reconfigurable equipment. The RA may be informed of improper operation by, for example, the SP, the OEM, other MD users or other system users.
(2) The RA signals the MD to cease its operations.
(3) The MD receives the instructions to cease the current operating mode.
(4) The MD ceases its improper functions. This may be, for example, through complete switch-off of the MD or by the MD's reversion to a known good operating mode such as a previous software version.

2.2.2.7 MD Discovery of Operational Database for Supporting Dynamic Reconfiguration of Equipment

In some cases the operation of the MD may be dependent on operational information obtained from an operational database (OD).

There are many methods by which the initial database discovery and contact may occur. By way of example here, an initial discovery use case is illustrated in Fig. 2.16. The following steps can be identified:

(1) The OEM or SW manufacturer embeds in the MD or the reconfiguration software a first link network address.
(2) For initial discovery the MD queries the first link network address and includes its certificate in the query.
(3) The first link network address replies with the appropriate operational database network address to the MD.
(4) The MD communicates with the operational network database.

In this procedure, note that in step 2, the MD includes its certificate in the query so that the first link network address can verify the OD requirements of the MD. In step 3, the response from the first link network address typically will include a means by which the MD can verify the authenticity of the response (to prevent the MD) being directed to a "rogue" OD.

Fig. 2.16 Use case for discovery of operational database

2.3 Scenarios for CR Deployment: Visions from Research Projects

Marja Matinmikko[1] and Jarkko Paavola[2]

[1]VTT Technical Research Centre of Finland, Oulu, Finland
[2]Turku University of Applied Sciences, Turku, Finland

This section presents a set of scenarios proposed within research projects that participated to COST-TERRA activities. Although the section will not aim to cover all the projects in the field of CR, it will describe a wide set of scenarios relevant to many projects, highlighting for each of them the opportunities and the corresponding research issues. Projects that are included in this review are:

- EU funded OneFIT (Opportunistic networks and cognitive management systems for efficient application provision in the Future Internet) project in 2010–2012;
- Finnish COGNAC (Cognitive and opportunistic wireless communication networks) project in 2008–2010;
- Finnish CORE and CORE+ (Cognitive radio trial environment) projects in 2011–2014;
- Finnish WISE/WISE2 (White space test environment for broadcast frequencies) projects in 2011–2014.

2.3.1 OneFIT Project

The main objective of the OneFIT project in 2010–2012 was to design, develop and validate the concept of applying opportunistic networks and respective cognitive management systems for efficient application/service/content provisioning in the Future Internet, see [9]. The motivation stemmed from the requirement to satisfy the increased demand for new applications and services through increased efficiency in resource provisioning and utilization. The project developed the concept of Opportunistic Networks that are operator-governed, temporary and coordinated extensions of the infrastructure. They are dynamically created—through operator spectrum, policies, and information—in places and at time instants they are needed to deliver multimedia flows to mobile users, in the most efficient manner. They can include network elements of the infrastructure and devices potentially organized in an infrastructure-less manner.

The project addressed and developed solutions for the following five scenarios [9]:

- **Scenario 1: Expanding infrastructure coverage**
  In this scenario, a device cannot connect to infrastructure due to lack of coverage or mismatch of radio access techniques. Opportunistic network is created to serve a device that has no connection to the infrastructure.
- **Scenario 2: Resolving cases of congested access to the infrastructure**
  A device cannot connect to infrastructure due to congestion. Opportunistic network is created to route traffic to decongested area.
- **Scenario 3: Local infrastructure-less service provision**
  Opportunistic network is created among devices without infrastructure components to serve local needs reducing the traffic load that has to go through the infrastructure.
- **Scenario 4: Cost-efficient bridging to the outside world via opportunistic networks**
  Opportunistic networks are created in specific service areas to optimise the resource use by allowing traffic aggregation.
- **Scenario 5: Opportunistic network using multiple access points for increased backhaul capacity**
  Opportunistic network is created across multiple access points to resolve congestion in the backhaul.

The project developed the concept of opportunistic network to address these scenarios with focus on the CRS based solutions for the efficient management of these networks and information exchange among the elements. To accomplish this, the project developed [10]:

- *Cognitive management systems*. Two types of systems were developed, namely "*Cognitive systems for Managing the Opportunistic Network*" (CMONs) and "*Cognitive management Systems for Coordinating the Infrastructure*" (CSCIs). They provide the means for determining the suitability, creating, modifying and handling forced terminations of an opportunistic network in the five scenarios;
- *Control Channels for the Cooperation of the Cognitive Management Systems* (C4MS). Control channels were developed to enable the delivery of information from the infrastructure towards the Opportunistic Networks and provide the means for the management of Opportunistic Networks.

2.3.2 COGNAC Project

The Finnish COGNAC project in 2008–2010 has developed scenarios for CRS which are summarized in [11]. COGNAC project built upon its predecessor, the Finnish CHESS (Channel State Estimation and Spectrum Management for Cognitive Radios) project in 2006–2007 summarized in [12]. The CHESS project focused on ad hoc type of operations with spectrum sensing as the key CRS technology. The COGNAC project shifted the focus towards applications of CRS technology to mobile communication systems instead of ad hoc systems and also considered other CRS technologies beyond spectrum sensing. The COGNAC project has predicted that evolution of CR systems is expected to start from optimising the use of spectrum and progressing ultimately towards sharing all mobile communication network resources in the global scale. The following four scenarios (steps) were identified in [11]:

- **Scenario 1: Optimizing the use of spectrum**
  First the focus is on optimizing spectrum use in the band including spectrum sharing between different systems in the same band or using the band more efficient by a single system exploiting CRS technology.
- **Scenario 2: Network optimization**
  Autonomous network optimization becomes more and more important as the number of base stations/access points of mobile systems and overlapping networks increases. The resource usage optimization of the resulting heterogeneous networks that partly operate in same frequency bands must be performed dynamically by the systems themselves. Networks will be able to support the adaptation to different environments, using several different radio interfaces. Intelligent selection of the radio interface or other available channel is made cognitively in order to select the best method. As the experience with the operation in heterogeneous environment is gained, it will become feasible to include aspects of e.g. database based machine learning features into the systems, allowing them to make predictions and pre-adaptations their behaviour to match future conditions.
- **Scenario 3: Access to all local resources**
  Flexible access rights are introduced to all local network resources, not only spectrum. Nodes of a network, e.g. wireless devices, not necessarily wish to communicate but they also bring their resources to the network. Different types of resources include radio resources (e.g. power, frequency, time and space), built-in resources (e.g. mass storages, processing units, and batteries), user interface resources (e.g. speakers, microphones, and cameras), social resources (e.g. individuals and groups), and connectivity resources (e.g. air interfaces). Resources are shared widely.
- **Scenario 4: Access to world-wide resources**
  Concepts of cloud computing become relevant for wireless subscribers with needs for computing and other services. Gradually all networks resources can be managed and shared in cognitive network. Evolution of CR systems and cloud computing leads to mobile clouds.

The investigations in the COGNAC project indicated that the CRS technology was applicable to several areas in mobile communication systems to improve the resource usage ranging from optimising the spectrum use to other resources as well.

2.3.3 CORE Project

The Finnish CORE project (2011–2012) in the Tekes Trial program continued the work of the COGNAC project and expanded the approach to cover also business aspects. The focus was returned back to spectrum, it being the key resource to be studied in the context of mobile communication networks. The CORE project developed four scenarios for future cognitive spectrum sharing networks by using value creation and capture as two defining axes. Six key ecosystem players were identified including spectrum regulator, dominating mobile network operator (MNO), challenger MNO, infrastructure vendor, equipment vendor, and content provider. The following scenarios were developed by taking into account the dimensions of the degree of licensing and source of customer attraction and lock-in (services versus devices) [13, 14]:

- **Scenario 1: Cruella de Vil**
  The regulator plays the key role and allocates available spectrum mostly to dominating MNOs, and thus does not enforce spectrum sharing. MNOs can act as a smart bitpipe that exploits its vast spectrum resources to offer mobile services and connectivity to its wide customer base.
- **Scenario 2: Snow White**
  The regulator allocates spectrum to both challenger and dominating MNOs and enforces sharing indirectly. This leaves the operators to compete with more diverse strategies as they have to become an efficient bit-pipe in order to exploit their limited spectrum resources more efficiently in the growing competition.
- **Scenario 3: Cheshire Cat**
  The regulator promotes sharing and opens up more unlicensed spectrum opportunities. Non-MNO service providers exploit these newly opened unlicensed spectrum opportunities entering the field currently dominated by MNOs, which reshapes the market dynamics, especially of dominating MNOs' business.
- **Scenario 4: Gyro Gearloose**
  The regulator promotes sharing. The MNOs' business is affected by new innovative and focused devices that directly connect to specified services. Also, there are cases where device vendors and platform providers are merging to create "Gyro Cats," such as Nokia-Microsoft or Google-Motorola.

The CORE project focused specifically on how mobile communication networks could gain access to new spectrum resources via spectrum sharing using the CRS technology. The studies concluded that a promising way for MNOs was to share spectrum from other type of incumbent spectrum users with rules and conditions that resemble current exclusive licensing to guarantee predictable quality of service (QoS) in the shared bands.

2.3.4 CORE+ Project

CORE+ project (2013–2014) [15] in the Finnish Trial program continued the work of the CORE project with a focus on a special scenario for mobile communication systems with spectrum sharing using the new Licensed Shared Access (LSA) concept that will be introduced in more detail in Sect. 2.6. The LSA concept allows a limited number of authorised users to operate in the same frequency band with sharing rules that provide certain QoS for all users. The LSA approach when applied to mobile communication systems sharing spectrum from other type of incumbent spectrum users refers to the Authorised Shared Access (ASA) concept [16]. ASA allows the MNO to obtain a license from the regulator to operate in the band with pre-determined rules and conditions agreed with the regulator and incumbent spectrum that resembles exclusive licensing and offer guaranteed QoS. To become successful, the ASA concept has to benefit all involved stakeholders, namely spectrum regulator, incumbent spectrum user, and the ASA licensee which can be a dominating MNO or a challenger MNO. Within the ASA scenario, the following business benefits have been initially identified for the key stakeholders:

- **Spectrum regulator** is responsible for issuing ASA licences and negotiating sharing conditions/usage requirements. The ASA concept could allow the regulator to maximise the value of the spectrum assets through sharing while still retaining control over the spectrum. It allows the balancing of the spectrum demands between different wireless systems by improving the spectrum availability via sharing.
- **Incumbent spectrum user** (non-MNO) can take advantage of the ASA concept by offering parts of its unused spectrum to be shared with the MNO and possibly get compensation for it.
- **Dominant MNO** could gain access to new low cost ASA bands to satisfy the growing traffic demand without strict coverage obligations. It could obtain significant savings in the network expenditure by acquiring new ASA spectrum and deploying base stations there using existing sites maintaining the dominant position.
- **Challenger MNO** could gain access to new low cost ASA spectrum bands which should be particularly appealing to Challenger MNOs, as they often suffer from the lack of sufficient amount of spectrum. They could gain access to new localised business opportunities challenging the dominating MNOs.

2.3.5 WISE and WISE2 Projects

WISE projects (2011–2014) focus on incumbent protection methods for shared spectrum. In the WISE project TV white space test (TVWS) environment was deployed in Turku, Finland. Two incumbent scenarios are considered for TVWS:

- **TV signals**: Spectrum occupancy for TV broadcasting is stable and the information on transmission sites can be obtained from the national regulator.
- **Wireless microphones**, or program making and special event (PMSE) equipment in general: this use case is hard to predict, and therefore the incumbent user is more complex to protect. For example in Finland, a radio license is required for wireless microphones. However, authorities have no possibility to guarantee that all users follow the registration process.

Detailed information can be found from the ECC Report 159 [17] with complementary studies in ECC reports 185 [18] and 186 [19].

In WISE2 project the utilization of test environment was extended for piloting following use case scenarios with TVWS technology:

- Rural broadband: TVWS can be used as a last mile connection to provide broadband access for rural areas, since TV frequencies have good propagation properties.
- Video surveillance: Wireless video surveillance systems utilize WLAN or cellular networks. These systems have issues with capacity and cost. TVWS could provide a cost efficient way for extra capacity.
- Smart grid: TVWS is tested for communication between control stations. Data rates are very low but delay requirements are strict. TVWS is considered for replacing satellite link. Typical distances considered for the transmission link are in the order of several tens of kilometres.
- ITS in public transport: Public transport has many wireless applications. For TVWS wireless ticket purchasing and connection between bus and information screen at the bus station are piloted before devices supporting mobility and handover are available.

Additionally, in WISE2 project, incumbent protection for ASA/LSA spectrum sharing model is investigated. Two incumbent scenarios are considered for ASA/LSA in 2.3–2.4 GHz frequency band [19]:

- **Wireless cameras**, or PMSE equipment in general: Several different use cases such as cordless, mobile and portable video links are included.
- **Military use**: From military side unmanned aircraft systems (UAS) use the frequency band. UAS use telecommand in the uplink and video links in the downlink direction.

It is assumed that the database approach is used to control spectrum sharing. Database must contain information on available frequencies, and the incumbent use on those frequencies. A device queries database with location and optionally operating parameters. The database sends a response containing frequency and transmission power information.

The approach in the project for the incumbent protection is two-fold. The PMSE manager has been developed to collect and control information on incumbent use. The manager communicates information to the database. A simple and easy-to-use user interface, which allows also automatic registration if supported by devices, will make it feasible to manage incumbent devices even when their use is hard to predict spatially or temporally.

On the other hand, RF measurements are performed in the project to verify defined protection ratios between incumbent use and new devices in the frequency band. Measurements are also carried out to validate database operations.

2.4 DSA Application Scenarios: Cases of Finland and India

Varadharajan Sridhar[1] and Arturo Basaure[2]

[1]Sasken Communication Technologies, Bangalore, India
[2]Aalto University, Espoo, Finland

2.4.1 Introduction

The paradigm of regulation has shifted from pure coordination and planning to the creation of a competitive and sustainable environment for various services, including telecommunications. Technologies have also evolved to accommodate flexibility in spectrum management. As a result countries are migrating in different degrees, and at different speeds to flexible spectrum management regimes. This is also augmented by the exponential increase in mobile data traffic that is expected to grow at a compound annual growth rate of 92 % from 2010 to 2015, reaching 6.3 Exabyte per month by 2015 [20]. In addition, this study shows that mobile internet will grow from 14 million at the end of 2010 to 788 million by the end of 2015. This phenomenon is happening in both developed countries such as Finland as well as in emerging countries such as India. For example, wireless Internet subscribers in India have reached about 143 million and continue to grow at an exponential rate [21].

This potential increase for mobile Internet services is pushing governments to better manage their spectrum resources. There is a trade-off before the policy maker, between the number of operators to be allocated spectrum and spectrum block allocated to each operator. In some emerging countries, the decision has been made in favour of competition and hence the associated maximal usage of allotted spectrum. On the other hand, in many advanced countries, such as those belonging to the European Union, the policies favour a limited number of operators with more spectrum blocks for each operator. In addition, these advanced countries had favoured technology harmonisation, while emerging countries favoured technology competition. Given this disparity in spectrum policies and market structure in different markets, it is interesting to analyse the future evolution path for spectrum management in these two extreme scenarios. Following Table 2.1 contrasts the spectrum allocation and market structure in the representative countries of Finland and India.

Table 2.1 Comparison of policies across India and Finland (adapted from [26])

Factor	In India	In Finland
Average spectrum allocation per operator per License Service Area	2 × 6 MHz in 900; 2 × 4.5 in1800 for 2G; 2 × 3 MHz in 800 for 2G/3G; 2 × 5 MHz in 2100 for 3G; 20 MHz unpaired in 2300 MHz for BWA	2 × 11.3 MHz in 900; 2 × 24.8 MHz in 1800; 2 × 15 MHz in 2100; 4.8 MHz unpaired in 2100; 2 × 20 MHz in 2600 MHz
Market Fragmentation (measured as HHI ranging from 0 to 1; 1 being monopoly)	0.15	0.33
Harmonization Policy of the Government	GSM in 900/1800; CDMA in 800; WCDMA in 2100; BWA in 2300 MHz; No unified view	GSM, WCDMA and LTE adopted in different spectrum blocks as per European harmonisation measures [69]

As is clear from the above Table, while Finland has taken the route of less number of operators, less market and spectrum fragmentation, increased spectrum assignment for each operator and a globally harmonized policy on spectrum allocation, India has taken a very different route, highlighted by intense competition, heavily fragmented market and spectrum allocation and no unified long term view of spectrum allocation policy.

2.4.2 Policy Options for Reacting to the Exponential Increase in Demand

Even though spectrum policy may differ depending on country and local circumstances, the national regulatory authorities (NRAs) have responded in general in the following ways:

(i) Allocate additional spectrum to mobile services. For example, release of the digital dividend spectrum in the 700 MHz;
(ii) Mandate refarming of the traditional 2G spectrum band, especially in 900 MHz and 1800 MHz for a more spectrally efficient (e.g. 3G and 4G) technologies so that the given spectrum is efficiently used;
(iii) The creation of secondary market that allows spectrum trading;
(iv) Consider allowing non-exclusive sharing of spectrum opportunistically using Dynamic Spectrum Access (DSA) technologies such as CR.

Out of these four options, the last two depend on the existing spectrum allocation policies and the market structure that vary across countries. Trading can be defined as a process in which spectrum changes its ownership, allowing spectrum transactions either geographically, by frequency or time. Many authors point out that such spectrum markets are viable if sufficient numbers of market participants exist and the amount of tradable spectrum is balanced to the demand [22, 23]. Trading can be mutually agreed between parties or performed in an open spectrum market. As opposed to trading, in an opportunistic spectrum access, the secondary user does not require a formal permission from the primary spectrum holder [24], but it accesses the available spectrum opportunistically (non-coordinated), making sure that it does not interfere with the primary user, e.g. through a spectrum database or using CR sensing technologies. However, the primary spectrum holder may still be willing to get an economical compensation from allowing secondary access to its spectrum [25]. The non-exclusive spectrum sharing may be operator induced or end-user induced. We discuss in the following section these two possibilities and their possible presence in different markets.

2.4.3 The Causal Model of Spectrum Policy

The model in Fig. 2.17 [26] summarizes the spectrum policy differences between a country practicing spectrum harmonization such as Finland and a country practicing market driven spectrum policy such as India. It uses "shifting the burden" archetype [27] which describes how the chosen path for solving a problem makes difficult to change to another alternative path and thus creating a path dependency.

As shown in Fig. 2.17, the increased disparity in capacity and coverage can be handled by enforcing a stronger harmonization policy leading to a more equal and efficient initial allocation and assignment of spectrum. This subsequently leads to the decrease of the disparity and a balancing loop 'B-Harmonization' which on a rough level corresponds e.g. to the Finnish spectrum policy as well as many other European markets.

On the other hand efficient centralized allocation means that operators do not need to conduct much market based sharing (or trading) and that end users do not have many options in terms of the different radio access possibilities which subsequently means that secondary market sharing or trading mechanisms between operators and CR type of capabilities in devices are not required. Operators may cooperate by infrastructure sharing or even by spectrum sharing, but they do not need a market based co-operative trading or opportunistic end user access. The inability of the market to redistribute the spectrum resources in turn leads to a reinforcing loop ('R-Efficiency through Centralized Planning') that possibly locks the market on a path of enforcing a harmonization policy. In this situation, it is expected that opportunistic access is made possible through operator centric devices.

On the other hand, the Indian market has followed the opposite dynamics. Since the initial spectrum allocation is inefficient, the increasing mobile service demand has been handled by the market in the form of co-operative trading between operators (i.e. national roaming) and opportunistic end user access (i.e. many data plans per user through multi-SIM devices). This in turn has led to what can be seen as a kind of a secondary market activity and subsequently to the decrease of the

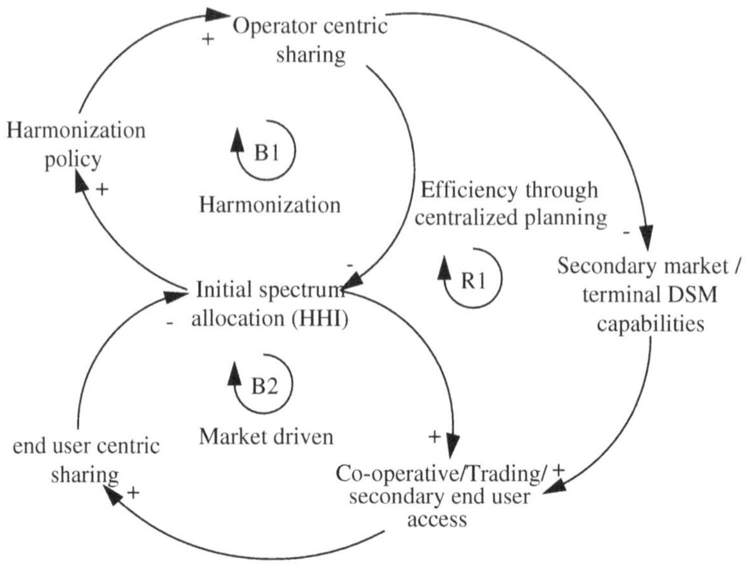

Fig. 2.17 Feedback model indicating the differences between spectrum policies (adapted from [26])

disparity and a balancing loop 'B-Markets'. In this case, when the market solves the disparities in coverage and capacity it leaves little space for a harmonization policy which in turn leads to a larger need of secondary market sharing and trading mechanisms between operators and CR capabilities of end-user terminals in order to efficiently use and redistribute the radio resources. This in turn leads to a reinforcing loop 'R-Efficiency through Centralized Planning' that works in the opposite direction when compared to harmonized markets such as Finland. The exponential increase in multi-SIM handsets is an example of this type where the end-user devices have CR like capabilities to access networks and hence spectrum of choice.

The fact that these two markets can be seen as being locked on two opposite paths have a significant impact on their future evolution, especially regarding the introduction of DSA technologies. While harmonized markets may favour mobile operator centric spectrum sharing, highly competitive markets may fit better with end-user centric spectrum sharing.

2.4.4 Conclusions and Future Evolution Scenarios

Based on the previous model, it appears that the Finnish market is set on a path of pursuing a harmonization policy where demand for wireless broadband in the future will be met by a centrally planned efficient initial allocation, the increasing adoption of spectrally efficient technologies and by releasing exclusive digital dividend spectrum. India on the other hand seems to be on a path of pursuing a more market based policy that could lead to higher activity in the secondary markets (as is already the case with national roaming and multi-SIM phones). The Indian market structure thus can be seen as having a better fit with the CR technologies which could lead to a more rapid diffusion of CR systems in India (and in other developing markets following the same market structure). Therefore when it comes to the diffusion of CR technologies, the emerging markets are in a good position to do technology leap-frogging much similar to the way they did in skipping analogue wireless technologies and adopting the second generation digital cellular systems.

2.5 Mobile Communications and Need for CR

Uwe Löwenstein[1] and Marja Matinmikko[2]

[1]Telefónica Germany, Munich, Germany
[2]VTT Technical Research Centre of Finland, Oulu, Finland

This section focuses on the spectrum matters of cellular mobile communication systems and the emerging role of CR in that regard. Starting from the mobile traffic growth predictions, the current status of mobile spectrum is presented. Future directions are depicted including the potential for new spectrum bands for the mobile service and the introduction of CR technology to facilitate spectrum sharing between a mobile communication system and incumbent spectrum users.

2.5.1 Mobile Traffic Growth

Back in 2006, the International Telecommunication Union Radiocommunication (ITU-R) sector published a comprehensive forecast of the global mobile telecommunication market in Report ITU-R M.2072 [28]. The report identified a number of new wireless applications and services and predicted a strong growth in the mobile traffic in the time span 2010–2020.

In preparation for the World Radiocommunication Conference in 2007 (WRC-07), the ITU-R developed a methodology [29] to estimate the spectrum requirements of IMT systems based on the global traffic forecast in 0 and technology developments summarized in [30], see also [31]. The results of the spectrum requirement calculation studies for IMT systems in the years 2010, 2015 and 2020 are presented in Report ITU-R M.2078 [32] including the spectrum requirements for two distinct radio access technology groups (RATGs):

(1) pre-IMT, IMT-2000 and its enhancements (RATG 1) and
(2) IMT-Advanced (RATG 2).

Table 2.2 IMT spectrum demand in MHz prior to WRC-07 from ITU-R M.2078 [31]

Market setting	RATG 1			RATG 2			Total		
	2010	2015	2020	2010	2015	2020	2010	2015	2020
Lower	760	800	800	0	500	480	760	1300	1280
Higher	840	880	880	0	420	840	840	1300	1720

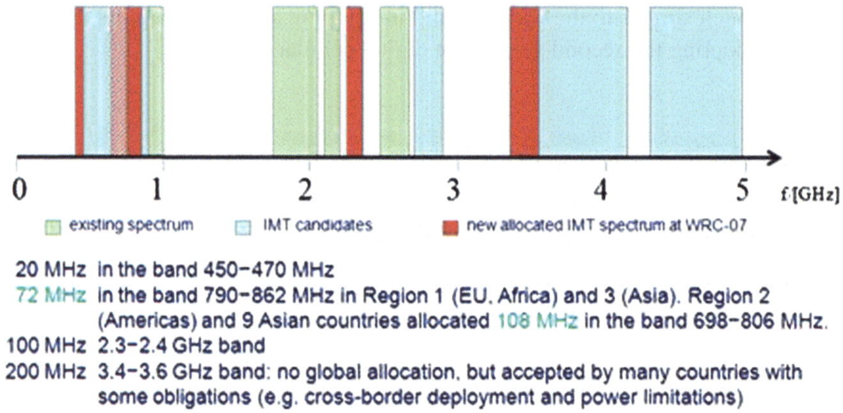

Fig. 2.18 Spectrum allocations at WRC-07

These two RATGs collectively cover the family of IMT systems. The study concluded that between 760–1720 MHz of spectrum needed for IMT systems depending on market setting and year as presented in Table 2.2. The study took into account the total mobile telecommunication traffic from [28] and considered the relevant radio systems that are capable of carrying this traffic.

The total mobile traffic was divided among four different RATGs where the first two covered IMT systems, while RATG 3 was defined as "existing radio LANs and their enhancements" and RATG 4 as "Digital mobile broadcasting systems and their enhancements". Thus, the study took into account all relevant radio systems and concluded the spectrum requirements for the IMT systems, i.e. RATGs 1 and 2.

Based on these studies, the ITU-R decided at WRC-07 to globally allocate additional spectrum between 392 MHz (EU, Africa, Asia) and 428 MHz (Americas, CHN, KOR, IND, JPN, NZL) to the mobile service with identification to IMT. The detailed allocation is shown in Fig. 2.18. It should be noted that this allocated spectrum did not become available immediately, but requires further collaborative work on international, regional and national level.

To underline the apparent fact of the increasing mobile traffic demand throughout the previous years, ITU recently published its annual MIS Report (Measuring the Information Society) 2013, showing two benchmarking tools to measure the information society: the ICT Development Index (IDI) and the ICT Price Basket (IPB). In the Top-Ten ranking of the IDI, 8 European countries

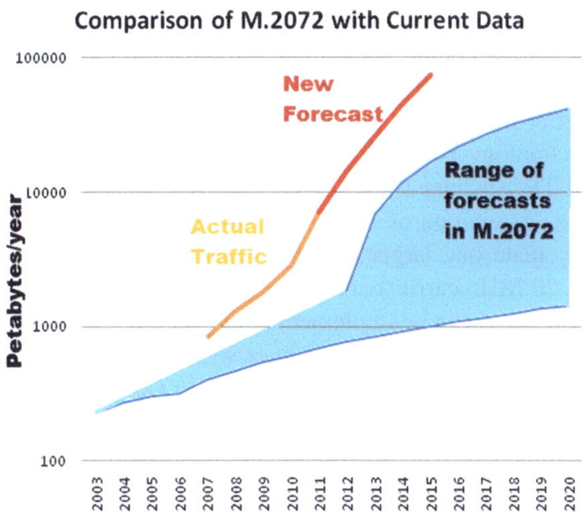

Fig. 2.19 Comparison of traffic from Report ITU-R M.2072 [28] prior to WRC-07 and Report ITU-R M.2243 [35] prior to WRC-12. Source: [35]

(including Sweden, Finland, United Kingdom) can be found beside Korea (#1) and Hong Kong (#10) [33].

There are several studies [32, 34] clearly showing that the trend of increasing traffic demand is also continuing in the future. In particular, the ITU-R has prepared a new report [35] on the market and traffic forecasts for IMT, collecting new traffic forecasts from a number of sources. This study calculated an increase in the mobile traffic resulting mainly from data, which exceeds the values of ITU-R M.2072 (WRC-07) by a factor of >5 already today as shown in Fig. 2.19.

To address the traffic growth which today is foreseen to be even stronger than previously anticipated, the ITU-R prepared an input for the previous WRC-12 to include a new agenda item for the next WRC-15 to review the spectrum availability for IMT, based on new studies presented in Report ITU-R M.2243 [35] on the spectrum requirements and additional allocations to the mobile service for broadband systems including IMT.

With the new traffic forecast summarized in Fig. 2.19 for the years 2015–2020 and further forecasts for the year 2025 from Report ITU-R M.2243 [35], it is obvious that the increasing traffic has to be accommodated somehow within the next decade. This can be done in different ways:

1. by allocating additional spectrum for the mobile service;
2. by re-using existing spectrum allocations more efficiently (e.g. refarming, carrier aggregation and other new technologies);
3. by using new methodologies to access spectrum (e.g. shared spectrum access).

To address the *first item*, the ITU-R will consider the allocation of additional spectrum for the mobile service at WRC-15. In preparation for WRC-15, the ITU-R has estimated the spectrum requirements of IMT systems as well as other mobile systems (such as RLAN).

For the *second item*, spectrum refarming is already implemented in various European countries by the majority of mobile operators (see also Table 2.4). This enables the operators to re-use existing spectrum with new, more efficient and better performing technologies. A very well suited and already widely deployed technology is LTE (Long Term Evolution) and in the future its enhancements "LTE-Advanced" (LTE-A). LTE-A—beside other new capabilities—also allows the aggregation of single carriers in different bands, called "carrier aggregation", to create one larger carrier (e.g.: 10 MHz @ 800 MHz + 10 MHz @ 900 MHz = 20 MHz carrier) and consequently enable higher data rates in given cell. LTE-Advanced can be implemented in existing LTE-networks with relatively low effort (e.g. SW-modification together with possible very small hardware extensions, but no site-reconstruction). According to GSA—The Global mobile Suppliers Association as of October 2013, 456 operators are investing in LTE and there are 213 commercially launched LTE networks deployed in 81 countries [36].

For the *third item*, new developments in the spectrum access methodology will further improve the spectrum use. There are different concepts under development such as Licensed Shared Access (LSA) and its application to mobile, i.e. Authorised Shared Access (ASA). The Radio Spectrum Policy Group (RSPG) prepared a response to the European Commission's request for an opinion on spectrum regulatory and economic aspects of LSA [37], concluding that "Licensed Shared Access (LSA) could provide new sharing opportunities on a European scale under a licensing regime, while safeguarding national current spectrum usages which cannot be refarmed. It is not intended that LSA will be an initial or temporary phase prior to the refarming of any band." More detailed information regarding ASA/LSA can be found in the following Sect. 2.6.

A mixture of all three approaches is envisaged to be required to successfully meet the growing traffic demand. Despite the above mentioned ways to accommodate the traffic, there is an increase in the needed/desired data-rates and higher data-rates need also wider carrier bandwidth. For that reason, the above mentioned methodologies have to be applied in higher frequency bands.

2.5.2 Current Status in Cellular Mobile Spectrum

Currently various frequency bands are allocated exclusively to mobile IMT (Table 2.3), with different detailed implementations across the three ITU-R regions.

Based on the spectrum allocations, different roll-out strategies have been implemented across Europe based on technology availability. The former existing strict separation of technologies and frequency bands has been changed and the same technology is available across different bands (mainly due to spectrum refarming), leading to the following technology usage across Europe as presented in Table 2.4.

Table 2.3 Globally allocated frequency bands for IMT

Band (MHz)	ITU radio regulations footnotes identifying the band for IMT
450–470	5.286AA
698–960	5.313A, 5.317A
1710–2025	5.384A, 5.388
2110–2200	5.388
2300–2400	5.384A
2500–2690	5.384A
3400–3600	5.430A, 5.432A, 5.432B, 5.433A

2.5.3 *Future*

As explained above, the currently available spectrum allocation for the mobile service is quite comprehensive and today sufficient to carry the current traffic load. However, recent ITU-R investigations summarized in this section have identified a strong increase in the mobile traffic in the future leading to increasing spectrum demand. This possible need for new spectrum for the mobile service is the topic of the next WRC-15.

In parallel to the process of possible new spectrum allocations, there is the process of actually making these bands available for mobile network operators to deploy their networks. From the IMT spectrum identifications summarised in Table 2.3, not all bands are available today in all European countries. Typically, a band allocated to the mobile service already encompasses some incumbent spectrum use prior to the allocation and clearing of the band for mobile is a time-consuming and costly process. For example, the 2.3–2.4 GHz band that was globally allocated to mobile service at WRC-07 encompasses a range of other incumbent users in Europe, such as military, programme making and special events (PMSE), etc.

To speed up the process of making these bands available while preserving the rights of the incumbents, a new sharing based approach has emerged taking advantage of the CR technology. Spectrum sharing under the recently introduced LSA framework can offer the benefits of traditional exclusive licensing while guaranteeing incumbents' rights. It facilitates spectrum sharing between an incumbent spectrum user and a new licensee with rules and conditions that can offer predictable QoS to both. This LSA approach can be more appealing for the mobile communication systems to share spectrum bands from other type of incumbent spectrum users in contrast to unlicensed sharing such as unlicensed access to the TV white spaces.

Table 2.4 Technologies deployed across the different frequency bands

Band (MHz)	Former technology	Current technologies	Remarks
450			Rarely used for cellular
700	–	–	Allocated in principle, but bandplan t.b.d.
800	–	LTE	
900	GSM	GSM & UMTS	
1800	GSM	GSM & LTE	
2100	UMTS	UMTS	
2600	–	LTE	
3500	WiMax	WiMax & LTE	Only partly used

2.6 Licensed Shared Access as an Example of Upcoming Implementation in Europe

Markus Mueck[1], Srikathyayani Srikanteswara[2], Graham Macdonald[3] and Mohamed El-Refaey[4]

[1]INTEL Mobile Communications Group, Neubiberg, Germany
[2]INTEL Labs, Hillsboro, Oregon, USA
[3]INTEL Global Policy Group, Swindon, UK
[4]INTEL Labs, Cairo, Egypt

2.6.1 Background

Wireless and mobile traffic growth is predicted to explode and grow exponentially by a paramount rate through 2025 [38]. On top of existing wireless applications, new technologies will require additional spectrum resources as they expand in new areas, in particular Machine-To-Machine communication, Wearable Devices communication, video streaming etc. will take their share while their presence (and thus the capacity share) in the market is negligible in 2013. Figure 2.20 illustrates this trend of mobile data traffic expected by 2017 as forecasted by Cisco Visual Networking Indexing (VNI).

It is obvious that existing spectrum for wireless and mobile communication will not be sufficient to satisfy the upcoming requirements. And worse, the traditional approach of re-purposing spectrum of other applications (such as TV Broadcast spectrum) to become exclusively licensed spectrum is reaching its limits. In the US, the National Broadband Plan [39] outlines requirements for 500 MHz of new mobile and wireless spectrum below 6 GHz by 2020. In Europe, the European Parliament and Council approved the first Radio Spectrum Policy Programme (RSPP) [40] with the concrete action that the European Commission together with all Member States will ensure that "*at least 1200 MHz spectrum are*

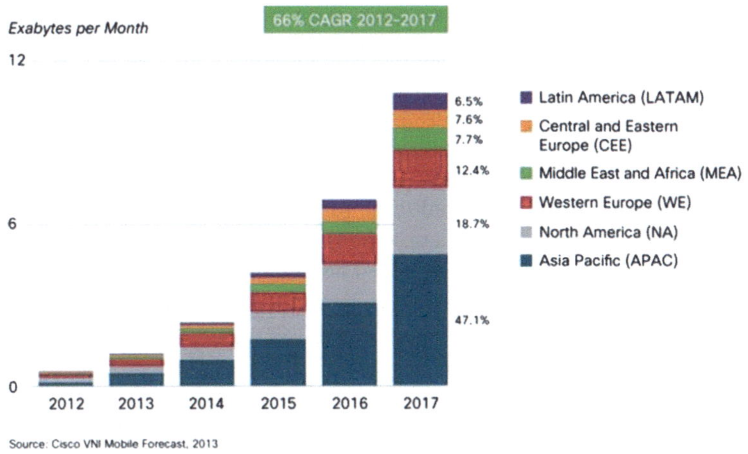

Fig. 2.20 Mobile data traffic expected by 2017—Cisco Visual Networking Indexing

identified to address increasing demand for wireless data traffic; and assessing the need for additional harmonised spectrum bands".

The industry is therefore considering feasible alternatives with a specific focus on:

- Small Cells;
- Use of bands above 6 GHz for short-range extensions of existing systems;
- Spectrum Sharing;
- Heterogeneous Networks.

The approaches of Small Cells and millimetre wave bands above 3 GHz rely on traditional thinking: the Small Cells idea is to reduce the cell sizes of Base Stations and Access Points such that the number of users per cell, Base Station and/or Access Point is reduced. As a result, the available spectrum is split over a reduced number of users and thus there is more capacity available to each user. In the mmWave approach, new frequency bands at 60 GHz or above are added to the existing mobile and wireless bands which are typically located below 6 GHz. Due to their high path loss characteristics, mmWave Base Stations and Access Points are complementing existing cellular infrastructure with high capacity, local cells of reduced size.

The Spectrum Sharing and Heterogeneous Networks approaches, on the other hand, rely on CR solutions as they have been extensively studied for over 10 years—large European Research Programmes such as IST-E^{2}R [41], IST-E^{2}R II [42] and ICT-E^3 [43] have substantially contributed to their development. A report of SCF Associates for the European Commission on "*Perspectives on the value of shared spectrum access—Final Report for the European Commission*" [44] has indeed identified for a given Scenario ("*Scenario 3*") a "*new spectrum capacity available from sharing*" of "*400 MHz*". Related standardization is actively driven in various SDOs including IEEE DySPAN with a focus on

Dynamic Spectrum Access (DSA) solutions and ETSI Reconfigurable Radio Systems Technical Committee (ETSI RRS) on Licensed Shared Access (LSA) [45] in 2.3–2.4 GHz and TV White Space (TVWS) secondary access in 470–790 MHz (see more on this in Sect. 1.3). DSA solutions are expected to play a role in the mid- to long-term since the inherent highly dynamic spectrum access characteristics are still uncertain from an economic and business feasibility perspective. LSA targets a rather short- to mid-term implementation building on static or quasi-static approaches which is supported by political and regulation activities—the European Commission has indeed issued a Standardization Mandate (EC Mandate M/512 [46] with a specific request to develop LSA standards with support from CEPT Working Group Frequency Management (CEPT WG FM) which develops the regulation framework for enabling an implementation of LSA in Europe in Project Teams PT52 [47] and PT53 [48]. Due to its immediate relevance, LSA will be discussed in further detail in the sequel.

Finally, the Heterogeneous Networks approach considers solutions for enabling the user Mobile Devices to exploit the entire available Radio Framework. Indeed, why shouldn't a Mobile Device exploit all available Radio Access Technologies (RATs) supported by the underlying modem platform? Even a simultaneous operation of multiple distinct RATs is possible with suitable solutions being proposed by IEEE 1900.4 [49] and other standards.

In the following, the current LSA activities are presented in further detail as an example for an upcoming implementation of CR in Europe.

2.6.2 Licensed Shared Access as a Use Case for Realizing CR

LSA is a technology enabling the secondary usage of spectrum based on a long-term license agreement between an LSA Licensee (typically a Cellular Operator) and an Incumbent (e.g. public safety). The European Commission's *Radio Spectrum Policy Group* (RSPG) currently defines LSA as follows [50]:

> A regulatory approach aiming to facilitate the introduction of radiocommunication systems operated by a limited number of licensees under an individual licensing regime in a frequency band already assigned or expected to be assigned to one or more incumbent users. Under the LSA framework, the additional users are allowed to use the spectrum (or part of the spectrum) in accordance with sharing rules included in their rights of use of spectrum, thereby allowing all the authorized users, including incumbents, to provide a certain QoS.

The licensing approach in combination with static or quasi-static availability of shared bands leads to a guaranteed level of Quality of Service (QoS) and a business case which is far more straightforward and obvious compared to the highly dynamic DSA case.

Indeed, LSA can provide additional resources to mobile operators and economic incentives to governments even if it is used for relatively static and long term spectrum sharing. Figures 2.21, 2.22, 2.23 illustrate the differences between the traditional, exclusively licensed, LSA and DSA based approaches. It becomes obvious that LSA will improve the exploitation of spectrum resources in the short- to mid-term but a quasi-optimum exploitation is expected to require a more dynamic DSA approach in the long term despite the technical, economic and business feasibility hurdles.

As mentioned previously, the European Commission has recently issued EC Mandate M/512 with a specific request to develop Standards enabling the implementation of LSA in Europe. ETSI is working on corresponding solutions in the ETSI RRS Technical Committee and has issued a first deliverable which details the intended first implementation in Europe in the 2.3–2.4 GHz band [51]. This band corresponds to 3GPP LTE Band 40 which was first made available to cellular communication in China. Thanks to this fact, the latest generations of Mobile Devices support this mode and are thus inherently "*LSA ready*" in the 2.3–2.4 GHz band – under the assumption, of course, that no further features are required in the Mobile Devices and the access to LSA bands is managed from the Network Infrastructure side. The high-level system design proposed by ETSI is further illustrated in Fig. 2.24. In this context, two new functions are introduced into the wireless and mobile ecosystem:

- The LSA Repository (that constitutes a geo-location database) interacts with Incumbents in order to gain information on LSA band availabilities and access conditions for LSA Licensees;
- The LSA Controller access the LSA Repository in order to derive LSA Spectrum access requirements for LSA Licensees. Typically, the LSA Controller interfaces with the network infrastructure via the cellular operators' OA&M system.

Building on the upper high-level, conceptual presentation of LSA, ETSI RRS currently further develops the LSA system specification. The definition of System Requirements is currently on-going in ETSI TS 103 154: *System requirements for LSA in 2300–2400 MHz* [52]. This activity is expected to be followed by the detailed definition of an LSA System Architecture and finally the definition of related interfaces.

In parallel to ETSI standardization activities, CEPT WG FM is working towards ensuring the readiness of LSA introduction to the market from a regulation perspective. Indeed, two project teams are developing deliverables with the following scope:

- **CEPT WG FM PT52** [47]: "… The Project Team shall: develop a draft ECC Decision, aimed at harmonising implementation measures for MFCN (including broadband wireless access systems) in the frequency band 2300-2400 MHz…"
- **CEPT WG FM PT53** [48]: "… The Project Team shall handle the following tasks: … Develop an ECC Report on general conditions, including possible

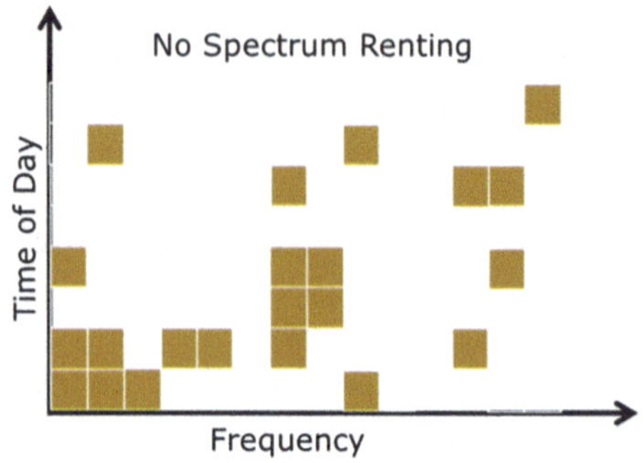

Fig. 2.21 Illustration—usage of spectrum capacity without spectrum sharing

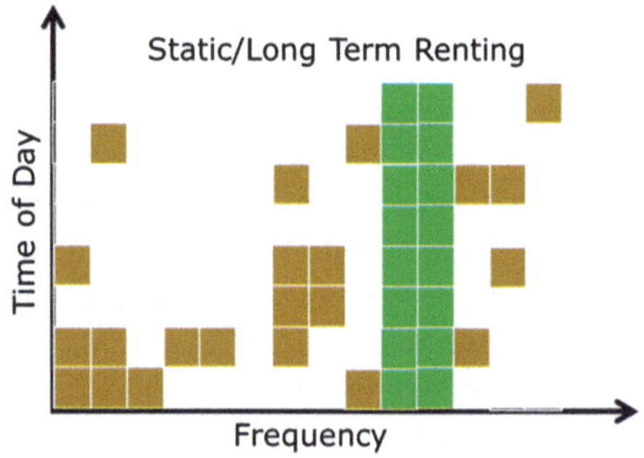

Fig. 2.22 Illustration—usage of spectrum capacity with licensed shared access

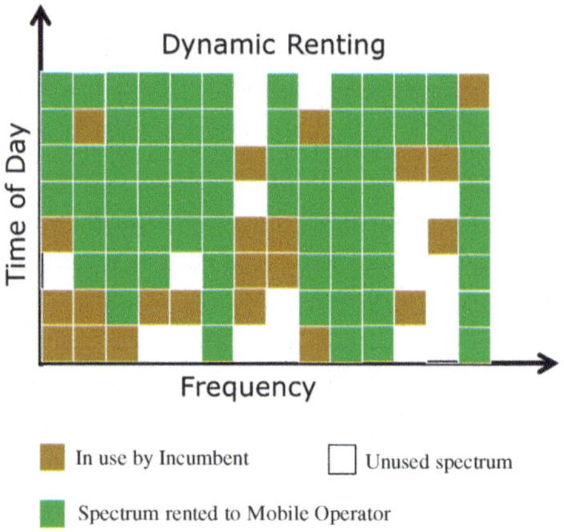

Fig. 2.23 Illustration—usage of spectrum capacity with DSA

sharing arrangements and band-specific (if not dealt with by a specific project team) conditions for the implementation of the LSA that could be used as guidelines for CEPT administrations.…" A corresponding ECC Report is about to be finalized under the title "ECC Report 205: Licensed Shared Access (LSA)".

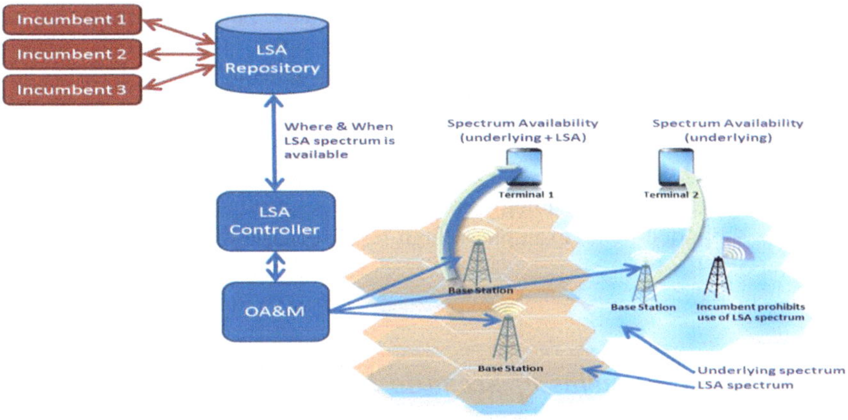

Fig. 2.24 Illustration of the licensed shared access approach [51]

The development of LSA is thus a brilliant example for an efficient interaction of the European Commission, CEPT and ETSI driving the introduction of a new technology to the market.

2.6.3 Conclusions

As a result of extensive research in the field of CR over more than 10 years, the technology is finally reaching market readiness, in particular with the imminent market introduction of LSA in Europe in the 2.3–2.4 GHz band. One reason of this late adoption of the technology is due to the complex interrelations between concerned stakeholders and the disruptive economic and business considerations. Investment in network infrastructure is only justified if there are guarantees on available mobile and wireless spectrum capacity—this simple fact still represents a key hurdle for further CR solutions which become mature from a technical perspective.

2.7 Scenario Planning Methodology[1]

Arturas Medeisis[1] and Dirk-Oliver von der Emden[2]

[1]Vilnius Gediminas Technical University, Vilnius, Lithuania
[2]Federal Office of Communications, Biel, Switzerland

[1] Dirk-Oliver von der Emden: the views expressed here are solely those of the author in his private capacity, and do not necessarily reflect the views of OFCOM.

This section outlines ideas for the application of scenario planning methodology to the realm of CR. It first reviews some basic principles involved with scenario planning. Then it proceeds with discussing possible avenues for classification and categorization of CR concepts as well as their inherent features in terms of enablers for or obstacles to their development. Such analysis would be helpful to better structure and organise the scenario planning. To conclude, the section outlines a particular methodology proposal for delineation of CR deployment scenarios, supported by an example of high-level scenario planning for the future evolution of CR.

2.7.1 Role and Benefits of Scenario Planning

As commercial CR applications begin to appear, it is required to back them with techno-economic modelling as means of proving economic viability of CR technology in various use cases. This could assist both in discovering of the profitable business models for CR as well as informing policy makers on shaping of regulatory regimes for radio spectrum access that would be favourable for CR innovation while not impairing on existing spectrum uses and users [53].

However before being able to conduct the meaningful techno-economic analyses, first it is important to understand and chart the general directions of possible evolutions of respective business opportunities. Obviously, this is not a trivial question in the case of as yet unproven CR technology, which until now existed mostly in the research labs or a few non-commercial pilots. So it is important to identify the general questions as to how and into which directions the overall development of CR will progress, before embarking on more detailed levels of operational planning and business forecasting.

It is in such instances of seeking broader views on overall trends and likely developmental paths of particular business and market segments, where the methodology of "scenario planning" comes into view [54–57].

In general, the essence and flow of scenario planning may be described by the following steps:

(1) Depicting sequences of events as *stories* that explain how we get from today to some future outcomes:

(a) what are present-day trends and influences?
(b) which forces may influence the evolution of the story?
(c) can the events be categorised as "highly likely", "uncertain", or "highly unlikely"?
(d) any events qualifying as milestones? I.e. would they have a decisive impact on directing overall development towards one specific outcome (with other words: must they happen or must they not happen if one or several outcomes is/are to occur)?
(e) attainability, desirability of particular outcomes?

(2) Identifying interactions between the scenarios, i.e. co-existence and/or relative progress of the scenarios:

(a) analysis of the linkages between events in order to identify the enablers and hurdles that will influence the relative speed of development of each scenario;

(3) Ensuring consistency and plausibility of scenarios through iterations:

(a) development of an integrated picture of the future, using elements of the identified scenarios.

(4) Based on the generated scenarios, outlining strategy options and developing most optimum strategy in order to achieve objectives.

Today several methods for building scenarios are known, such as the *Intuitive Logics* (IL), *Probabilistic Modified Trends* (PMT) and "*La Prospective*" [54]. Applied to the realm of CR business forecasting, perhaps the IL would be the most conducive methodology in comparison with the two other ones because of its inherent proclivity to addressing essential features of any CR planning exercise:

- absence of quantifiable historical data to characterise the development of future scenarios;
- the equal probability or, in other words, uniform uncertainty of the developed scenarios;
- orientation towards the learning and improved understanding of the involved processes rather than the reliability of the end products—the scenarios itself.

It is suggested [54] that while the inherent flexibility of IL method allows applying it in a broad variety of situations, the more deterministically oriented nature of PMT and "La Prospective" methods makes them optimal for use in "once-only problem solving" type of applications. In the case of evolving CR technology we would be interested in the long-term anticipative exploratory analysis, which thus naturally lends itself to the domain of IL [53].

Nevertheless, it is possible and sometimes useful bridging the differences between IL methodology and deterministic scenario planning of PMT kind, such as may be seen embodied by the Schoemaker's school of scenario planning [58]. Examples of recent works bringing the latter types of scenarios into the field of CR may be found in [56]. This bridging between the IL and PMT/Schoemaker's methods may be done by building the IL-derived scenarios along the clearly identifiable and quantifiable axes of uncertainty factors. This allows the clear identification of the most influencing factors and their analysis in order to discuss strategic development directions. Moreover, it gives the ability to detect warning signals as regards anticipated future developments once (one of) the identified uncertainties become more certain and/or certain milestone had been passed.

In the following subsections we shall address the issue of categorisation of CR with reference to its spectrum access modes, followed by the identification of respective enablers and hurdles for future development. This analysis would be helpful in carrying out the first steps of scenario planning.

2.7.2 Categorisation of CR Spectrum Access Modes

If CR is considered as disposing of user-independent intelligence, one possible approach when establishing a catalogue of CR employing Dynamic Spectrum Access (DSA) methods is to categorize the CR devices/networks according to the features of their "decision-making" processes with respect to spectrum access and usage mode. This decision making may be concerned with answering many questions such as "when" (moment & duration), "where" (location), "how" (frequency band, modulation, etc.), at which cost/price, etc. With this logic in mind, the Fig. 2.25 below proposes a categorisation of CR use with regards to spectrum access behaviour.

It is believed that these categories are not mutually exclusive. Actually, it may be presumed that many CR devices/networks will certainly have the ability to display several of such behaviours thanks to their inherent flexibility and re-configurability.

Examples as to how some presently envisaged CR deployment models relate to these three categories:

- Category A: master–slave;
- Category B: ad-hoc/mesh networks;
- Category C: autonomously sensing and deciding devices.

For instance, if a CR system intends to employ a Cognitive Pilot Channel (CPC), the category (or categories) retained to characterise the system will depend on the functionalities and features of the CPC [59].

2.7.3 Identification of Associated "Hurdles" and "Enablers"

It is proposed to develop a "matrix" for the identification of obstacles to and stimuli for the uptake and proliferation of CR in relation with the categories identified above. This identification has many similarities with the PEST analysis [56], however, the point here is really to propose identifying certain specific multi-dimensional axes for the Political-Economic-Societal-Technological (PEST) type analysis. An example of such matrix for identifying obstacles and enablers is offered below in Fig. 2.26.

It is proposed here to consider three categories of enablers/drivers (there may be others, or the categorization may be different), which may be identified as follows:

- "Technical enabler": (future) technological achievements, which will drive, give further incentive to research, development, production and exploitation of CR technology;

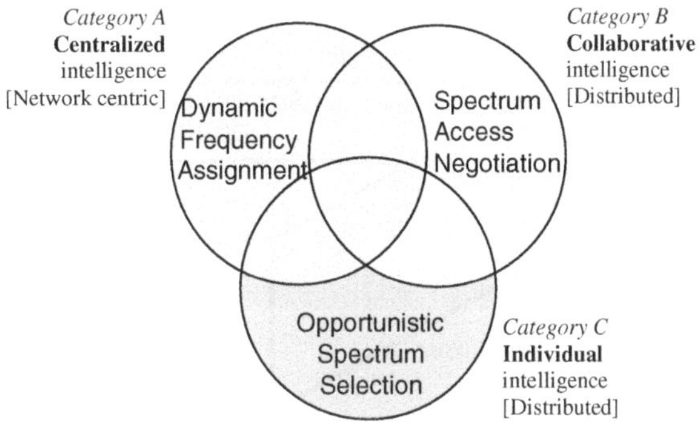

Fig. 2.25 Categorisation of CR use inspired by "spectrum use scenarios" in ETSI TR 102 802 [37]

- "Socio-economic enabler": social and/or economic factors that are incentives for the uptake and/or adoption of CR technology;
- "Regulatory enabler": regulatory measures, such as removal of some existing regulations, or enactment of some new ones, both of which acts would create incentives for investments in and/or adoption of CR technology.

Some examples for enablers[2] are given in the following Table 2.5.

The enablers, as well as hurdles, eventually identified as relevant for one CR category must not necessarily be relevant for another one. The relevance needs to be examined on a case-by-case basis.

In a similar vein, below the three categories of hurdles are proposed (and again, there may be others, or the categorization may be different), which may be identified as follows:

- "Technical hurdle": engineering challenges in relation to CR, whose solutions are not conceivable for production on a large industrial scale in the short to medium term;
- "Socio-economic hurdle": social and/or economic factors that inhibit the production and/or adoption of CR;
- "Regulatory hurdles": pieces of (present-day) regulation, compliance to which is disproportionately costly or inhibit uptake of CR concepts.

Some examples of hurdles for CR development are offered in the following Table 2.6.

Not surprisingly, it may be noted that the number of hurdles that spring to mind is larger than the number of enablers. The issue of balancing the situation by promoting incentives for CR innovation will be addressed later in this book.

---

[2] Some examples were taken from [60], noting that in this reference the drivers are associated with radiocommunications applications ("markets") that are more specific than the categories discussed in this section.

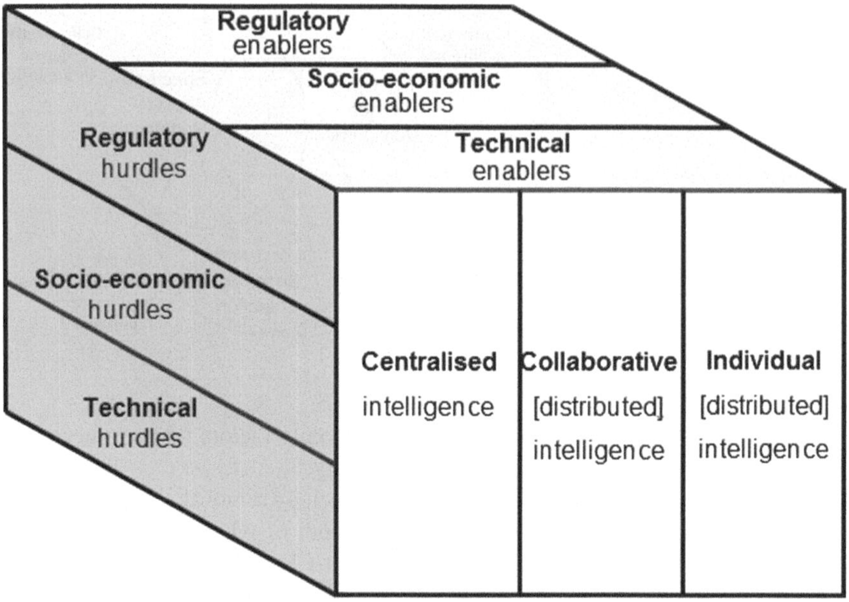

Fig. 2.26 Matrix of enablers and hurdles for CR implementation

Table 2.5 Examples of CR enablers

Enablers	Type
Simplified post sale service and bug fixing	Technical
Easier upgrade routes and clear path of migration to newer SDR based platform	Technical
Access to more waveforms	Technical
Potential for lower development cost due to more generic hardware platform and more flexible software architectures	Economic
Deregulation in the field of formal requirements when marketing R&TTE products	Regulatory

2.7.4 Importance of CR Scenario Planning for Radio Spectrum Management

The scenario-based approach could clearly assist in better understanding of realities and perspectives of CR evolution and proceeding towards developing a comprehensive techno-economic regulatory framework of radio spectrum access rules for CR/SDR-based wireless applications.[3]

[3] It has been recognized that scenarios were of considerable value in spectrum planning as they could provide useful insights into how spectrum management may be affected by future events having implications for demand for spectrum [61].

Table 2.6 Examples of hurdles impending development of CR

Hurdles	Type
Protection (e.g. detection) of incumbents spectrum users	Technical
Availability of CPUs: • In replacement of nowadays ASICs	Technical
Power consumption of mobile devices: • Technology of electric accumulators (batteries)	Technical
Irreversibility of allocation of spectrum to license-exempt CR operation	Economic
Transaction costs of "spectrum trades"	Economic
Spectrum and technology fragmentations: • Unfavorable for amortization of non-recurring costs • Hinders emergence of economies of scale	Economic
Inflexibility of provisions of primary radio licenses: • Inhibition of spectrum trading	Regulatory

The key would be to analyse and clarify the identified technological, economic and regulatory hurdles and enablers impacting future evolutions of CR, which would become the basis for building scenarios of CR development. Paths within the scenarios will depend, amongst other things, on chronological order when the various barriers identified will be overcome or the enablers become effective.[4] In that regard it is worth giving special focus to regulatory aspects. Because the peculiar nature of the regulation is that its shortcomings (when identified) can be reduced or eliminated overnight, simply by the (political) will of the competent governmental authority responsible for its enactment. Thus regulation can easily be both a hurdle (if status quo with shortcomings is upheld) and enabler (adoption of corrective regulation), as illustrated in Fig. 2.27.

But when completed, the scenarios are likely to generate different outcomes in respect of the strength of the demand for additional spectrum for CR and/or requirement for modifications to the radio regulatory environments. Thus the regulatory establishment will be in itself affected by future evolutionary developments of CR, see Fig. 2.28.

In fact, it may be observed that it is only natural to expect that the regulation be adaptive to any developments of wireless technology, both before and after the take-off of CR. More discussion on the impact analysis is offered in Chap. 5 of this book.

At the end, an important result of scenario planning could be evaluations with respect to the need for additional and/or specific spectrum made available for CR, its timing, the assumptions concerning the need re-allocations of spectrum or the adoption of completely novel approaches to spectrum management.

[4] Amongst other things, "there are differences between the scenarios in terms of the pace of change and the demand for additional spectrum" [57].

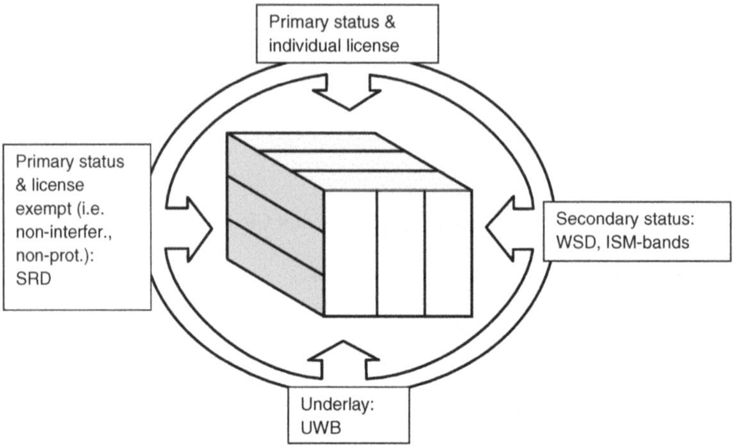

Fig. 2.27 The regulation as impacting factor of CR development

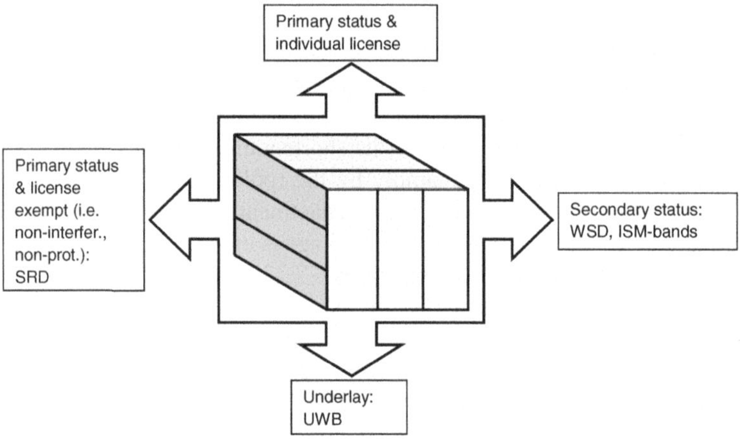

Fig. 2.28 CR development scenarios affecting the future regulations

2.7.5 Example of Scenario Planning for CR as DSA Technology

In this subsection we discuss scenarios for CR-based DSA market development. These scenarios were developed with the highest conceivable level of abstraction, in order to provide a broad look at possible directions for future development of CR technologies as business-changing technological innovation. Innovation is an important defining element in this context as it refers to the practical implementation of the new technology rather than solely theoretical understanding.

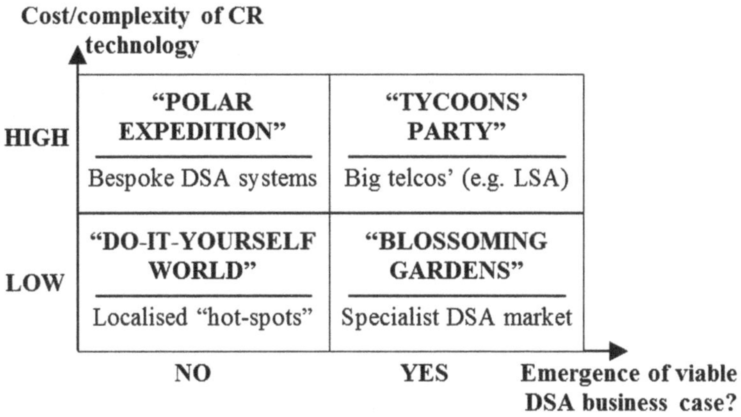

Fig. 2.29 Example of high-level scenarios for CR-based DSA business, adapted from [53]

It is thus proposed to build the high-level scenarios with reference to two most prominent uncertainties of CR-based DSA applications, namely:

- resolving the challenge of technical complexity of CR DSA technology, and
- finding a viable business proposition for DSA applications.

These two uncertainties are offered as top uncertainties affecting the pace of development of CR technologies with reference to both an intuitive logic (the IL method) and some real-life developmental facts. The first uncertainty is backed by the observation that, as of today, a solid commercially available prototype DSA application is still to appear. And as regards the second point, it may be surmised that the absence of significant industry efforts in promoting the CR DSA applications, seem to suggest that industry is, as yet, is not convinced there is any real revenue generating viable business case in exploiting of DSA concepts.

By accepting the identified uncertainties as defining axes for building scenarios for future evolution of CR-based DSA business, the four scenarios could be mapped accordingly against those uncertainties, as illustrated in Fig. 2.29.

Of the proposed four scenarios, the "Polar Expedition" appears to represent the current status quo, where the business prospects are uncertain and technological complexity is very high. Thus the name of scenario alluding to singular deployments of pilot systems, or military users with dedicated custom-built applications.

The "Tycoons' Party" scenario would represent the changing ecosystem characterised by appearance of clear business value proposition for DSA. Yet while the cost/complexity of CR technology would remain high, this scenario would likely to benefit a select few big players, be it large established telecommunications operators or some ambitious heavyweight newcomers such like Google. An obvious example springs to mind, namely that if and when the LSA concept is implemented, this would mean the industry transition into that scenario.

The "Blossoming Gardens" is an intriguing scenario whereas the falling costs and complexity of CR technologies would lower the barriers for market entry, while the DSA would be an attractive business proposition. This situation would be not unlike what happened with cellular mobile, which quickly turned from high complexity premium service technology to ubiquitous applications. So this previous example leads us to suggest that something similar would happen to CR DSA in such environment, whereas high profit margins would mean building certain "walls" that would keep the overall number of specialised players at some stable level.

On the other end, the "Do-It-Yourself World" scenario would represent the very different kind of developmental transition, with the CR technology becoming affordable, yet the business case for DSA remaining elusive. In such environment it could be envisaged that the CR DSA applications would spread out into a large number of local deployment islets, much like the Wi-Fi Hot Spots.

In summary it may be noted that the above example represents but example of very high-level considerations. Nevertheless, it already offers certain valuable insights and exploratory understanding of the environmental conditions steering the development of particular niche application of CR. Further insights might be obtained by applying such methodology in a more refined and multi-dimensional manner to analyse the various categories of issues discussed earlier in this section.

2.8 Taxonomy of CR Use Cases and Applications

Arturas Medeisis[1], Oliver Holland[2] and Luca De Nardis[3]

[1]Vilnius Gediminas Technical University, Vilnius, Lithuania
[2]King's College London, London, UK
[3]Sapienza University of Rome, Rome, Italy

As shown in the preceding sections, there already exist many different CR deployment scenarios and their number constantly grows. These scenarios are being presented either as a description of different technological solutions or a set of business and use cases. To add to this somewhat bewildering picture, there also exist many definitions of CR itself. Thus the absence of a single classification system for the entire range of CR applications and use cases persists to this day. This brings to mind an "elephant in the room" metaphor, when blind men attempt to describe what they do not know by touching different parts of the colossus [62].

It seems important to start a discussion on definitions of CR and its applications by stepping back to the origins. The original definition of CR has seen many alterations and derivations since it was coined by Mitola and Maguire back in 1999 [63–67]. All of those definitions, however, use various permutations of the three core features of CR: some kind of environmental awareness (such as spectrum sensing, geolocation database, etc.), functional re-configurability, and "intelligent adaptive behavior" [65]. A comparative glossary of CR definitions is available in [65], p. 375.

The most formal definition of CR seems to be offered by ITU-R [68], and it will be assumed as a basis for further discussion here. The ITU-R sees CR as a system that employs technology allowing it: "*to obtain knowledge of its operational and geographical environment, established policies and its internal state; to dynamically and autonomously adjust its operational parameters and protocols according to its obtained knowledge in order to achieve predefined objectives; and to learn from the results obtained*". It is important to note that such a definition avoids the issue of purpose of all those functionalities and adaptive operation, so it appears to be "application-agnostic".

On the matter of CR applications, the original vision postulated at least four use cases [63]:

- Spectrum pooling;
- Network management protocols;
- Personalised services delivery;
- Stability of type-certified downloads.

The ensuing decade of active R&D effort in this field again resulted in numerous alterations and expansion of that original vision. A good recent example of those alterations may be seen in an ITU-R report, being drafted at the time of writing this book, on the subject of CR use in land mobile services. It attempts to depict a dozen or so various CR applications from frequency agile autonomous systems to CR-assisted reconfiguration of user terminals for supporting heterogeneous radio access networks, see Sect. 2.1.

This not only illustrates the futility of fixing a complete definitive set of CR uses, it also shows intricate interlacing of CR methods and features at different layers of wireless technologies.

It might be argued that this absence of a clear and homogeneous classification of CR applications contributes to the protracted ambiguity with the standard definitions of CR and inhibits efficient codification of CR innovation ideas. To address this shortcoming, below a single universal taxonomy of CR use cases and applications is described.

It is proposed to start by considering two separate yet complementary planes on which the development of CR could be mapped. One plane would be based on the applications (utility) and the other one on a set of fundamental CR technical features to be implemented. The end target is a fully inclusive classification framework that is compatible with the need to adhere to the regulatory reality (through the description of the application hence allowing the link to underlying rules) and technological capabilities which must be satisfied to achieve the application. It is important that such an ultimate single taxonomy is broad and flexible enough as to also include possibility of future expansion by adding as-yet unknown prospective CR use-cases.

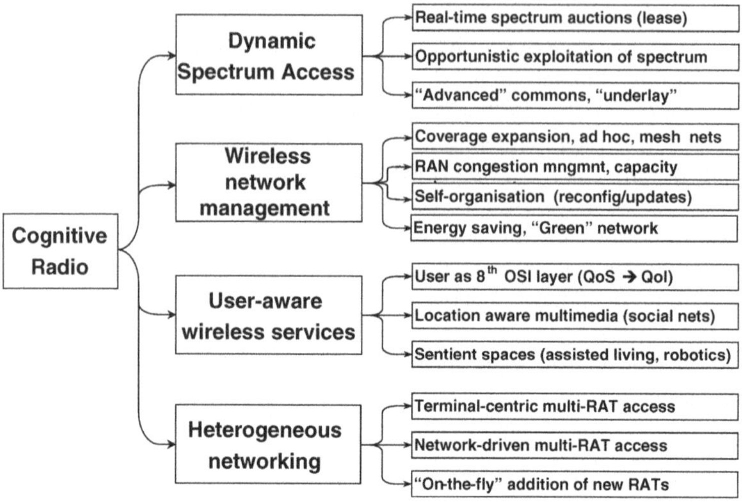

Fig. 2.30 CR utility strands in the application plane

Within the application plane, the various strands of CR evolution could be defined by considering the distinctive utility functions of various applications. This is the same dimension as used originally by Mitola [63] to describe the envisaged set of CR use scenarios. This seems to be held valid today and in the future, as it does not have hard limit on the number of distinct utility strands, nor their length or degree of branching. Another important aspect of this utility plane is that the classification could be directly mapped to the radio spectrum access types and, hence, to spectrum regulations.

Figure 2.30 depicts the proposed outlook of the application plane of CR classification framework. It is based on initial branching of CR into four main utility classes-strands, broadly corresponding to the ones envisaged by Mitola [63]. Then, each of these strands may be branched again and again into sub-layers corresponding to the deeper degrees of utility differentiation for respective family of applications.

When looking at this figure it becomes obvious that only some (mostly one) of the possible CR development strands had seen sufficient attention and R&D activities. It is namely the DSA, which had stolen most of the lime light recently and shadowed other possible development strands. As a result, the other three strands had been seeing less or no attention from the CR research community. Granted, some of those alternative strands, such as heterogeneous networking or wireless network management, had seen significant developments. However mostly they grow from within the respective interest communities as incremental process innovations, without realising the bigger picture of relevance to the CR development.

It should be noted that the vision of application-plane CR taxonomy, as proposed in Fig. 2.30, accommodates well all types of CR use cases and applications discussed elsewhere in this book and in the literature (e.g. sources quoted in this section). In fact, all of those different cases and applications could be mapped directly to one or another case-box in the proposed scheme [69]. But it may be also deduced that different authors may speak of cases in applications being on different levels (i.e. DSA application on level two vs. some more specific applications on level 3 or deeper). Such inadvertent de-focusing of discussion leads to the fuzziness and ambiguities of the aforementioned "elephant in the room" phenomenon.

When analysing the classification of CR applications offered in Fig. 2.30, it might be noted that this representation can be further enriched by bringing into view the additional angle of technology. The point here is that other than purely utilitarian functions performed by a given CR device or network, it would also be useful to consider at which technological capability they could operate. Here it is worth reminding that all CR technology rests on several fundamental premises (essentially, technological abilities), i.e., having awareness of the radio environment, and being able to adapt radio parameters based on that environment, with or without the cognition aspect. So, the technical capabilities represented in the taxonomy should take into account these key elements of the CR. Following this approach, it may be therefore assumed that the root of the technology path comprises of the four core features: radio information detection, radio parameter adaptation, learning and re-orientation. This is illustrated in Fig. 2.31 as four evolutionary levels of CR development in technological plane.

This figure helps to convey a simple message that it is unrealistic to expect all-singing-all-dancing CR applications to appear at once. To the contrary, it is likely that fully technologically capable CR applications will be developed through a gradual phased implementation. So it is logical to speak of 1st generation CR (such as TV White Space devices, for instance) and so on.

Inside each generational phase a further micro-classification might evolve by delving into more detailed aspects of the realised technical capabilities, such as specifying the types of radio awareness (e.g. autonomous sensing methods vs. Geolocation databases, etc., including their combinations), the allowed range of parameter adaptations (e.g. frequency ranges covered, modulations that can be achieved, or perhaps pre-defined full RATs), the learning means (e.g. reinforcement learning, artificial neural networks), etc. This might be taken down to the levels of highest resolution necessary to allow complete specification of the CR requirements. However these levels of details are not shown in Fig. 2.31 to keep it illustrational and conceptually simple.

While it is proposed to differentiate classification of CR uses in two separate planes, it may be also noted that in most cases they will be complementary whereas any given CR use case or application may be and should be characterised through combination of the utility and technology aspects. It is therefore suggested that such combined dual characterisation is essential in order to avoid confusion and ambiguity in specifying any past or future CR uses and applications. Once established, such classification system would by itself provide important practical value and uses, as discussed further.

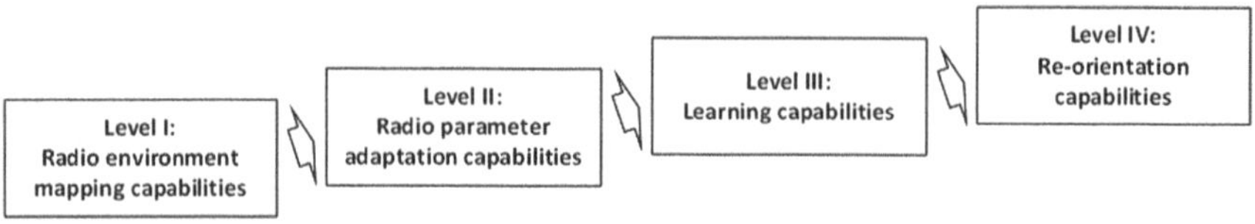

Fig. 2.31 CR technology evolution path

One important practical aspect of having a universal taxonomy would be its contribution to the codification of information on CR use cases and applications, which is being constantly produced by numerous R&D efforts. More specifically, the accepted taxonomy should make possible unambiguous classification within an information structure that devices/networks could use (communicate to each other) to understand which forms of CR are already used in given area and, as relevant, which other forms can be used based on observation of any regulatory constraints and technical capabilities of the devices that are going to communicate.

A practical implementation of this concept would be to use the taxonomy tree as a basis for "genetic" codification of CR applications. For instance, taking the application-driven scheme proposed in Fig. 2.30 as a reference, each branch at a given level could be described by letters A, B, C, and so on. Then each application class could be described by the unique "genetic code-word" where the position and value of each letter define the respectively branching level and particular branch. For example, GDB-based white space devices could be described as belonging to CR class "A-B-B-A", see Fig. 2.32, whereas Terminal-centric Multi-RAT CR device could be defined as belonging to class "D-A", and so on.

An important inherent feature of such genetic codification would be its unlimited growth in terms of number of branching levels by simple extension of code-word, and the possibility to accommodate new future branching at certain levels by adding new letters at given position in the code-word.

The same principle of "genetic coding" could be applied to the technology-driven classification illustrated in Fig. 2.31 if each respective generational level (and further detailed internal branching) is given the appropriate code-letters. Then any conceivable CR application and use case could be uniquely defined by a combination of "application" and "technology" class codes.

An illustration of why this is important and how it may be used in practice is illustrated below in Fig. 2.33.

As shown in the illustration, other than providing explicit referencing, such codification might be useful for future automation of CR operations, such as being used as part of the necessary peer recognition algorithms and the processes for CR "frequency rendezvous".

In conclusion, this section outlined a uniform system for comprehensive classification of all conceivable CR use cases and applications by using two planes: applications and technology. Such broad taxonomic view allows characterising specific CR applications and placing them contextually on the overall big picture of various strands and generations of CR development.

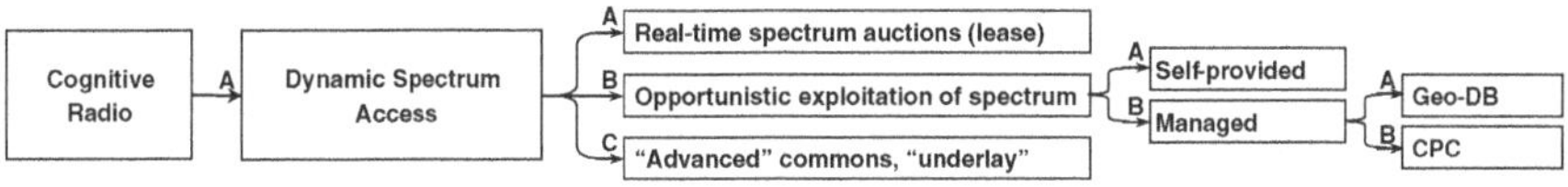

Fig. 2.32 Example of codifying a specific CR application

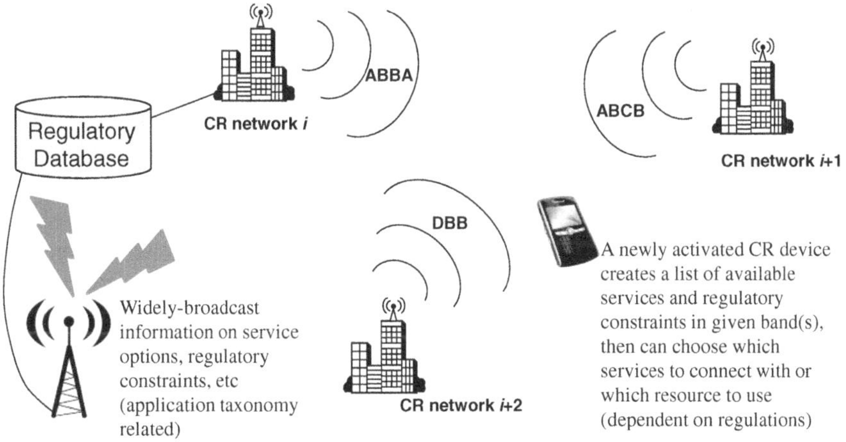

Fig. 2.33 Illustration of using application type coding for CR rendezvous

More specifically, it may be observed that up to now, the popular association of CR solely with OSA/DSA utility concepts is in fact a misconception, as it puts a shadow on several other equally important strands of development of CR technology and applications.

An important practical effect of having a single universal taxonomy would be its contribution to the codification of information on CR use cases and applications.

2.9 EN 301 598: A European Harmonised Standard for Deployment of TV White Space Devices

Cesar Gutierrez and Hamid Reza Karimi

Office of Communications (Ofcom), London, UK

2.9.1 Introduction

The European regulatory regime for the use of wireless devices was designed some time ago with the aim of removing the need for national type approval of devices. In this regime, manufacturers who wish to place equipment on the European market are required to self-declare conformance to the "essential requirements" of the Radio and Telecommunications Terminal Equipment (R&TTE) Directive. Such self-declaration can be achieved via a number of routes.

The primary route is through compliance with harmonized standards developed by European standards organizations. Harmonized standards that address the requirements of Article 3.2 of the R&TTE Directive (effective use of the spectrum to avoid harmful interference) are usually produced by ETSI. These typically include radio frequency (RF) requirements such as limits on radiated wanted and unwanted powers.

The European harmonized standard EN 301 598 [70] has recently been developed by ETSI committee BRAN to address white space devices operating in the UHF TV band. EN 301 598 is different from past harmonized standards for other bands, in that it includes many requirements which relate to control functionalities—in addition to the more traditional RF requirements. This is because unlike other radio devices, the regulatory limits on the RF characteristics (e.g. radiated signal frequency and power) of white space devices are not fixed and cannot be hard-coded into the device, but are rather communicated to the device by a database. Consequently, in addition to the usual RF requirements found in other standards, EN 301 598 specifies a number of control and monitoring functions (along with the relevant compliance tests) to ensure that WSDs behave in accordance with the regulatory parameters set out by databases. In this section we describe some of the key elements of the harmonized standard.

2.9.2 Masters, Slaves and Databases

As illustrated in Fig. 2.34, the specifications in EN 301 598 are based on a framework for access to TV white spaces (TVWSs) which involves the following four entities (see also the compatible UK framework described in Sect. 1.5.2):

(1) White space databases (WSDBs)—provide location-specific and device-specific information on TVWS availability (as well as other instructions) to white space devices (WSDs);
(2) WSDB web-listing—identifies the WSDBs that are approved by a given national regulatory authority (NRA) to provide service in the relevant jurisdiction;
(3) Master WSDs—geolocated radio devices that are capable of accessing a web-listing, and directly communicate with a WSDB (via means other than the UHF TV band);
(4) Slave WSDs—radio devices that do not communicate directly with a WSDB, but operate subject to the control of a master WSD.

WSDBs and WSDs are required to exchange the following parameter sets in order for WSDBs to determine and communicate the available TVWS radio resource to the WSDs:

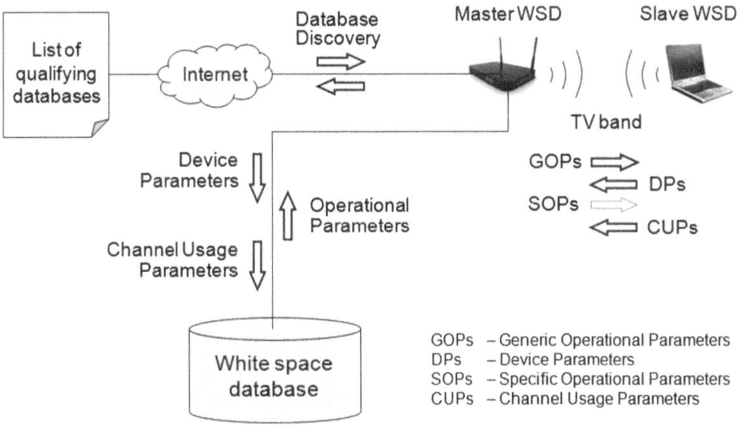

Fig. 2.34 Framework for access to TV white spaces

(a) Device Parameters»CPöThese are parameters that are communicated from a WSD to a WSDB, and identify specific characteristics of the WSD. These include WSD location (and location uncertainty), type (A/B), category (master/slave), unique ID, technology ID, emission mask class, emission mask enhancements, and reverse intermodulation enhancements.

(b) Operational Parameters»CPöThese parameters are generated by a WSDB and communicated to WSDs. They specify the TV channels that a WSD can use, the emission limits in each channel, and a number of other ancillary parameters. There are two types of Operational Parameters:

- Specific Operational Parameters, and
- Generic Operational Parameters.

Specific Operational Parameters are specific to an individual WSD, while Generic Operational Parameters apply to all slave WSDs that are served by a particular master WSD. A WSDB communicates Generic Operational Parameters to a master WSD, which in turn broadcasts these to all slave WSDs in its coverage area. The Generic Operational Parameters account for certain characteristics of the serving master WSD (e.g. location, power, and hence coverage area), but assume default values for the Device Parameters of the slave WSDs.

(c) Channel Usage Parameters—These parameters are reported by a WSD to inform a WSDB of the *actual* radio resources that are used by the WSDs. These include the indices of the TV channels used and the emission levels in each channel.

2.9.3 The Operation of a White Space Device

Here we set out the sequence of procedures considered by EN 301 598 for the exchange of information between WSDBs, master WSDs, and slave WSDs. These procedures reflect the following three high level regulatory requirements:

(1) A WSD must only operate in compliance with the Operational Parameters provided by a WSDB that is approved by the relevant national regulatory authority.
(2) A WSDB must receive certain essential information (Device Parameters) from a WSD in order to generate and communicate Operational Parameters to that WSD.
(3) A WSDB must maintain a record of the actual usage of the TV white spaces. To this end, each WSD must report back to the WSDB the actual digital terrestrial TV (DTT) channels and the powers that it uses.

Requirement (3) arises from a need for a cautious approach to the deployment of WSDs. Specifically, the circumstances for secondary use in the UHF TV band—namely the widespread primary use of the band, and the new paradigm of radio resource allocation by WSDBs—mean that the impact of interference must be managed carefully, making it important for regulators to be able to monitor the use of TV white spaces. For this reason, it is required that WSDBs collect detailed information about TVWS usage.

A typical sequence of procedures may be described in the context of four separate phases A to D as presented next, and illustrated in Fig. 2.35. The Specifications in EN 301 598 are not prescriptive in relation to these four phases and, in practice, different wireless technologies may implement these in a variety of ways so long as the necessary information is exchanged correctly and WSDs radiate in the UHF TV band subject to the correct Operational Parameters.

2.9.3.1 Phase A: Specific Operational Parameters for a Master WSD

Phase A relates to the generation and communication of Specific Operational Parameters for master WSDs, and involves the following steps:

(1) The master WSD must access a list of approved WSDBs via the internet. This web-list may be hosted by the relevant national regulatory authority (as is the case in the UK) or by a trusted party;
(2) The master WSD will then select a WSDB from the list, and request Specific Operational Parameters from the WSDB for its own transmissions. In this process, the master WSD must first communicate its Device Parameters (including its location) to the WSDB;
(3) The WSDB will then generate the Specific Operational Parameters that the master WSD must comply with for its transmissions. To perform this, the WSDB will use a) the Device Parameters provided by the master WSD, and b)

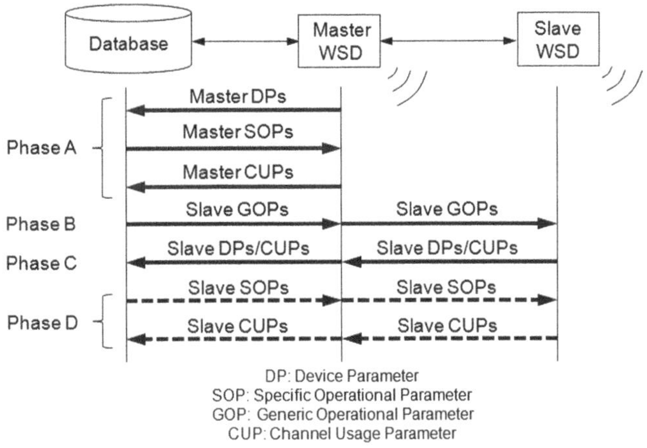

Fig. 2.35 Illustration of operational phases

TVWS "availability data" which it holds, and which indicate the maximum power that a WSD is permitted to radiate within each DTT channel at a particular location in order to ensure a low probability of harmful interference to the incumbent primary users. The WSDB will communicate the Specific Operational Parameters to the master WSD;

(4) The master WSD must communicate back to the WSDB the actual channels and powers which it intends to use (its Channel Usage Parameters). These usage parameters are likely to be different from the Specific Operational Parameters provided by the WSDB. This is because the master WSD may not be capable of transmitting in all DTT channels indicated by the WSDB, or at the channel-specific emission limits, or there may be a network control function that restricts the emissions of several master WSDs to ensure that they do not interfere with each other;

(5) The master WSD can then start transmissions in the UHF TV band according to its reported Channel Usage Parameters.

2.9.3.2 Phase B: Generic Operational Parameters for Slave WSDs

Phase B relates to the generation and communication of Generic Operational Parameters. These parameters describe the DTT channels and maximum powers that any slave WSD within the coverage area of a given master WSD can use for its transmissions.

Generic Operational Parameters primarily describe the radio resources that a slave WSD may use in order to associate with a master WSD. We use the term "association" to refer to the process whereby a slave WSD initially identifies itself to its serving master WSD. This is a usual process in many wireless technologies.

A networked element—the base station or access point—broadcasts information to indicate to the non-networked elements—the terminals—the radio resources (typically frequencies) that the latter may use in order to identify themselves to the network and to request further access to the medium. In the case of TV white spaces, it is envisaged that (following association) some technologies will continue to use the radio resources specified by the Generic Operational Parameters for on-going transmission of data, while other technologies will only use these as a means to submit a subsequent request for additional radio resources (see phase D). Phase B involves the following steps:

(1) The master WSD must contact the serving WSDB and request Generic Operational Parameters for the transmissions of any slave WSDs that might be located within its coverage area.
(2) The WSDB will then use the TVWS "availability data" and the information that it holds about the master WSD (see Phase A) to calculate the master's coverage area. The WSDB will then calculate the Generic Operational Parameters by assuming (a) that slaves may be at any location within the master's coverage area, and (b) default conservative values for the Device Parameters of the slaves. Note that at this stage no slave Device Parameters are available at the master WSD or at the WSDB. This is because no slave WSDs will have yet associated with the master WSD. The WSDB will communicate the Generic Operational Parameters to the master WSD.
(3) The master WSD must then broadcast the Generic Operational Parameters to slave WSDs within its coverage area. The actual Generic Operational Parameters broadcasted will normally be a subset of those communicated by the WSDB, and may even correspond to a single channel only. This is because the master WSD may not be able (or willing) to receive transmissions from slave WSDs in all the channels identified by the WSDB.
(4) Slave WSDs must comply with the channel-specific powers limits specified in the broadcasted Generic Operational Parameters when they transmit in the UHF TV band for purposes of association with the master WSD.

2.9.3.3 Phase C: Association of a Slave WSD with a Serving Master WSD

Phase C relates to the association of slave WSDs with master WSDs. Any slave WSD wishing to transmit in the UHF TV band, must undertake the following:

(1) A slave WSD must associate with a master WSD by identifying itself to the master. A slave WSD may submit its full set of Device Parameters for this purpose.
(2) To perform the above, the slave WSD must transmit in compliance with the Generic Operational Parameters broadcasted to it by the master WSD.
(3) The master WSD must forward the identities, or the full set of Data Parameters, of its associated slave WSDs to the WSDB.

(4) Slave WSDs which have already associated with a master WSD may continue to use Generic Operational Parameters for subsequent transmissions. Alternatively, they may request Specific Operational Parameters in order to benefit from increased TVWS availability (see Phase D).

2.9.3.4 Phase D: Specific Operational Parameters for a Slave WSD

Phase D relates to the generation and communication of Specific Operational Parameters for individual slave WSDs. Specific Operational Parameters describe radio resource availability that is greater than that described by Generic Operational Parameters. This is because, absent the required data, WSDBs make cautious assumptions regarding the Device Parameters of slave WSDs when they generate Generic Operational Parameters, and these results in somewhat restrictive radio resources in terms of available DTT channels and channel-specific emission limits. A slave WSD that is able to accurately determine its location or whose Device Parameters are more favourable than those assumed by the WSDBs in generating Generic Operational Parameters (e.g. cleaner spectrum emission masks), will be able to gain access to greater radio resources if it communicates its Device Parameters to a WSDB in order to receive Specific Operational Parameters. The above is described as phase D and involves the following steps:

(1) A slave WSD will provide its Device Parameters to its serving master WSD and request Specific Operational Parameters. The master WSD will forward this request to the WSDB. An alternative implementation may be one where the Device Parameters of the slave WSDs reside in the master WSD, and it is the master WSD which requests Specific Operational Parameters for the slave WSDs from the WSDB;
(2) The WSDB will generate Specific Operational Parameters by using the TVWS “availability data” that it holds and the slave Device Parameters provided by the master WSD;
(3) The WSDB will communicate the Specific Operational Parameters for a slave WSD to the master WSD. The master WSD will then communicate the Specific Operational Parameters to the associated slave WSD;
(4) The slave WSD will communicate to the master WSDB the actual channels and powers that it intends to use (its Channel Usage Parameters). The DTT channels described by the Channel Usage Parameters may be a subset of those identified by the Specific Operational Parameters. By definition, the powers described by the Channel Usage Parameters must be lower than the emission limits specified by the Specific Operational Parameters. The master WSD will forward the Channel Usage Parameters to the WSDB. An alternative here is that all intelligence resides in the master WSD, in which case the master makes decisions on behalf of the slave WSD regarding the channel(s) and powers(s) to be used by the slave, and the master itself generates the Channel Usage Parameters on behalf of the slave.

2.9.4 *Requirements and Specifications*

The requirements in EN 301 598 are intended to mitigate harmful interference by ensuring that the wanted emissions (inside the band) and unwanted emissions (both inside and outside the band) do not exceed specific limits. The limits outside the UHF TV band are specified in EN 301 598 in the same manner as existing harmonized standards, and include limits on Tx/Rx spurious emissions and transmitter inter-modulation. On the other hand, the limits inside the UHF TV band are specified in a novel way. This is for two reasons:

(a) A WSD may operate in a single DTT channel, or it may operate simultaneously in a mixture of contiguous and non-contiguous DTT channels.
(b) The limits on the wanted emissions are not fixed, but are location-specific, channel-specific, and are specified by a WSDB.

Given the above, the specifications in EN 301 598 are divided in two groups: namely RF requirements and control/monitoring requirements. The latter ensure that the WSDs communicate the necessary Device Parameters to a WSDB, and that they operate in compliance with the Operational Parameters that they receive from the WSDB. EN 301 598 requires that the Device Parameters, Operational Parameters, and Channel Usage Parameters include a minimum set of information. Their detailed specification (such as the format and size of the data) is mostly up to design of the WSD-WSDB protocols (via proprietary, *de facto*, or industry-defined standards such as IETF PAWS). The requirements are summarized below.

2.9.4.1 Nominal Channel

A Nominal Channel is defined as one or more contiguous DTT channels that are used by a WSD for its wanted transmissions. Its lower and upper edge frequencies must coincide with the European harmonized DTT channel raster ($470 + 8k-1$ and $470 + 8k$ MHz, respectively). The bandwidth of a Nominal Channel (and indeed the total bandwidth of wanted emissions) must not exceed the value specified by the WSDB.

2.9.4.2 In-Block Power and Power Spectral Density

The total radiated in-block RF power of a WSD must not exceed a specified level P. In the case of WSD operation in a single DTT channel, the value of P is equal to P_1 in dBm/(8 MHz) as communicated by the WSDB for that channel. In the case of simultaneous operation in multiple DTT channels, the value of P must be the lowest of the P_1 values in dBm/(8 MHz) specified for each of the channels being used. The radiated in-block RF power spectral density of a WSD within any DTT channel must not exceed a level P_0 dBm/(100 kHz) as specified by the WSDB for that DTT channel.

2.9.4.3 Emission Masks Inside the UHF TV Band

EN 301 598 specifies a total of five spectrum emission masks (or *classes*) inside the UHF TV band. The manufacturer must declare the emission class with which the WSD complies. Class 1 devices have the most stringent emission mask and will benefit from increased TVWS availability.

2.9.4.4 Database Discovery

At start up, and before initiating any transmissions, a master WSD must locate and consult a web-listing of approved WSDBs relevant to its geographical domain. A Master WSD must not transmit if it cannot consult a web-listing. Furthermore, a master WSD must not request Operational Parameters from a WSDB that is not on the web-list.

2.9.4.5 Master and Slave Updates

It may be necessary in certain circumstances (e.g. to protect incumbent primary users) to instruct a WSD to cease transmissions within a short time interval. As a result, there is a requirement that master WSDs must support an update function, through which a WSDB can inform that the Operational Parameters of the master WSD and its served slave WSDs are no longer valid. For this, a master WSD must either a) be able to receive an update from the WSDB within T_{Update} seconds (push update), or b) send an update request to the WSDB every T_{Update} seconds (pull update). A master WSD must support a minimum T_{Update} value of 60. A master WSD must cease transmission, and must instruct the slaves attached to it to cease transmission, if it receives an update from the WSDB that the relevant Operational Parameters are no longer valid. The actual value of T_{Update} will be specified by the relevant national regulatory authority and communicated to the WSDs by the WSDBs.

Similar requirements apply to slave WSDs, but here the update interval is fixed. A slave WSD must cease transmission within 1 s when updated to do so by its serving master WSD. Furthermore, a slave must cease transmission within 5 s of discovering that it can no longer receive updates from its serving master WSD.

2.9.4.6 Loss of Connection with the Database

A master WSD must at all times be reachable by a WSDB within a time interval T_{Update} (via pull or push update). However, this will not be possible if the connection with the WSDB is lost. In such cases, the master WSD must cease transmissions, and instruct its associated slave WSDs to also cease transmissions.

2.9.4.7 Geolocation

A key element in the operation of WSDs is the ability of the WSDB to provide Operational Parameters on the basis of the location of the WSD. However, not all WSDs are required to geolocate. For example, the broad location of slave WSDs can be derived from an estimate the coverage area of the serving master WSD. WSDs which do geolocate, however, must observe certain requirements. A WSD, whose location is more than a defined threshold distance away from the location it reported for obtaining Operational Parameters, must stop using those Operational Parameters. This threshold is itself an Operation Parameter, and its value can be set by the national regulatory authority. Furthermore, a geolocated WSD must check its location at least every 60 s.

2.9.4.8 Device Types

EN 301 598 defines two types of WSDs. Type A WSDs are intended for fixed use only, and can have integral, dedicated or external antennas. Type B WSDs are not intended for fixed use and can have an integral or a dedicated antenna. The requirement for an integral or dedicated antenna is to mitigate the risk of users tampering with the antenna.

2.9.4.9 User Access Restrictions and Security Measures

An important concern from the perspective of interference to incumbent primary services is the risk of users tampering with the WSDs. If a WSD user is capable of bypassing the process of receiving Operational Parameters from a WSDB, or is capable of inputting bogus Operational Parameters into the WSD, then serious interference could result. For this reason, EN 301 598 contains strict requirements to prevent users from modifying the configuration of the WSD, and to ensure that communications with a WSDB are secured and authenticated.

2.9.5 Conclusions

Database-assisted operation of white space devices in the UHF TV band is an example of a new paradigm in spectrum management, and relies on the real-time communication of regulatory emission limits and instructions to the wireless devices. However, just like other wireless devices, in order to be placed on the European market a white space device must comply with the Radio and Telecommunications Terminal Equipment (R&TTE) Directive.

EN 301 598 has been produced by ETSI as a harmonized standard for white space devices in the UHF TV band as a means to prove compliance with the requirements of the R&TTE Directive. It contains a number of traditional requirements, such as restrictions on emission masks, as well as many new requirements to account for database-assisted operation.

Since the database-assisted approach to spectrum management is still in its infancy, the current version of EN 301 598 is quite likely to be revised as white space technologies develop, and as regulators, manufacturers and users gather information on the operation of these devices in the field.

One area in particular where we expect to see further development relates to the issue of interference among TV white space devices. Mitigation of such interference can be achieved in a variety of ways, for instance by databases taking on a coordination role, or by means of polite protocols implemented by the devices themselves. At this stage it is too early to include requirements for such mitigation techniques in the harmonized standard—as the number of TV white space devices in the field is likely to remain low in the short term—but such requirements may become increasingly necessary as usage of TV white spaces becomes widespread.

References

1. ITU-R Report M.2225: Introduction to cognitive radio systems in the land mobile service (2011)
2. ITU-R Report M.2242: Cognitive radio systems specific for international mobile telecommunications systems (2011)
3. ITU-R Report M.2134: Requirements related to technical performance for IMT-advanced radio interface(s) (2008)
4. ETSI TR 102 802: Reconfigurable Radio Systems (RRS); Cognitive Radio System Concept. http://www.etsi.org
5. ETSI TR 102 907: Reconfigurable Radio Systems (RRS); use cases for operation in white space frequency bands. http://www.etsi.org
6. Review of the Directive 1999/5/EC, Commission proposal for the revision of the R&TTE Directive. http://ec.europa.eu/enterprise/sectors/rtte/documents/legislation/review/
7. ETSI TR 102 967: Reconfigurable Radio Systems (RRS); use cases for dynamic equipment reconfiguration. http://www.etsi.org
8. Mueck, M. D., Ivanov, V., Choi, S., Ahn, C., Yang, H., Kim, J., Jin, Y., Baldini, G., Steer, D., Alavi, H., Piipponen, A.: Reconfigurable radio systems as enabler for exploiting the future heterogeneous wireless communications landscape. ETSI Reconfigurable Radio Systems Workshop, Dec 2012
9. OneFIT Deliverable D2.1: Business scenarios, technical challenges and system requirements, p. 72 (2010)
10. OneFIT Deliverable D2.2.2/D6.4: Functional and system architecture version 2.0, p. 73 (2012)
11. Matinmikko, M., Bräysy, T. (eds.): Towards cognitive radio systems: main findings from the COGNAC project. VTT Research Notes 2575, p. 80 + app. 23 (2011)
12. Matinmikko, M., Höyhtyä, M., Mustonen, M., Sarvanko, H., Hekkala, A., Katz, M., Mämmelä, A., Kiviranta, M., Kautio, A.: Cognitive radio: An intelligent wireless communication system. Research report, VTT-R-02219-08. http://www.vtt.fi/inf/julkaisut/muut/2008/CHESS_Research_Report.pdf, March 2008
13. Ahokangas, P., Matinmikko, M., Myllykoski, J., Okkonen, H.: Future scenarios, ecosystems and business models for cognitive radio systems. VTT Technology 55, p. 54 (2012)

14. Ahokangas, P., Matinmikko, M., Yrjölä, S., Okkonen, H., Casey, T.: "Simple rules" for mobile network operators' strategic choices in future cognitive spectrum sharing networks. IEEE Wirel. Commun. **20**(2), 20–26 (2013)
15. Cognitive Radio Trial Environment + (CORE+) project web site http://core.willab.fi/
16. Report on ASA concept. CEPT Working Group document FM(12)084 Annex 47
17. ECC Report 159: Technical and operational requirements for the possible operation of cognitive radio systems in the "white spaces" of the frequency band 470–790 MHz. http://www.erodocdb.dk/docs/doc98/official/Pdf/ECCRep159.pdf (2011)
18. ECC Report 185: Complementary Report to ECC Report 159 Further definition of technical and operational requirements for the operation of white space devices in the band 470–790 MHz. http://www.erodocdb.dk/docs/doc98/official/Pdf/ECCRep185.pdf (2013)
19. ECC Report 186. Technical and operational requirements for the operation of white space devices under geo-location approach. http://www.erodocdb.dk/docs/doc98/official/Pdf/ECCRep186.pdf (2013)
20. Cisco Visual Networking Index: Global mobile data traffic forecast update, 2010–2015. [http://newsroom.cisco.com/ekits/Cisco_VNI_Global_Mobile_Data_Traffic_Forecast_2010_2015.pdf]. Accessed 24 October 2013
21. Sridhar, V., Krishna Kumar,G.: Realizing India's broadband dream. Financial Express (24 September 2013)
22. Caicedo, C.E., Weiss, M.B.: The viability of spectrum trading markets. IEEE Commun. Mag. **49**(3), 46–52 (2011)
23. Crocioni, P.: Is allowing trading enough? making secondary market in spectrum work. Telecommun. Policy **33**(8), 451–468 (2009). doi:10.1016/j.telpol.2009.03.007
24. Chapin, J.M., Lehr, W.H.: Cognitive radios for dynamic spectrum access-the path to market success for dynamic spectrum access technology. IEEE Commun. Mag. **45**(5), 96–103 (2007)
25. Holland, O., De Nardis, L., Nolan, K., Medeisis, A., Anker, P., Minervini, L.F., Velez, F., Matinmikko, M., Sydor, J.: Pluralistic licensing. In: IEEE International Symposium on Dynamic Spectrum Access Networks (2012)
26. Sridhar, V., Casey, T., Hämmäinen, H.: Flexible spectrum management for mobile broadband services: How does it vary across advanced and emerging markets? Telecommun. Policy **37**(2–3), 178–191 (2013)
27. Senge, P.M.: The fifth discipline: the art & practice of the learning organization. Doubleday, New York (1990)
28. Report ITU-R M.2072: World Mobile Telecommunication Market Forecast (2006)
29. Recommendation ITU-R M.1768: Methodology for calculation of spectrum requirements for the future development of the terrestrial component of IMT-2000 and systems beyond IMT-2000 (2006)
30. Report ITU-R M.2074: Radio aspects for the terrestrial component of IMT-2000 and systems beyond IMT-2000 (2006)
31. Takagi, H., Walke, B. (eds.): Spectrum Requirement Planning in Wireless Communications: Model and Methodology for IMT-Advanced, p. 248. Wiley, NJ (2008)
32. Report ITU-R M.2078: Estimated spectrum bandwidth requirements for the future development of IMT-2000 and IMT-Advanced (2006)
33. ITU "MIS Report 2013" (Measuring the Information Society). http://www.itu.int/en/ITU-D/Statistics/Pages/publications/mis2013.aspx. Accessed 07 October 2013
34. Ericsson "Mobility Report". http://www.ericsson.com/mobility-report (08, 2013)
35. Report ITU-R M.2243: Assessment of the global mobile broadband deployments and forecasts for international mobile telecommunications (2011)
36. GSA. Fast facts: Global mobile suppliers association. http://www.gsacom.com/news/statistics . Accessed 07 October 2013
37. RSPG: Draft Opinion on licensed Shared access", European Commission, Radio Spectrum Policy Group, RSPG13-529 rev1, (2013)
38. UMTS Forum Report 44, Mobile traffic forecasts 2010–2020 report. http://www.umts-forum.org/component/option,com_docman/task,doc_download/gid,2537/Itemid,213/, January 2011

39. The National Broadband Plan. http://www.broadband.gov/plan/
40. First Radio Spectrum Policy Programme (RSPP): European Parliament and Council http://ec.europa.eu/digital-agenda/en/rspp-roadmap-wireless-europe (2012)
41. E2R IP in Europe—Overview, Didier Bourse. http://www.ofcom.org.uk/static/archive/ra/topics/research/topics/converge-new-emerging/sdr/10-bourse.pdf (2003)
42. EC FP6 E2R II Achievements and EC FP7 E3 Ambitions: Didier Bourse. http://www.itu.int/dms_pub/itu-r/oth/0A/06/R0A060000420001PDFE.pdf (2008)
43. End-to-End Efficiency (E3): Didier Bourse. ftp://cordis.europa.eu/pub/fp7/ict/docs/future-networks/projects-e3-factsheet_en.pdf (2008)
44. Perspectives on the value of shared spectrum access—Final Report for the European Commission, SCF Associates for the European Commission. http://ec.europa.eu/information_society/policy/ecomm/radio_spectrum/_document_storage/studies/shared_use_2012/scf_study_shared_spectrum_access_20120210.pdf, February 2012
45. ETSI Reconfigurable Radio Systems (RRS) Technical Committee. http://portal.etsi.org/portal/server.pt/community/RRS/329
46. European Commission, Document M/512 EN: Standardisation Mandate to CEN, CENELEC and ETSI for reconfigurable radio systems.www.http://ec.europa.eu/enterprise/standards_policy/mandates/database/index.cfm?fuseaction=search.detail&id=515(2013). Accessed 31 October 2013
47. CEPT WG FM PT52, Terms of References. http://www.cept.org/ecc/groups/ecc/wg-fm/fm-52/page/terms-of-reference
48. CEPT WG FM PT53, Terms of References. http://www.cept.org/ecc/groups/ecc/wg-fm/fm-53/page/terms-of-reference
49. IEEE 1900.4 Working Group on Architectural Building Blocks Enabling Network-Device Distributed Decision Making for Optimized Radio Resource Usage in Heterogeneous Wireless Access Networks. http://grouper.ieee.org/groups/dyspan/4/
50. Radio Spectrum Policy Group.: Report on CUS and other spectrum sharing approaches: Collective Use of Spectrum. http://rspg-spectrum.eu/_documents/documents/meeting/rspg26/rspg11_392_report_CUS_other_approaches_final.pdf (2011)
51. Electromagnetic compatibility and Radio spectrum Matters (ERM): Mobile broadband services in the 2300 MHz–2400 MHz frequency band under Licensed Shared Access regime—System Reference Document, ETSI (2012)
52. ETSI TS 103 154: Reconfigurable Radio Systems (RRS); System requirements for operation of Mobile Broadband Systems in the 2300 MHz–2400 MHz band under Licensed Shared Access (LSA) regime, to appear
53. Medeisis, A., Delaere, S.: High-level scenarios for the future of cognitive radio business. In: 2011 IEEE 22nd International Symposium on Personal Indoor and Mobile Radio Communications (PIMRC) (2011)
54. Bradfield, R., et al.: The origins and evolution of scenario techniques in long range business planning. Futures **37**, 795–812 (2005)
55. Karlson, B., et al.: Wireless Foresight: Scenarios of the Mobile World in 2015. Wiley, UK (2003)
56. Casey, T.: Analysis of radio spectrum market evolution possibilities. Commun. Strat. **75**, 109–130 (2009)
57. Mapping the Future of … Convergence and Spectrum Management: A Scenario-Based Review of Spectrum Management Strategies in the Converged Era, Study Report for UK Radiocommunications Agency by NerveWire, Indepen Consulting, and Intercai Mondiale. http://www.ofcom.org.uk/static/archive/ra/topics/convergence/document/mappingfuture/may2000/conv_rpt.pdf, May 2000
58. Schoemaker, P.J.H.: Multiple scenario development: its conceptual and behavioral foundation. Strateg. Manag. J. **14**(3), 193–213 (1993)
59. ETSI TR 102 683 V1.1.1 (2009-09): Reconfigurable Radio Systems (RRS); Cognitive Pilot Channel (CPC)

60. Roadmap for Software Defined Radio—A technology, market and regulation perspective, Danish Technological Institute and GateHouse, pp.22–23. http://www.teknologisk.dk/_root/media/38349_Roadmap%20for%20Software%20Defined%20Radio_dec2009_public.pdf, December 2009
61. ECC Report 016: Refarming and Secondary Trading in a Changing Radiocommunications World. http://www.erodocdb.dk/Docs/doc98/official/pdf/ECCREP016.PDF, September 2002
62. Buddhikot, M.: Understanding dynamic spectrum access: models, taxonomy and challenges. In: Proceedings of IEEE DySPAN 2007, Dublin, Ireland, 17–21 April 2007
63. Mitola, J.: Cognitive radio: An integrated agent architecture for software defined radio. Ph.D. Dissertation, Royal Institute of Technology (KTH), Kista, Sweden (2000)
64. Mitola, J., Maguire, G.: Cognitive radio: making software radios more personal. IEEE Pers. Commun. **37**(10), 13–18 (1999)
65. Wyglinski, A.M., Nekovee, M., Hou, Y.T.: Cognitive Radio Communications and Networks: Principles and Practice. Elsevier, Amsterdam (2009)
66. Chen, K-C., Prasad, R.: Cognitive Radio Networks. Wiley, West Sussex (2009)
67. Hossain, E., Niyato, D., Han, Z.: Dynamic Spectrum Access and Management in Cognitive Radio Networks. Cambridge University Press, Cambridge (2009)
68. ITU-R Report SM.2152: Definitions of Software Defined Radio (SDR) and Cognitive Radio System (CRS) (2009)
69. Medeisis, A., Holland, O., De Nardis, L.: Taxonomy of cognitive radio applications. In: 6th IEEE Symposium on New Frontiers in Dynamic Spectrum Access Networks, Washington, USA, October 2012
70. Draft ETSI EN 301 598 V1.0.0 (2013-07): White space devices (WSD); Wireless access systems operating in the 470–790 MHz frequency band; Harmonized EN covering the essential requirements of article 3.2 of the R&TTE Directive
71. ETSI TR 102 802 V1.1.1 (2010-02), Reconfigurable Radio Systems (RRS); Cognitive Radio System Concept
72. European Commissions Office (ECO): ECO report 03—the licensing of mobile bands in CEPT. [http://www.cept.org/eco/deliverables/eco-reports] (2011). Accessed 24 October 2013

Chapter 3
Technical Approaches for Improved Spectrum Sharing

Fernando J. Velez and Marja Matinmikko

Abstract The aim of this chapter is to showcase several important contributions towards identification of techniques for spectrum sharing and coexistence. It is envisaged that such novel techniques may be of value in various regulatory considerations. They might assist in shaping technical conditions that will govern the access to radio spectrum by CR technologies. Section 3.1 reviews the recommended principles of Geolocation Databases' operation in European regulatory environment, as well as the envisaged structural composition of technical solutions for their implementation. Section 3.2 is composed of several contributions that offer different angles of looking at CR spectrum sensing algorithms and implementation techniques. It also considers the possibilities of dynamic re-configurability through beam forming capabilities. This is followed by discussion in Sect. 3.3 of possibilities for spectrum aggregation from non-contiguous frequency bands, as made possible by the DSA capabilities of CR. This opens up possibilities for significantly increasing the available operational bandwidth—and hence the data throughput—of the radio transceivers. It also allows pursuing energy efficiency objectives. Section 3.4 looks at the possibilities of developing unsynchronised CR networks using Filter-bank multi-carrier, an alternative type of modulation that offers a superior performance and reduced out-of-band emissions compared with other traditional types of modulation. The detection of malicious users is addressed in Sect. 3.5, by employing a statistical approach. This approach allows reliable detection of users even when the system does not have a priori information about primary channel activity and characteristics of users. The following Sect. 3.6 looks at the spectral efficiency of CR systems and the related possibilities of using Iterative Water-filling method, which may be highly

F. J. Velez (✉)
Universidade da Beira Interior - DEM, Instituto de Telecomunicações, Covilhã, Portugal
e-mail: fjv@ubi.pt

M. Matinmikko
VTT Technical Research Centre of Finland, Oulu, Finland
e-mail: Marja.Matinmikko@vtt.fi

A. Medeisis and O. Holland (eds.), *Cognitive Radio Policy and Regulation*, Signals and Communication Technology, DOI: 10.1007/978-3-319-04022-6_3,

beneficial for broadband wireless channels under static or slowly varying conditions. The final Sect. 3.7 contains two contributions that present different aspects associated with the issue of assessing the amount of white spaces, or in other words—spectrum resource available for DSA access. It first looks at the principles of Radio Environment Mapping, which is then complemented by an example of evaluating amount of TV band white spaces in Italy.

3.1 Geo-Location Databases

Alexander Kholod

Federal Office for Communications, Biel/Bienne, Switzerland

3.1.1 Background

An important issue when getting the access to spectrum by a cognitive radio system [1] is the protection of incumbent users of this spectrum. The geo-location database is one of the cognitive techniques used to assess the current usage of the spectrum and to assist accessing this spectrum without placing undue constraints on incumbent users of the spectrum.

The principles and requirements to the cognitive technique employing the geo-location database have been developed in the course of the CEPT studies on white space devices (WSDs). WSDs are defined as devices that can utilise White Space spectrum without generating harmful interference to incumbent services by making use of cognitive techniques [2][1].

To this end, ECC Report 159 [3] sets initial principles and operation requirements to WSDs under the geo-location database technique, which are further developed in ECC Report 186 [4].

3.1.2 Introduction

An increased interest to the geo-location technique is outlined by its flexibility and scalability in addressing various cognitive wireless systems: from local television

---

[1] *White Space spectrum is defined by the CEPT* [2] *as "a label indicating a part of the spectrum, which is available for a radio communication application (service, system) at a given time in a given geographical area on a non-interfering/non-protected basis with regard to other services with a higher priority on a national basis".*

WSDs [3] to industrial level geo-location solutions to support Licensed Shared Access operations of cellular mobile networks [5].

Though the concept of White Space spectrum was first used in connection to the band 470–790 MHz, it can be translated to other frequency bands, which are considered underutilized either in terms of location or time. Furthermore, the principles for the geo-location database operation are generic and can be considered without application to a particular frequency band and related protection of a particular incumbent user.

This section presents the general concept of the geo-location database operation and discusses different issues in this regard. The requirements to two integral parts of the geo-location database system, namely the database and the WSD, are considered as well.

3.1.3 General Principles

The geo-location technique is the process, where a WSD measures its geographical location and queries a geo-location database in order to get information on available frequencies at this location. As discussed later in this section this process requires some exchange of information to occur between the WSD and the geo-location database.

The determination accuracy of the WSD geographic location is one of the crucial elements in the whole process. In particular, any errors in location determination may alter the fundamental purpose of using the geo-location database, which consists in ensuring the operation of incumbent users free from WSD interference. The WSD should maintain this accuracy while in operation.

A closely related aspect is the mesh used by the geo-location database to store information on spectrum usage across a given geographical area. Under a large scaled mesh, the information on spectrum availability might be inaccurate. Too fine mesh will likely result in elevated computing efforts and an increased data transfer.

3.1.3.1 Master-Slave Concept

A radio communication link between the devices of the same application will always involve at a minimum two transmitting/receiving units. Therefore, any operational scenario with WSDs can be described as a communication between a master device and a slave device.

A master WSDs is a WSD that can establish a direct communication channel with a geo-location database to obtain operational parameters pertinent to its geographical location. A slave WSDs cannot establish such a communication channel and obtains the relevant operational information directly from its serving master WSD.

To this end, one distinguishes between two cases of master/slave WSD deployment: one for so-called non-geo-located slaves and another for geo-located slaves. Both cases are illustrated schematically in Fig. 3.1.

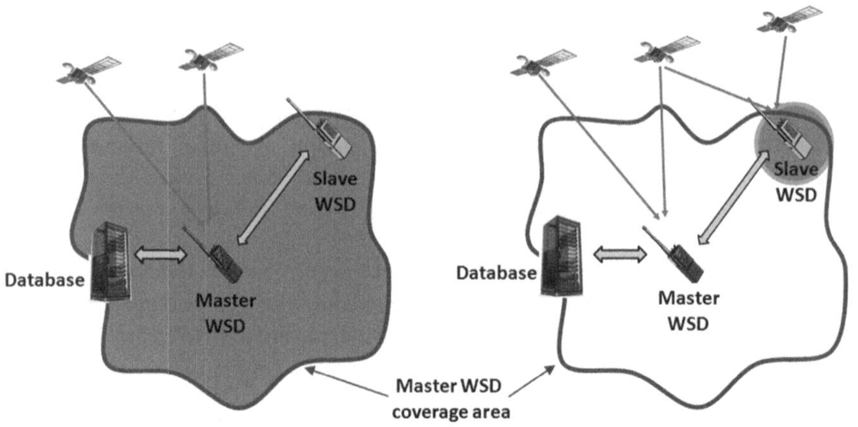

Fig. 3.1 Master/slave concept with non-geo-located (*left*) and geo-located (*right*) WSDs. The *shadow regions* show the area within which the slave WSDs can be determined to be located

In the case of non-geo-located slaves, the master does not know the exact geographical locations of the slaves within its coverage area. Therefore, the operational information obtained by the master WSD from the database and communicated further to the slave WSDs is valid for the whole coverage area of the master WSD. The coverage area can be determined either by the master WSD itself or by the geo-location database on the basis of information received from the master.

In the case of geo-located slaves, the master knows the exact geographical locations of its slaves. It queries, therefore, the database for its own operational parameters and then sequentially for each slave. The received information is then communicated to appropriate slaves.

The requirements set to the geo-location database operation are differentiating between those applied for master WSDs and those applied for slave WSDs.

3.1.3.2 Technical and Operational Requirements

The baseline for the transmission allowance for master WSDs in the territory of a country is the discovery of a geo-location database, which has been approved by the regulator of that country. This requirement is sometimes called 'no-go requirement'. If it is met successfully, an exchange of information between the master WSD and the geo-location database as well as between the master WSD and the slave WSDs will need to take place.

In particular, the master WSD shall communicate to the database its device parameters including the geographical location, the accuracy this location has been measured, device type (fixed installation or portable/mobile), device emission characteristics (e.g., spectrum mask, emission technology, etc.), device model (to enable solving interference problems), device unique identifier (to allow tracing individual devices) and device category (master or slave). The master WSD shall also communicate to the database the device parameters of its associated slaves. The master WSD may optionally communicate to the database the detailed antenna characteristics, including the antenna height, antenna angular discrimination, and antenna polarisation, for itself and associated slave WSDs provided such information is available.

The geo-location database will then communicate back to the master WSD the list of available frequencies within the device's location (for both the master and slaves WSDs), the associated maximum permitted transmit power levels for each of available frequencies, and the validity time of parameters provided. The geo-location database may optionally communicate to the master WSD a sensing threshold for the detection of incumbent services as well as a so-called "cease operation" message that can be used in exceptional cases to request the master WSD and its associated slave WSDs to stop instantaneously their transmissions.

It stands to reason that the master WSD shall only operate in accordance with the above technical and operational parameters received from the geo-location database. The following needs to be added here. After having received the above instructions from the geo-location database and before commencing any transmission, the master WSDs shall communicate to the database the channel usage parameters, which include the selected frequency, intended transmit power and, optionally, its coverage area (in case of non-geo-located slave WSDs). These parameters will be used by the database to assess the spectrum usage by WSDs in a given geographical area and, if required, take account of cumulative interference.

In a similar way as it is done between the master WSD and the geo-location database, the slave WSD shall communicate to its serving master WSD the list of device parameters that includes the device type, device emission characteristics, device model and device unique identifier. For the geo-located slave WSDs, the list of device parameters sent to the database will also include the slave WSD antenna location and the antenna location accuracy. Optionally, the detailed antenna characteristics may also be communicated by the slave WSD to its serving master WSD.

The slave WSD shall be capable of receiving from the master WSD the information on available frequencies, the associated maximum permitted transmit power levels, and the validity time of parameters provided. Optionally, the master WSD may send to its slave WSD a sensing threshold for the detection of incumbent services as well as the "cease operation" message. The slave WSD may only operate in accordance with the instructions received from the master WSD. It shall stop immediately its transmission if instructed to do so by its serving master WSD or when no communication with the master WSD can be established after the validity time.

3.1.3.3 Combined Sensing and Geo-Location

It is also possible to combine the geo-location database technique with the spectrum sensing. Such a combined technique presents two potential advances compared to the geo-location database only; however, these advances appear to be mutually exclusive.

One benefit consists in the improved detection of incumbent users of the spectrum compared to the cases when either spectrum sensing alone or geo-location database alone are used. A favorite decision on the spectrum availability is taken only when both the geo-location database and the spectrum sensing confirm the absence of an incumbent spectrum user at the WSD location.

Another benefit is that the combined technique may also allow the detection of incumbent spectrum users, which are not registered in the database. In particular, such incumbent users are detected by means of the spectrum sensing without support with information from the database. The combined detection and associated decision is only then employed for the incumbent users, which are registered in the database.

It needs to be mentioned here that the implementation of reliable spectrum sensing has a number of challenges [6]. Therefore, the potential benefits of the combined approach mentioned above may not be achievable in practice.

3.1.3.4 Database Management Issues

A general approach to the geo-location database management is presented in Fig. 3.2. The goal of this process is to ensure protection of incumbent users of the spectrum while providing WSDs with access to this spectrum. There are a number of key elements associated with the geo-location database management.

In order to ensure protection of incumbent users of the spectrum, the database should be loaded with the information on these users. This could be achieved by recording transmitter parameters (like antenna location, height and pattern, output power, etc.) and receiver characteristics (like sensitivity level, protection requirements, etc.) for each of the protected services/systems. Another possibility could be to associate each point of the database mesh with the received signal characteristics of the incumbent users.

The database is to contain the latest information on incumbent users. An important parameter in this regard is the database update delay, which defines the time period set to update the database once incumbent users provide a notice of a change in their assignments. Another important parameter related to the information on incumbent users is the database update frequency. This parameter sets the periodicity of database update required to keep the information in the database valid.

The information received by the database from the WSD and the information contained in the database on incumbent spectrum users would have to be converted into the list of available frequencies at the WSD location and associated maximum permitted transmit power levels for each of available frequency. Hence, a translation process is to be performed to bridge these two ends. This is a critical element in the whole geo-location database architecture indeed and is to be implemented properly. If it is not, there is a risk either of interference into incumbent users, or of the WSD access to the spectrum being limited without cause.

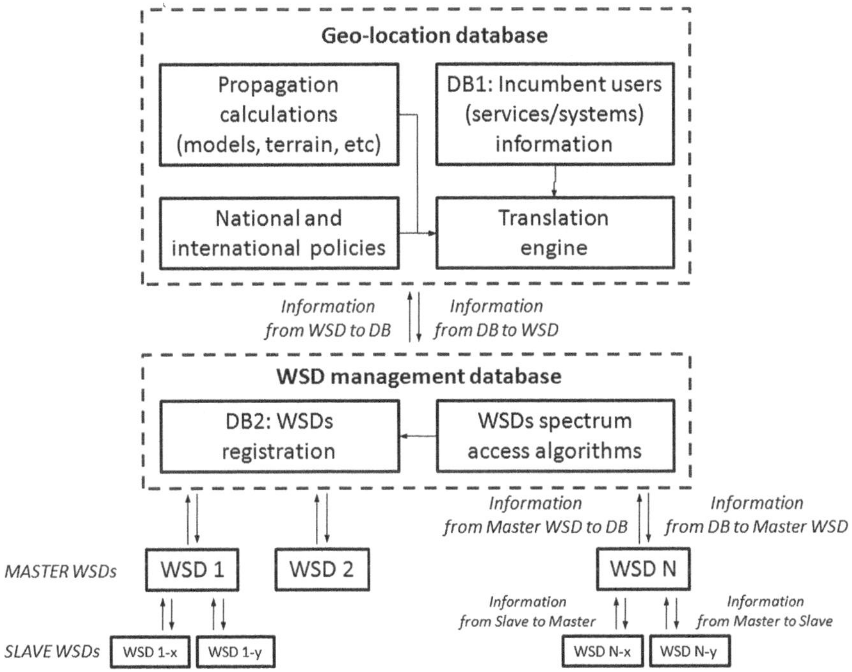

Fig. 3.2 Geo-location database management

The algorithms implementing the translation process depend primarily on the type of the incumbent service/system. Propagation loss calculations constitute an inherent part of these algorithms. It needs to be highlighted here that the way the translation process is implemented must be in conformity with related national and international policies. In particular, international policies cover the cases when a WSD in one country could potentially interfere harmfully with an incumbent spectrum user in a neighbouring country.

More than one WSD may seek for spectrum access in a given geographical area. In order to ensure that different WSDs access the spectrum in a coordinated manner (e.g. without mutual interference) special spectrum access algorithms need to be implemented. These algorithms are also governed by national policies with regard to spectrum sharing between WSDs and incumbent users and between WSDs themselves.

Furthermore, the WSDs that were already granted spectrum access need to be registered in the database. This would help the geo-location database to correctly assess and coordinate the spectrum usage by WSDs in a given geographical area. The knowledge about WSDs operating in the area is also required in order to consider properly cumulative interference from WSDs into incumbent spectrum users.

3.1.4 Concluding Remarks

The geo-location technique is a powerful tool to ensure a controlled access to underutilized spectrum. Development of a regulatory framework for this technique raises some issues and sets a number of requirements. The treatment of these issues and the implementation of these requirements require coordinated efforts by both the regulatory and standardization organizations.

3.2 Cooperative CR Spectrum Sensing and Beam Forming

Agapi Mesodiakaki[1], Ferran Adelantado[2], Luis Alonso[1] and Christos Verikoukis[3]

[1]Technical University of Catalonia (UPC), Barcelona, Spain

[2]Open University of Catalonia (UOC), Barcelona, Spain

[3]Telecommunications Technological Centre of Catalonia (CTTC), Barcelona, Spain

3.2.1 A Novel Energy-Efficient Contention-Aware Channel Selection Algorithm for CR Networks

In recent years, due to the ever increasing traffic demands and the limited spectrum resources, it is very likely that several cognitive radio ad hoc networks (CRAHNs) will coexist and opportunistically use the same primary user (PU) resources. In such scenarios, the ability to distinguish whether a licensed channel is occupied by a PU or by another CRAHN can significantly improve the spectrum efficiency of the network, while the contention among the CRAHNs already operating on the licensed channels with no PU activity, may further affect the performance of the network. Therefore, the proper selection of the frequency bands, already being used by other CRAHNs, could result in notable throughput and energy efficiency gains for the network under study.

To that end, in [7] we have proposed a novel energy-efficient contention-aware channel selection algorithm in a scenario where the CRAHN under study coexists with other non-cooperating CRAHNs based on a specific coexistence scheme. According to the algorithm, the CRAHN: (i) initially locates the spectrum holes by exploiting cooperative spectrum sensing (CSS) (ii) categorizes the idle licensed channels based on their contention level (i.e., number of secondary users (SUs) belonging to other non-cooperating CRAHNs that are operating on the licensed channels) and (iii) selects the less contended licensed channel to access first.

During CSS, feature detection [8, 9] is combined with a technique that estimates the number of SUs in each licensed channel. On the one hand, feature detection enables the distinction between different types of signals (e.g., PUs' and SUs' signals) at the expense of higher complexity and longer sensing time. Moreover, it requires prior information about the PU waveforms, which however, it is typically known for most standard technologies that operate on licensed channels [10]. Hence, the use of feature detection enables the distinction of the licensed channels into: (i) those with PU activity, (ii) those with SU activity and iii) those with no activity at all and it is used as the reference algorithm for the performance comparison of our approach; Please also note that the use of feature detection is fundamental, as a simpler technique (e.g., energy detection) would result in very low spectrum efficiency, as all the idle licensed channels being used by other CRAHNs, would be considered busy and thus would be avoided. On the other hand, the second technique estimates the contention in each licensed channel with SU activity. Specifically, it estimates the number of SUs belonging to the other non-cooperating CRAHNs that are already operating on the channel. Thereafter, the algorithm categorizes the idle licensed channels based on their contention level and then selects the less contended channel to be accessed first by the CRAHN under study.

In [7], it has been shown that the proposed algorithm manages to achieve notable performance gains without inducing any significant additional overhead in comparison with the reference algorithm. This is due to the fact that the performance of the algorithm only depends on the comparison between the estimated values of SUs in each licensed channel and not on their exact value. Specifically, the performance of the proposed algorithm has been evaluated by means of simulations and it has been shown that it can present gains up to 70 % in throughput and up to 68 % in energy efficiency for high traffic in the licensed channels in comparison with the reference algorithm.

In such coexistence scenarios, achieving fairness among the coexisting CRAHNs (i.e., equal transmission opportunities among their SUs) constitutes another key issue. Generally, achieving fairness among the SUs that share the same PU spectrum is a research topic that has received a lot of attention. However, most proposals assume that a licensed channel that is occupied by an SU cannot be accessed by another SU. In particular, the channel appears as being busy to the SU and thus it is avoided. To overcome the aforementioned problem, in [11], the authors propose FMAC, a MAC protocol that utilizes a three state sensing model. FMAC uses a spectrum sensing algorithm [12] to distinguish whether a busy channel is occupied by a PU or by an SU and, in the latter case, gives the option to the SU to share the channel with the SUs of other CRAHNs that are currently using it. Nevertheless, in [11] a simple system model consisting only of one licensed channel is considered, while more importantly, the scheme employs a constant back-off window. As a result, unlike the secondary network coexistence scheme (SNCS) proposed in [7], it shows low adaptability to any changes in the number of contending SUs in a licensed channel.

To that end, in [13] we evaluated the performance of SNCS in terms of fairness among the coexisting CRAHNs in comparison with FMAC by means of simulations and it has been shown that it can achieve significant throughput and energy efficiency gains, while maintaining or even achieving better fairness among the

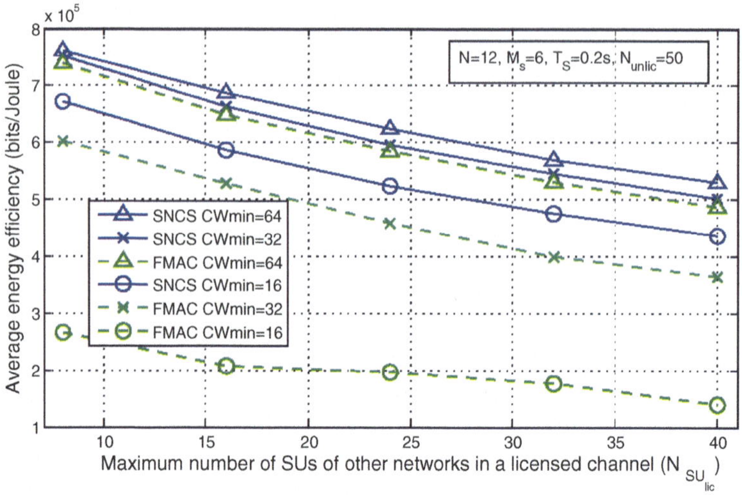

Fig. 3.3 Average energy efficiency of the CRAHN under study versus the maximum number of SUs of other networks in a licensed channel, N_{SUlic}, for different minimum back-off window values, $CW_{\min}$

coexisting CRAHNs. Please note that for the channel selection the algorithm proposed in [7] is used for both coexistence schemes.

Then, in Fig. 3.3, the average energy efficiency of the CRAHN under study is depicted for both SNCS and FMAC versus the maximum number of SUs of other networks in a licensed channel, N_{SUlic}, for different minimum back-off window values, $CW_{\min}$.

As it can be observed, the energy efficiency of the CRAHN under study is decreased as the contention in the licensed channels increases, due to the increased number of collisions among the SUs. However, notice that the SNCS achieves better performance than FMAC for all the considered values of $CW_{\min}$. This is due to the fact that, in FMAC, the SUs use a constant back-off window every time a collision takes place, while the SNCS employs an exponential one and thus it manages the collisions more efficiently. Therefore, the maximum performance gain of SNCS is achieved for the lowest minimum back-off window value ($CW_{\min} = 16$) and for high contention in the licensed channels ($N_{\mathrm{SUlic}} = 40$).

Moreover, as far as the fairness of the considered approaches is concerned, we employ in our study the well-known Jain's fairness index [14], which is given by:

$$J(x_1, x_2, \ldots, x_n) = \frac{\left(\sum_{i=1}^{n} x_i\right)^2}{n \sum_{i=1}^{n} x_i^2} \tag{3.2.1}$$

where n is the number of contending SUs and x_{i} denotes the number of transmission opportunities of the SU i. We consider that an SU has a transmission opportunity every time it transmits a packet on the channel independently of whether a successful transmission or a collision occurs.

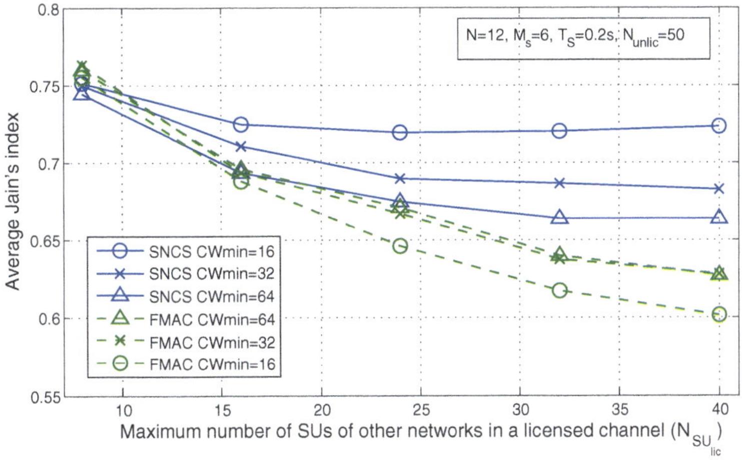

Fig. 3.4 Average Jain's index versus the maximum number of SUs of other networks in a licensed channel, N_{SUlic}, for different minimum back-off window values, $CW_{\min}$

Then, the average Jain's index of all the SUs, that contend to gain access to a licensed channel, is depicted for both approaches in Fig. 3.4, for different minimum back-off window values, $CW_{\min}$.

As it can be noticed, SNCS can achieve up to 25 % better fairness than FMAC ($CW_{\min} = 16$) for high contention in the licensed channels ($N_{\mathrm{SUlic}} = 40$). This stems from the fact that for short contention periods, i.e., in the case that the PU resumes its activity in the licensed channel shortly after the CRAHN under study has hopped to it, SNCS achieves much better fairness among the SUs than in FMAC, as an SU that is involved in a collision defers its transmission for a longer time, and thus the transmissions opportunities are more equally distributed among the contending SUs.

3.2.2 Cooperative CR Beam Forming in the Presence of Location Errors

Auon Muhammad Akhtar[1], Luca De Nardis[2], Mohammad Reza Nakhai[1], Oliver Holland[1], Maria Gabriella Di Benedetto[2] and A. Hamid Aghvami[1]

[1]King's College London, London, UK

[2]Sapienza University of Rome, Rome, Italy

3.2.2.1 Background and Purpose

Thus far, most of the research on cognitive radio has been focused on single-hop scenarios, tackling physical (PHY) layer and/or medium access control (MAC) layer issues. However, recent research finding have highlighted the potentials of multi-hop cognitive radio networks [15]. The cognitive paradigm can be applied to different scenarios of multi-hop wireless networks, one such scenario being the cognitive radio ad hoc network which consists of CR nodes which communicate with each other in a peer to peer fashion through ad hoc connections [16]. To fully realize the potential of such networks, cross-layer design issues must be addressed, for example, the routing decisions at the network layer should be made in conjunction with the PHY layer characteristics.

This chapter, summarising work first presented in [17], introduces a transmit beamforming strategy taking into account the positions of primary, secondary victim and intended secondary receivers, to achieve underlay secondary access in multihop cognitive radio networking. The transmit beamforming strategy defines a novel path optimization scheme that deviates from a preselected path given by the routing module, based on local information and according to a relay selection metric. This metric is designed to improve both coexistence with primary/secondary victim receivers and performance of the secondary cognitive network. Simulations compare the proposed strategy with a baseline solution that does not adopt beamforming, and with a strategy that applies beamforming on each hop without modifying the original path. Results show that the proposed strategy is capable of improving coexistence with primary/secondary victims, and highlight that a trade-off exists between the meeting of coexistence constraints and maximisation of secondary network performance.

3.2.2.2 Transmit Beamforming Strategy

A first key element of this work is the transmit beamforming strategy between hops in the path. This strategy aims to minimize the total transmitted power subject to certain constraints. These constraints are the interference margins of the primary nodes and minimum signal-to-noise ratio/signal-to-interference-and-noise ratio (SNR/SINR) requirements of the secondary nodes. Instead of using instantaneous channel state information (CSI), we use the second order statistics of the channel state information at the transmitting nodes. It is assumed that the secondary transmitters have knowledge about the locations of all the primary nodes within their transmission range. Instead of using a large number of samples to obtain the second-order statistics, we use the expression derived in [18] to obtain these statistics directly. Based on this, an optimisation problem is derived to minimise the transmit power at the beamforming secondary node, subject to constraint of a minimum required SINR at the secondary receiver and an interference threshold at all primary receivers [17]. Since the resulting problem is non-convex, it is converted into convex sequential dynamic programming (SDP) form, which can solved by an SDP solver like SeDuMi. More information on the procedure for this is available in [17].

3.2.2.3 Routing Algorithm Design

We next utilize transmit beamforming to design a routing algorithm for multi-hop cognitive radio networks. The objective of the algorithm is three-fold: (i) To minimize the end to end power consumption, (ii) To minimize the co-channel interference generated within the secondary network, and (iii) To minimize the number of primary interference constraint violations.

To achieve the goals set above, we adopt a centralized approach whereby the optimal power saving route is initially calculated through Dijkstra's algorithm by using the point to point link costs without beamforming. After this initial step, the algorithm modifies the selected route by using a novel cost metric. To ensure that the modified route does not deviate too much from the optimal power saving route, the cost metric is used only on alternate hops, for example, for every odd numbered hop of the optimal route, the hop destination is selected based on the proposed cost metric, while the destinations of the even numbered hops remain unchanged.

We now propose a cost metric which is used to select the node which is most suitable to act as a relay. The proposed metric takes into account the potential impact of the selection of a relay on the primary receivers and other secondary nodes, within the transmission range of the source and the candidate relay node. In the following, we refer to the source, relay and destination nodes as *S*, *R* and *D*, respectively. A terminal *R* will only be eligible as a relay if it meets all the following conditions:

(1) *S* does not violate the interference constraint of any of the primary users when it transmits data to *R* using beamforming;
(2) *R* does not violate the interference constraint of any of the primary users when it transmits data to *D* using beamforming;
(3) The position of *R* is such that the distance between *R* and *D*, indicated as $dist_{\mathrm{RD}}$, is not larger than the distance from *S* to *D*, $dist_{\mathrm{SD}}$. This condition ensures physical connectivity between the selected relay and *D* and it ensures that the algorithm remains loop free. The above description translates into the cost *Cost(S,D,R)* associated to the generic terminal *R* as a potential relay between *S* and *D*.

Further information on the associated cost function can be obtained in [17]. However, for the purpose of this chapter, Fig. 3.5 gives a demonstration of the proposed cost metric in action. Here, green nodes are the secondary source and destination nodes, the yellow nodes are the primary nodes while the nodes labelled as *R1,…,R6* are the candidate relay nodes. Amongst the candidate relay nodes, the nodes in blue are the eligible candidates while the nodes in red, i.e., *R4* and *R6* are

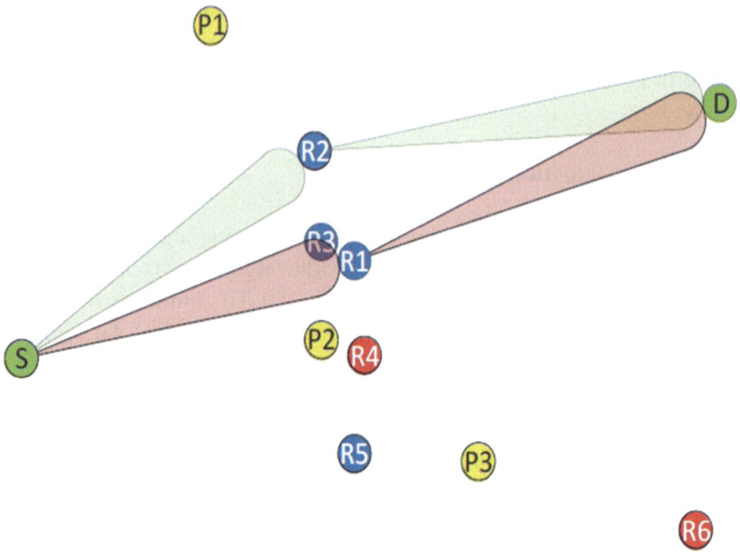

Fig. 3.5 Demonstration of the proposed cost metric

not eligible to act as relays. *R4* is rejected because *S* cannot transmit to *R4* without violating the interference constraint of *P2* while the distance between *S* and *R6* is larger than the distance between *S* and *D*. The Dijkstra's algorithm selects *R1* as the best option to act as a relay. However, it can be seen that if the data is transmitted to *R1*, a lot of interference is exerted upon *R3*. On the other hand, if *R3* is selected as relay, interference will be exerted upon *R1* when *R3* forwards the data to *D*. Using the proposed cost metric, the best option here is to select *R2* as relay.

3.2.2.4 Performance Evaluation

The performance of the proposed solution is evaluated by means of computer simulations executed by combining MATLAB and OMNeT++ as follows:

(1) MATLAB was used to implement the transmit beamforming strategy introduced in Section III and to analyse the performance of the route optimization approach defined in Section IV by measuring the interference generated towards each secondary node as well as the average number of constraints set by primary receivers that are met.
(2) OMNeT++ was used to test the proposed strategy in presence of actual packet transmissions in order to measure the impact of the proposed solution on throughput, moving from results generated in MATLAB.

3.2.2.5 Simulation Scenario and Setup

The MATLAB code was used to simulate a network of secondary nodes equipped with a ULA with $N = 8$ antenna elements and a spacing between adjacent elements $d = 0.0625$ m, corresponding to half a wavelength for a carrier frequency $f_c = 2.4$ GHz, and capable thus to perform DOA estimation and beamforming. An angular spread $\Delta\theta = 2^\circ$ was introduced around the exact angle for each measurement. A noise power $P_n = -101$ dBm was assumed at each receiver, while the path loss exponent for propagation was set equal to $\alpha = 2$. MATLAB was used to solve the beamforming optimization problem of taking advantage of the SeDuMi solver provided by the cvx package, imposing an upper bound φ_m on the allowed interference towards the primary nodes and a minimum SNR level of $\gamma_r = 10$ dB for all the secondary nodes.

The following steps were executed in MATLAB for each run:

(1) Generation of a topology composed of N_S secondary nodes and N_P primary nodes randomly deployed in an area of $X_{max} = 50$ m by $Y_{max} = 50$ m^2;
(2) Generation of N_{conn} connection requests in the secondary network with random source and destination nodes, random duration uniformly distributed between *minDuration* and *maxDuration* and random delay from the previous connection request from same source node uniformly distributed between *minDelay* and *maxDelay*; then, for each connection request: (a) Selection of the best path according to the minimum power routing strategy defined previously in this chapter; (b) Optimization of the path according to the proposed metric, described previously and in detail in [17]; (c) Measurement of interference generated towards secondary nodes not involved in the connection with and without optimization; (d) Measurement of number of primaries for which the constraint on the maximum interference value is met with and without optimization.
(3) Export to file of the data required by OMNeT++, consisting of: (a) primary and secondary network topology; (b) the list of the N_{conn} generated connection requests, including source, destination and duration; (c) original and optimized paths for each connection; (d) the reduction in the interference $I(x, y, z)$ perceived in y guaranteed by the introduction of beamforming in the link from x to z, for all x–z pairs involved in any connection, for both original and optimized paths.

The inputs generated in MATLAB were used in a simulated secondary network built in OMNeT++, with each secondary node characterized by the architecture shown in Fig. 3.6. With reference to such architecture, it should be noted that:

- The mobility and clustering modules were not activated, as a static network with flat organization was assumed.
- The positioning module was configured so to provide perfect position information about all network nodes.
- The application module for a generic node x was in charge of reading from file connection requests having x as source, and generate for each connection

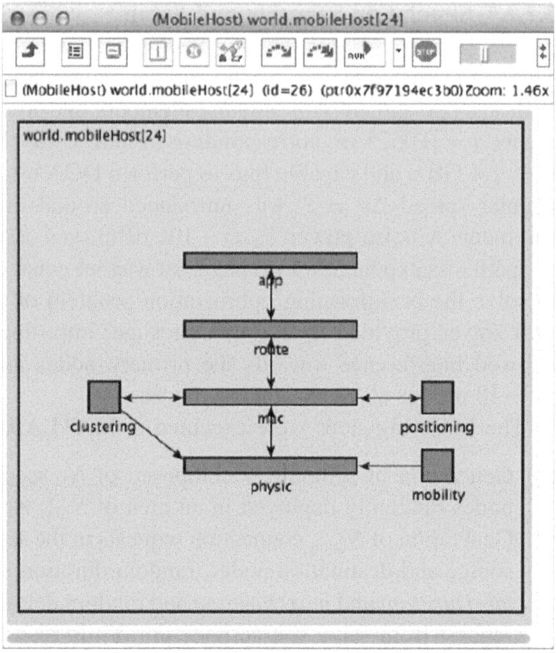

Fig. 3.6 Secondary node architecture implemented in OMNeT++

packets of size *appPacketSize* bits spaced in time by a constant delay set to *applicationRate/appPacketSize* (modeling thus a Constant Bit Rate (CBR) packet stream) for a time equal to the connection duration read from file;

- The routing module for a source node x, upon receiving from the application module the first packet of a connection, was in charge of (a) loading from file the corresponding end-to-end path determined in MATLAB, (b) record such path in each packet; the routing module of intermediate nodes took care of forwarding the packet towards the destination by reading the next hop from the packet itself, while routing module of a destination node simply forwarded the packet to application module.
- The MAC module implemented a simple Aloha protocol without retransmission, taking care of immediately forwarding packets received from the routing module to the physical layer module and vice versa.
- The physical layer module had the responsibility of transmitting and receiving packets taking into account path loss, propagation delays and interference generated by packet collisions.

The impact of interference, in particular, was modelled with accuracy significantly higher than what currently found in existing OMNeT++ frameworks, such as INET and MixiM, in order to ensure a correct analysis of the impact of the proposed optimization on network performance. The simulator is in fact able to keep track of all transmitted packets and, for each packet reception, determines the interference level on a symbol by symbol basis (note that, as binary modulation was considered in all simulations, in the following bits will be considered in place of symbols). Consecutive bits subject to the same interference are grouped into so called bit regions. Next, for each bit region the average Bit Error Probability (BEP) is evaluated by adopting the Standard Gaussian Approximation for the interference power, and the number of bit errors is randomly determined according to the BEP. Finally, the total number of bit errors generated is evaluated by summing up errors introduced in each bit region, and compared with the maximum number of errors admitted for the packet as determined by the adoption of a Reed-Solomon code with a coding rate $RS_{\text{rate}} = 0.835$ (corresponding to a correction capability roughly equal to 10 % of the packet bits) in order to decide if the packet is correctly received or discarded. The following steps were executed in OMNeT++ for each run: (1) Loading of primary and secondary network topologies from file; (2) Loading of connection requests from file and for each request: (a) Generation of packets; (b) Forwarding of packets along the end-to-end path read from file; (c) Measurement of end-to-end throughput and other relevant metrics; (3) Averaging of measured metrics.

Table 3.1 presents the values for the simulation parameters.

3.2.2.6 Simulation Results: Matlab

Figure 3.7 shows the average interference imposed on the secondary nodes when the data is routed between the secondary source and destination nodes. To ensure continuity of the simulations, the constraint on primary interference is relaxed if the cost of is $+\infty$ for all the secondary nodes within the transmission range of the transmitter for a specific hop. As can be seen from the figure, the optimized routing with beamforming, i.e., routing with the proposed cost metric, gives the best performance in terms of interference imposed within the secondary network. As expected, routing without beamforming gives the worst performance. Furthermore, it must be mentioned here that to compare the performance of routing with beamforming and optimized routing with beamforming, one must also consider the number of primary constraint violations, since we relax the primary interference constraint when none of the secondaries is able to satisfy this constraint. To make this comparison, Fig. 3.8 shows the number of primary constraint violations for different number of primary nodes. From Figs. 3.7 and 3.8, it can be seen that the difference in performance between the optimized and non-optimized routing with beamforming in Fig. 3.7 is large when the corresponding difference in primary violations in Fig. 3.8 is relatively small, e.g., the performance when the number of primary nodes is 30. Otherwise, when the difference in performance in Fig. 3.7 is small, the difference in the number of primary constraint violations is relatively large. In order to have a fair comparison between the two, the number of primary

Table 3.1 Simulation parameters

Parameter	Value(s)
Number of Secondary nodes N_S	50
Number of primary nodes N_P	From 10 to 50
Number of connection requests per run N_{conn}	1000
Minimum connection duration *minDuration*	25 s
Maximum connection duration *maxDuration*	75 s
Min delay between connection requests *minDelay*	50 s (High Traffic)/500 s (Low Traffic)
Max delay between connection requests *maxDelay*	100 s (High Traffic)/750 s (Low Traffic)
Transmission rate at physical layer	1 Mb/s
Maximum transmission power for secondary nodes	1 μW
Application packet length *appPacketSize*	512 bits
Application source rate *applicationRate*	320 kbit/s

constraint violations for routing with beamforming and optimized routing with beamforming should force to be the same.

3.2.2.7 Simulation Results: OMNeT++

OMNeT++ simulations considered the following four different scenarios, obtained by varying the traffic load in the secondary network and the number of primary nodes:

- *Low traffic, free network*—Low traffic (obtained by setting the *minDelay* and *maxDelay* variables to the corresponding values in Table 3.1) and no primary nodes;
- *Low traffic, constrained network*—Low traffic and $N_P = 10$ primary nodes;
- *High traffic, free network*—High traffic (obtained by setting the *minDelay* and *maxDelay* variables to the corresponding values in Table 3.1 and no primary nodes;
- *High traffic, constrained network*—High traffic and $N_P = 10$ primary nodes.

The throughput, defined as the ratio between end-to-end received packets and generated packets was measured in the four scenarios above for the three strategies previously introduced in the paper. Figure 3.9a presents the throughput in the case of the Low traffic, free network scenario. The figure shows that in this scenario the optimization in the routing path leads to an increase in throughput, as on each other hop the strategy is able to select the node that provides the lowest amount of interference to neighbouring nodes, thus increasing the probability of correct packet reception throughout the network. Moving to the Low traffic, constrained network scenario, Fig. 3.9b shows that the introduction of constraints determined by the presence of a significant number of primaries has the impact of reducing the gap between the two BF-based strategies, due to the fact that in several cases potential relays that would lead to lower interference in the secondary network are

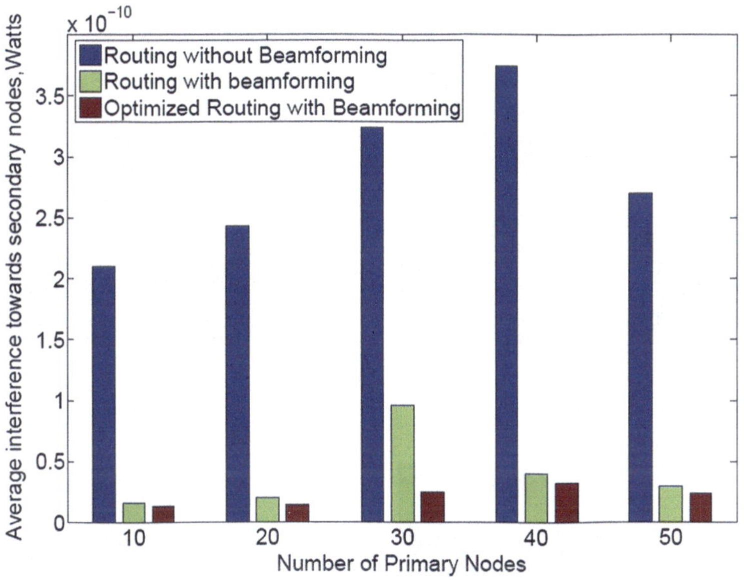

Fig. 3.7 Average total interference exerted upon the secondary nodes between source and destination

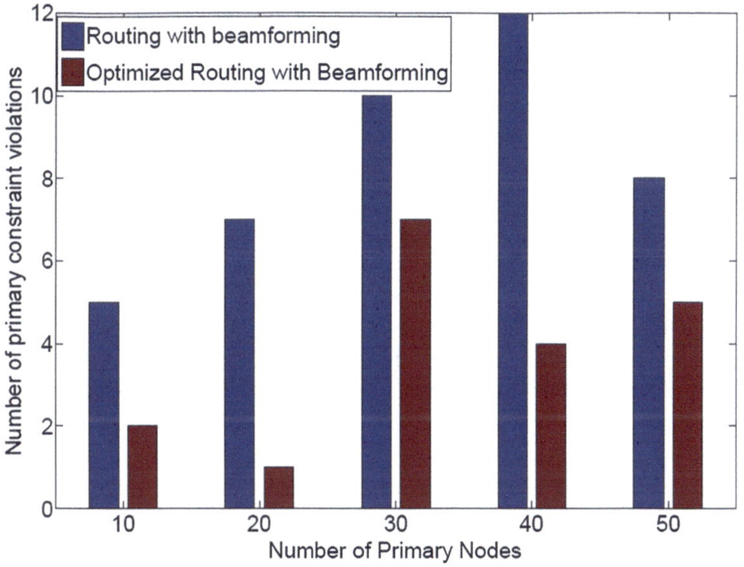

Fig. 3.8 Number of primary constraint violations

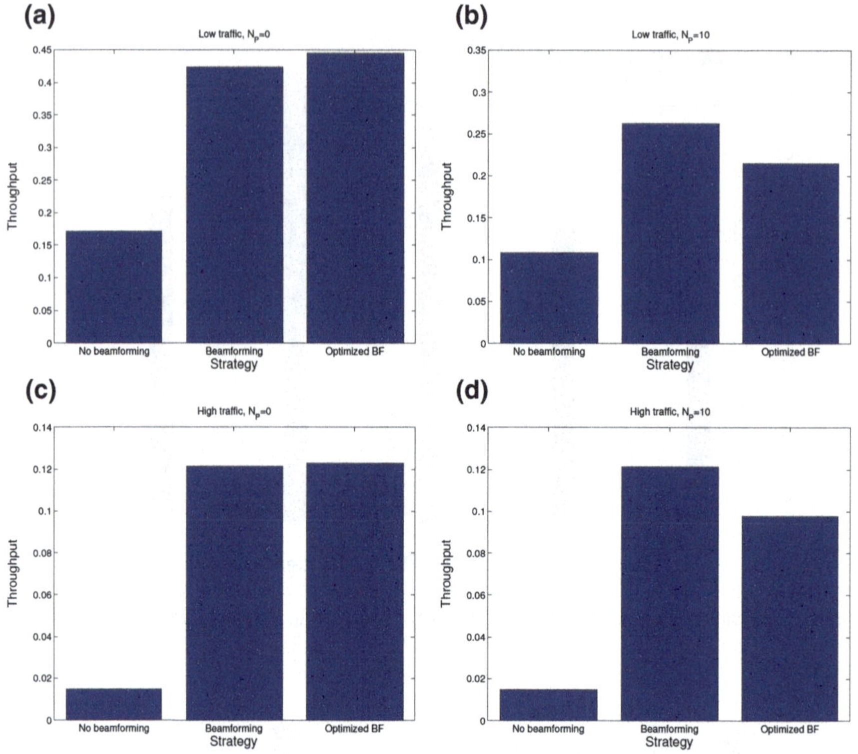

Fig. 3.9 Number of primary constraint violations. **a** Throughput in the Low traffic, free network scenario for the three considered routing strategies **b** Throughput in the Low traffic, constrained network scenario for the three considered routing strategies **c** Throughput in the High traffic, free network scenario for the three considered routing strategies **d** Throughput in the High traffic, constrained network scenario for the three considered routing strategies

discarded as they do not satisfy the hard constraint on the level of interference towards one or more primary receivers. Figure 3.9c shows how the throughput is affected in the High traffic, free network; results show how for all strategies performance is significantly reduced due to the higher number of collisions, and the corresponding higher average value of the interference power during packet reception. Finally, Fig. 3.9d shows results in the High traffic, constrained network, that introduces again the presence of the primary nodes; interestingly, results highlight that in this case the Optimized Routing with Beamforming leads to slightly worse results compared to simple Routing with Beamforming. However, as shown by Fig. 3.8, this comes together with a better coexistence capability with primary receivers, highlighting the presence of a trade-off between coexistence and secondary network performance.

3.2.2.8 Conclusions

This chapter has focused on transmit beamforming and routing in a multihop, ad hoc cognitive radio network. After introducing the transmit beamforming strategy, we proposed a new cost metric which was used to design an optimized, beamforming based routing algorithm with three-fold objective: to minimize the end-to-end path power consumption; to minimize the co-channel interference imposed within the secondary network and to minimize the number of primary interference constraint violations. Simulation results from MATLAB have confirmed that the optimized routing algorithm outperformed the original routing algorithm in terms of both, the interference generated within the secondary network and the number of primary interference constraint violations. The simulations carried out in OMNeT++ confirmed the improved throughput of the secondary network when no constraints from primary nodes are imposed, while they highlight a trade-of between coexistence capability and secondary network performance when the presence of primary nodes is taken into account.

Acknowledgements This work has been supported by the ICT-ACROPOLIS Network of Excellence, FP7 project no. 257626, www.ict-acropolis.eu, COST Action IC0902, COST Action IC0905 "TERRA", and the UK's Mobile VCE Consortium.

3.3 Spectrum Aggregation Over Non-contiguous Frequency Bands

Filippo Meucci[1], Orlando Cabral[1], Fernando J. Velez[1], Jessica Acevedo Flores[1], Daniel Robalo[1], Albena Mihovska[2], Neeli R. Prasad[2] and Oliver Holland[3]

[1]UBI-Instituto de Telecomunicações, Aveiro, Portuga

[2]University of Aalborg, Aalborg, Denmark

[3]King's College London, London, UK

Dynamic spectrum access (DSA) techniques are promising to enable spectrum aggregation with intra-operator multi-band scheduling [19]; hence, alleviating the spectrum scarcity problem.

Spectrum Aggregation (SA) consists in aggregating several (and possibly) fragmented bands, exploiting multiple small spectrum fragments simultaneously to yield to a (virtual) single larger band and ultimately deliver a wider band service. Consequently, SA allows that new high data rate wireless communication systems can coexist while reusing the spectrum of legacy systems. The non-contiguous approach may provide larger flexibility as well as diversity.

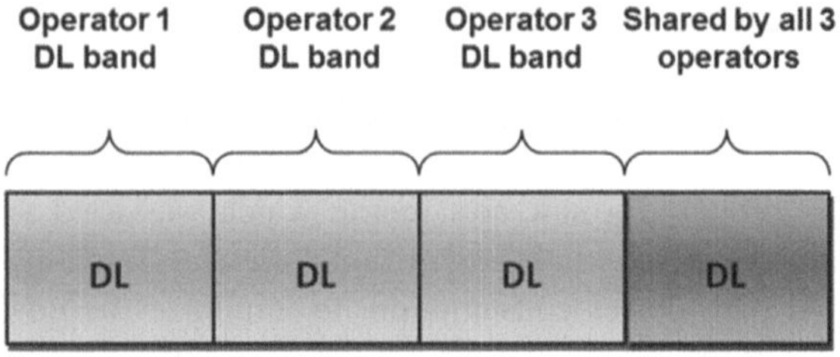

Fig. 3.10 Scenario of common frequency pool

3.3.1 Spectrum Aggregation with Multi-band User Allocation Over Two Frequency Bands

Affordable, high-bandwidth mobile access improves the quality of experience for users, enabling them to get more out of existing services and opens up opportunities for new mobile broadband services. Supporting additional system capacity and higher data rates will improve the value of these services. However, this is impeded by the existing highly fragmented radio frequency spectrum that does not match the actual demand for transmission and network resources. Therefore, mobile operators might be forced to aggregate spectrum of two or more separated sub-bands for downlink (DL) and uplink (UL) bands.

3.3.1.1 Problem Statement

The objective is to determine the best user allocation for a single operator over two (or more) High Speed Downlink Packet Access (HSDPA) frequency bands in order to maximize the total network throughput. The operator has exclusive usage of the 2 GHz band and can access the shared frequency pool at 5 GHz. The quantity of radio resources available at 5 GHz is determined by spectrum trading (or bargaining) among all the operators that have been granted access to the frequency pool, as shown in Fig. 3.10.

The problem of scheduling the users into two bands can be formulated as an Integer Programming optimization problem [20, 21]. The total throughput of the operator is the profit function to be maximized. The profit function depends on the channel qualities and on the number of users in each band. The number of users that can be allocated in the primary band (i.e., 2 GHz) are determined based on the load thresholds; then the multi-band scheduling is determined as a General Assignment Problem (GAP) [20] where the number of users in each band is upper bounded.

3.3.1.2 General Multi-band Scheduling: Spectrum Sharing as a General Assignment Problem

Profit Function. Solving Multiple-Objectives GAP (MO-GAP) can be very difficult and usually the objectives are combined together via a linear combination, called "scalarization" [20].

The profit function (PF) for the maximum throughput in the General Multi-band Scheduling (GMBS) is the following:

$$(\text{PF}) \quad \max \sum_{b=1}^{m} \sum_{u=1}^{n} R_{bu} x_{bu} \tag{3.3.1}$$

where R_{bu} is the throughput for user u on band b and x_{bu} is a 0−1 integer variable $x_{bu} \in \{0, 1\}$ defining the user allocation over the bands that is mathematically represented by:

$$x_{bu} = \begin{cases} 1 \text{ if user u is allocated on band b} \\ 0 \text{ otherwise} \end{cases}$$

For a single transceiver User Terminals (UTs), the GMBS has two constraints:

1. Each user can be allocated only to a single frequency band (assumption of single transceiver UT), which is represent by the integrality of $x_{bu} \in \{0, 1\}$;
2. The total number of users on each band is upper bounded; this is the bandwidth constrain (BC) on each frequency band, which is reported in Eq. 3.3.2.

$$\sum_{u=1}^{n} x_{bu} \leq N_b^{max} \, for \, b = 1 \ldots m \tag{3.3.2}$$

In general, the throughput R_{bu} is a function of the channel quality of user u on band b, $R_{bu} = f(CQI_{bu})$, where Channel Quality Indicator, CQI_{bu} is dependent on the Signal to Interference Ratio (SIR) that depends on the propagation conditions, on the interference from other cells and on power per code available:

$$CQI_{bu} = f(SIR) \tag{3.3.3}$$

The higher the number of users allocated in the band, the lower the CQI. R_{bu} can be changed ahead to depend on other variables, however, for simplicity in this work, it is dependent only on the CQI.

$R_{bu} - x_{bu}$ interrelation: two step GMBS procedure. The GMBS aims to determine x_{bu} (∀u = 1…n ; ∀b = 1…m) or the allocation of the users over the available frequency bands in order to maximize the total throughput. As such, the interdependence of R_{bu} with x_{bu} needs to be solved.

The GMBS problem is formulated in two steps. First, the maximum number of users that can be allocated on each frequency band is determined, then R_{bu} is defined as the available throughput considering that N_{max} users are allocated in the band. The number of maximum users in each band is calculated depending on the load, as $N_b^{max} = L_{normalized}^{-1}$, where $L_{normalized}$ is estimated based on the resources available for the cell, estimated as follows:

$$L_{normalized}(i) = \frac{\sum_{u=1}^{N_b} Load(u)}{R_{HSDPA}} \qquad (3.3.4)$$

where: N_b is the number of HSDPA user in band *b*, R_{HSDPA} is the number of High Speed Downlink Shared Channel (HS-DSCH) allocated in the cell, and *Load*(*u*) is the average number of HS-DSCH required by user *u* to support its service rate, *R*(*u*). This number is given by the following Eq. 3.3.5:

$$Load(u) = \frac{R(u)}{R(CQI_{bu}) \bullet N_{HS-PDSCH}(CQI_{bu})} \qquad (3.3.5)$$

where: the average propagation condition determines the channel quality indicator ID of user *u* on band *b*, CQI_{bu}, $R(CQI_{bu})$ is the achieved bit rate when one CQI_{bu} block is allocated in every frame and $N_{HS\text{-}PDSCH}(CQI_{bu})$ is the number of HS-DSCH associated to CQI_{bu}.

Once N_b^{max} has been determined, R_{bu}, or the expected data rate for each user should be dependent on CQI_{bu}, the effective packet error rate (*PER*) experienced (or predicted based on the position in case of a first transmission), and the number of users:

$$R_{bu} = \frac{R(CQI_{bu})}{N_b} \times (PER_{bu})^{-1} \qquad (3.3.6)$$

The direct mapping between CQI_{bu} and the throughput can be expressed as:

$$R(CQI_{bu}) = \begin{cases} 188.5 \text{ if } CQI_{bu} = 5 \\ 396.0 \text{ if } CQI_{bu} = 8 \\ 1659.5 \text{ if } CQI_{bu} = 15 \\ 3584 \text{ if } CQI_{bu} = 22 \end{cases} \qquad (3.3.7)$$

Suboptimal Multi-Band Allocation Algorithm. UTs are able to use both of the frequencies. The Common Radio Resource Management (CRRM) entity keeps track of the CQI in both frequencies by making use of the pilot channel. By keeping the load at the same level, users may be scheduled on one frequency or another, depending also on the CQI available. When a user arrives to the system, the UT is allocated to the 2 GHz band. The load is checked in both bands. If the load is higher in the 2 GHz, the user with the highest CQI in the 2 GHz will be moved to the 5 GHz band. The load is estimated based on the resources available for the cell. The procedure is the same as proposed in [22].

Results. The performance of the algorithm is assessed by using the service throughput which is the total number of bits transmitted and correctly received by the all users in the cell:

$$Serv\_thr_{[s^{-1}]} = \frac{b_{serv}[p]}{k \cdot T} \tag{3.3.8}$$

where: $b_{serv}[p]$ is the number of bits received in given period p, T is the transmit time interval, k is the number of steps. $k \cdot T$ is the simulation duration. Users are displaced in the cell within a distance from the BS from 300 to 3000 km with uniform distribution. The Near Real Time Video (NRTV) calls are modeled by a Poisson distribution, the call duration is exponentially distributed with an average of 180 s.

Figure 3.11a shows the results without multi-band scheduling (MBS). The operator has the availability of two frequency bands and each one is managed separately; calls requests are divided in the two bands and it is not possible to switch a service from one band to the other. The "*Overall Serv*" is the sum of the service throughput in both bands, the traffic requirement is the traffic required to satisfy all the users.

Figure 3.11b shows the results with the MBS, the operator can decide on which band to allocate the user. It is shown that, even for a single-band UT, both bands have a higher throughput due to the switching of users between them. Figure 3.12 compares the results with and without MBS. A 200 Kbps improvement is evident over a wide range of UTs in the cell.

By use of MBS a constant throughput gain over a wide range of active services in the cell can be achieved. MBS is able to support a higher number of NRTV users due to the ability of scheduling users considering their respective radio channel quality in different parts of the radio spectrum. The achieved improvement is relative to a scenario where users are randomly displaced in the cell. Because the 5 GHz bandwidth has much lower coverage then the 2 GHz band, the results shown here are a first step towards the analysis of the attainable gain.

3.3.2 Integrated Common Radio Resource Management with Spectrum Aggregation

Radio Resource Management (RRM) plays an important role in wireless system design, due to the scheduling algorithm which decides among packets that are ready for transmission, allowing certain quality of service (QoS) levels be achieved. CRRM refers to the set of functions that are devoted to ensure an efficient and coordinated use of the available radio resources in heterogeneous networks scenarios, as shown in Fig. 3.13.

A non-contiguous SA (from an upper layer point of view) and an integrated CRRM (iCRRM) entity are proposed, where SA and CRRM functionalities are handled simultaneously. The proposed iCRRM enables the integration of spectrum and network resource management functionalities leading to higher performance

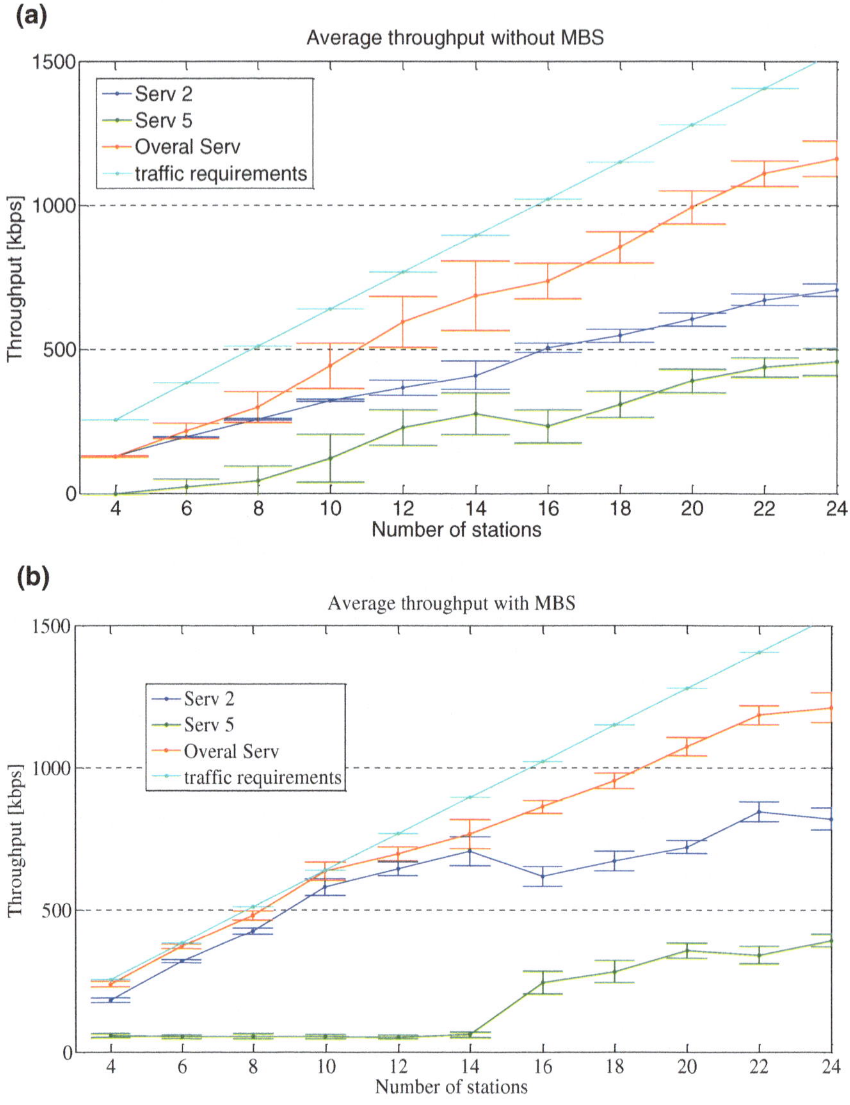

Fig. 3.11 Average throughputs: **a** without MBS, **b** with MBS

and system capacity gains, performing the scheduling via the optimal solution of a GMBS problem. The integration of dynamic spectrum use and SA, achieved with the use of iCRRM techniques is shown to provide significant throughput gain compared to a system where the iCRRM is not used. A formulation is proposed for the average Signal to Interference-plus Noise Ratio (SINR) that allows for setting

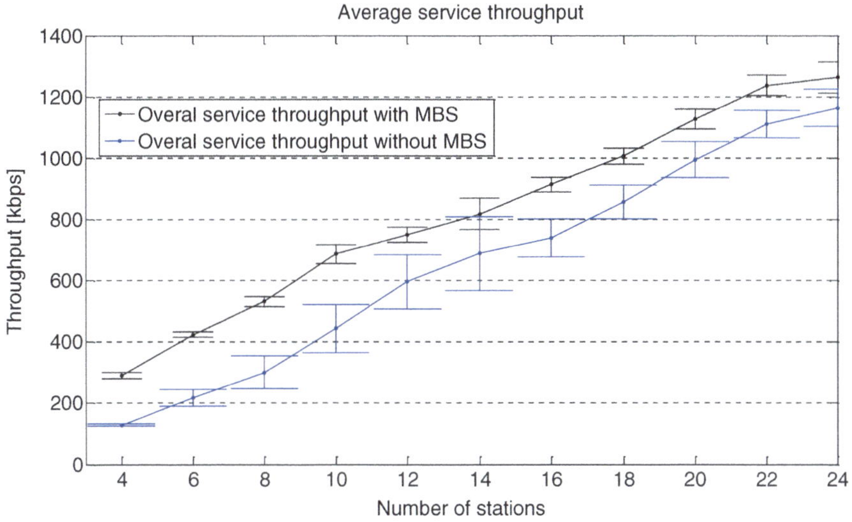

Fig. 3.12 Service throughput with and without MBS

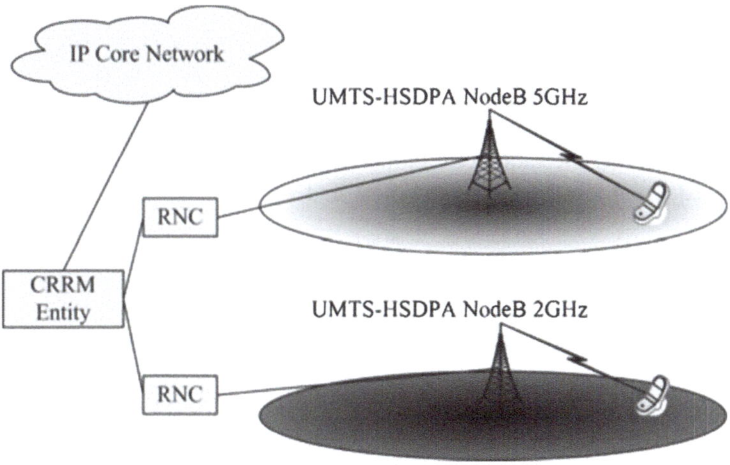

Fig. 3.13 CRRM in the context of SA with two separated frequency bands

the basic limits for the dependence of SA with GMBS on the coverage distance with an optimal solution. This section addresses radio resource management with SA. It discusses innovative results for communications using multiple bands to improve overall throughput and reduce packet delay for future mobile communication systems.

Table 3.2 Parameters and models used for 2 GHz and 5 GHz bands

Carrier frequency	2 GHz	5 GHz
Bandwidth BW	5 MHz	5 MHz
Path loss model	$L_{2GHz} = 128.1 + 37.6 \cdot \log_{10}(R)$	$L_{5GHz} = 141.5 + 28 \cdot \log_{10}(R)$

3.3.2.1 Objective and System Model

The inclusion of the iCRRM in the system model is used to facilitate the best user allocation and maximize the total network throughput in HSDPA systems, considering two frequency bands b are considered 2 GHz and 5 GHz with frequency reuse pattern one.

The radio channel follows the ITU radio propagation models, as summarised in Table 3.2. The interference in the Mobile Station (MS) is calculated with the signal strength received from the first ring of neighboring Base Stations (BSs) and the thermal noise [23].

Considering u as the user index, the CQI is a mapping of the averages of the CQIs recorded over time; a direct mapping between CQI_{bu} and the available rate at the physical layer, in kbps, is shown in "$R_{bu} - x_{bu}$ interrelation: two step GMBS procedure".

3.3.2.2 General Multi-band Scheduling

The Profit Function (PF) maximizes the total throughput of the operator, and is defined considering the ratio between the rate available on the single DL channel and the requested rate by the service flow and is expressed as follows:

$$(PF) \sum_{b=1}^{m} \sum_{u=1}^{n} W_{bu} x_{bu} \tag{3.3.9}$$

where: x_{bu} is the allocation variable, n the number of users and the normalized metric W_{bu} is:

$$W_{bu} = \frac{[1 - PER(CQI_{bu})] \cdot R(CQI_{bu})}{S_{rate}} \tag{3.3.10}$$

where: S_{rate} is the NRTV service rate, $PER(CQI_{bu})$ is the average Packet Error Rate (PER) occurred in previous transmissions that the DL channel for user u on band b is suffering for the Modulation and Coding Scheme (MCS) supported (0 in the case no transmissions has ever occurred), and $R(CQI_{bu})$ is the DL channel throughput for user u on band b, as a function of the MCS supported.

The constraints for GMBS vary, depending on the ability of the MSs to simultaneously transmit and receive in multiple frequencies (multiple transceivers at the MS), or just over a single band at the time. HSDPA physical layer [24] provides a set of orthogonal codes available for data transmission within a sub-frame. Codes maybe allocated to service flows/users in a flexible manner. More than one code can be assigned to a single user or a single code can be assigned to more than one user. The users on the same code adopt a time-division multiple access which is managed by the packet scheduler. The allocation variable x_{bu} reflects the code allocation per users and is either a Boolean value in the case of single code allocation, or a positive integer, $x_{bu} \in \{0, 1, \ldots, \max(N_{codes})\}$, in the case of multi-code allocation.

In the following explanation, a single-frequency single-code allocation will be explored with a Round Robin (RR) scheduler. The RR scheduler was selected since it is the one with least interference on upper layer algorithms.

In single-frequency single-code allocation, the GMBS presents the following constraints:

1. Allocation Constraint (AC): each user can be allocated only to a single frequency band with a single orthogonal code:

$$(AC) \sum_{b=1}^{m} x_{bu} \leq 1, x_{bu} \in \{0, 1\}, \forall \, \text{user} \, u \tag{3.3.11}$$

2. Bandwidth Constraint (BC): the total number of users on each band is upper bounded by the maximum normalised load that can be handled in the band, $L_b^{max} \in [0, 1]$:

$$(BC) \sum_{b=1}^{m} \frac{S_{rate} \cdot (1 + R_{Tx} \cdot PER(CQI_{bu}))}{N_{codes} \cdot R(CQI_{bu})} \cdot x_{bu} \leq L_b^{max}, \forall \, \text{b and} \, b \tag{3.3.12}$$

where: the first term is the requested serviced at a rate for user u, including the packet loss, normalized with the maximum data rate that the network can offer to the user u on band b which is $N_{codes} \cdot R(CQI_{bu})$ where N_{codes} is the maximum number of parallel codes available in HSDPA. $R_{Tx} = 2$ is the number of H-ARQ retransmissions. BC accounts for the user traffic requirement, DL capacity and overhead caused by packets lost. The load constraint for each band, L_b^{max}, is lower than one because of the padding caused when the packets from upper layers are fragmented to fit the MAC Packet Data Unit (MPDU), and signaling overhead.

3.3.2.3 Average SINR Analysis with Unitary Frequency Reuse Pattern

The SA gain is going to be evaluated for several inter-cell distances with a frequency reuse pattern K, equal to one. The average SINR was obtained following a similar method to the one described in [25]. In order to have comparable results, SA needs to be analyzed at constant average SINR, which is achieved only by tuning the BSs transmitter power.

SINR at a given position. Since our topology contemplates a MS in a given position (*x*,*y*), and given a BS transmitter power P_{Tx}, the MS at position (*x*,*y*) can be expressed by:

$$SINR(P_{Tx}, x, y) = \frac{P_{ow}(P_{Tx}, x, y)}{(1-\alpha)\cdot P_{ow}(P_{Tx}, x, y) + P_{nh}(P_{Tx}, x, y) + P_{noise}} \quad (3.3.13)$$

where: P_{ow} is the power received from the own cell, α is the orthogonality level of the codes according to [25], P_{nh} is the total amount of interfering power coming from the neighbor cells (6 cells in the case of hexagonal cell deployment), p_{noise} is the thermal noise power. The interference power received by a MS from the first ring of six neighbor cells is given by:

$$P_{nh}(P_{Tx}, x, y) = \sum_{i=1}^{6} I_i(P_{Tx}, x, y)_{\text{ with, } I_1=I_6, I_2=I_5, I_3=I_4} \quad (3.3.14)$$

where: $I_i(P_{Tx}, x, y) = P_{Tx}G_{Tx}G_{Rx}10^{\frac{-PL(x,y)^i}{10}}$.

Average SINR in the cell. The average SINR within a cell is the SINR measured by a MS with uniform probability density function for its deployment over the cell area. It depends on the cell radius, *R*, and on the BS transmitter power, P_{Tx}, as follows:

$$\overline{SINR}(P_{Tx}, x, y) = \frac{\overline{P}_{ow}(P_{Tx}, x, y)}{(1-\alpha)\cdot \overline{P}_{ow}(P_{Tx}, x, y) + \overline{P}_{nh}(P_{Tx}, x, y) + P_{noise}} \quad (3.3.15)$$

The average interference generated by a neighbor cell can be calculated by integrating each fraction of the interfering power over the area of the affected cell. Figure 3.14 shows one cell affected by interference in the origin of the coordinates and one interfering cell, at (x_O, y_O). By integrating over the hexagonal cell area, the average level of the received power from a neighbor cell, $\overline{I}$, may be calculated as:

$$\overline{I}(R, P_{Tx}) = \int_y \int_x f_I(P_{Tx}, x, y)dxdy = \int_y \int_x \frac{P_{Tx}G_{Tx}G_{Rx}}{A_{cell\,nh}} PL(x, y)dxdy \quad (3.3.16)$$

where: A_{cellnh} is the total affected cell area. As the surrounding interfering neighbors are all at the same distance, $\sqrt{3}R$, $\overline{P}_{nh}(P_{Tx}, x, y) = 6\cdot\overline{I}(P_{Tx}, R)$, as it is shown in Fig. 3.14.

$\overline{P}_{ow}(P_{Tx}, x, y)$ is the average signal power within a cell and it is constant for the same frequency model, no matter what value of *K* is used.

Transmitter Power Normalization Procedure. The goal of the average SINR analysis is to determine a set of transmitter powers, P_{Tx}, to be considered in future simulations, in order to have a constant average SINR for all the cell radii. Figure 3.15 shows the average SINR results.

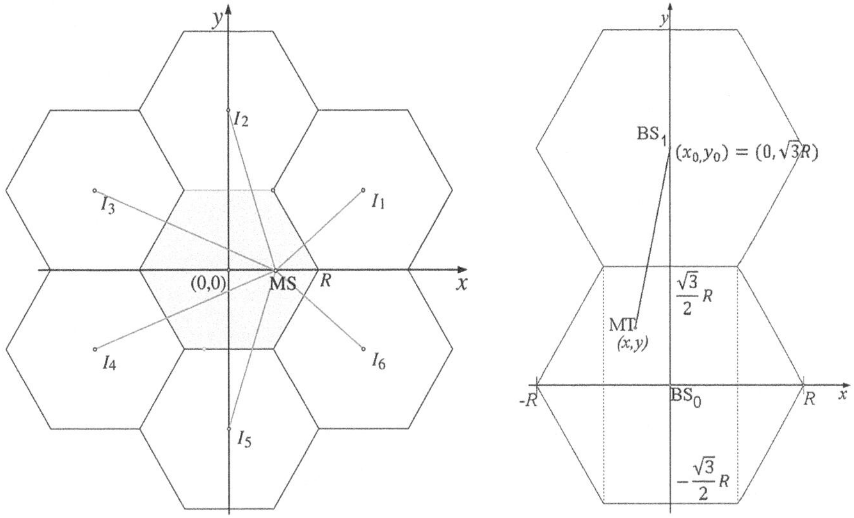

Fig. 3.14 Topology considered in our formulation with $K = 1$, the inter and intra cell interference for the HSDPA LTE network

Results. The performance of the SA user allocation is assessed by using the total Service Throughput (*Serv_thr*) metric which was discussed in "Results" of Sect. 3.3.1.2.

The system can achieve better performance if MSs are allowed to be switched between bands. Figure 3.16 shows the results in the presence of the iCRRM with the proposed GMBS algorithm, for several cell radii with normalized power. Due to the power normalization procedure, the performance of the iCRRM is almost constant for all the cell radii. In the saturation point, a gain of 30 % is achieved compared to the absence of iCRRM.

Figure 3.17 presents the throughput gain in percentage and the absolute gain as a function of the cell radius under a constant average SINR: the power normalization procedure proposed in the previous section allows for fair results comparison with variable cell radius. The almost constant gain demonstrates the potentiality of iCRRM over a wide range of cell coverage distances.

Figure 3.18 shows the load variation depending on the number of active MSs in the cell for both frequency bands and $R = 300$, 900 and 1,500 m. Since the path loss is lower at 2 GHz compared to the 5 GHz band, in the case of low cell load, the iCRRM entity will mainly allocate the MSs to the 2 GHz band.

Figure 3.19 presents the average *PER* as a function of the MS-BS distance for three cell radii: $R = 300$, 900 and 1,500 m; the use of iCRRM shows a significant reduction of *PER*, especially for the 5 GHz band.

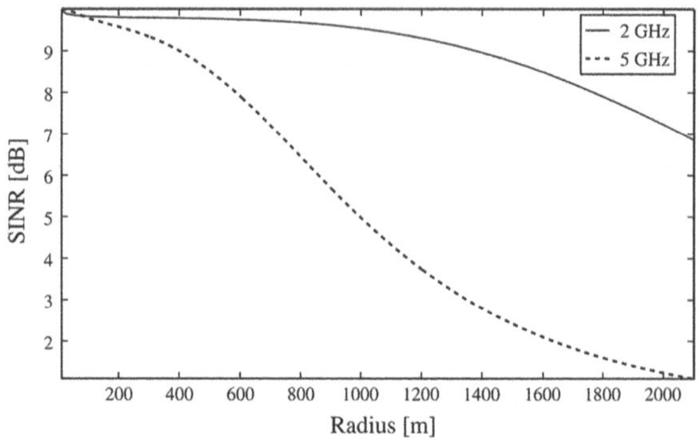

Fig. 3.15 Average SINR (dB) as a function of the cell radius [m]

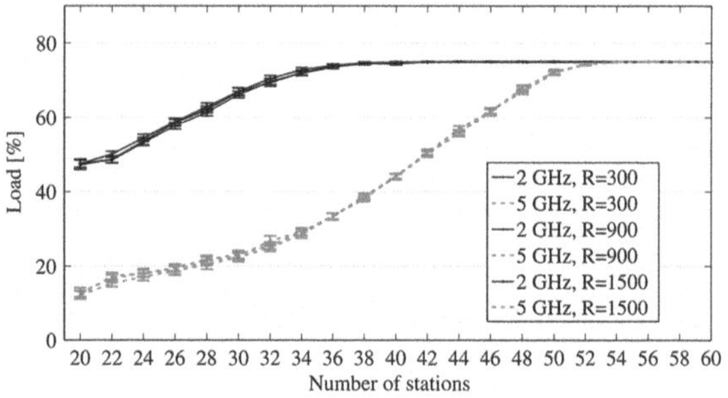

Fig. 3.16 Average service throughput with the iCRRM with normalized power

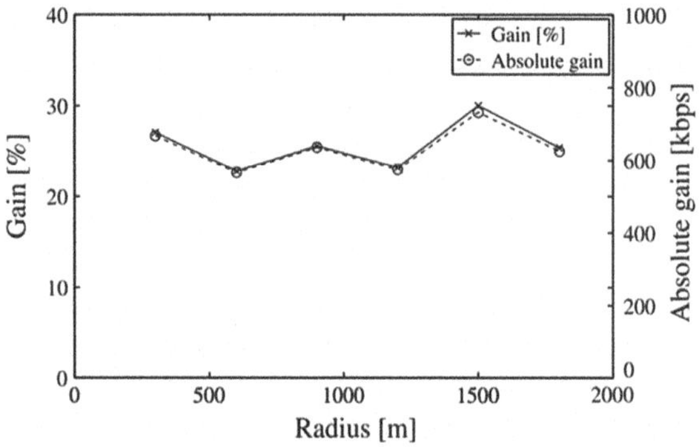

Fig. 3.17 Gain between the presence and absence of the iCRRM as a function of the cell radius

For all the cell radii, the GMBS is able to reduce the *PER* from 0.01 to 0.005 via an accurate selection of the MSs to be moved between 2 and 5 GHz, based on their channel quality in each frequency band.

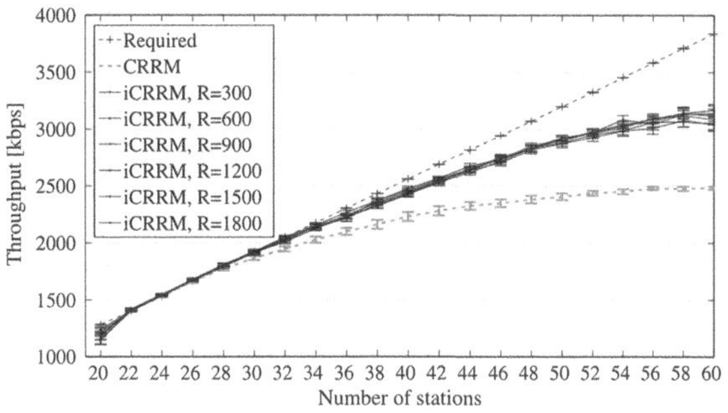

Fig. 3.18 Variation of the load with the number of SSs for 60 users for both frequency bands for $R = 300$, 900 and 1,500 m

3.3.3 Transmitted Power Formulation for the Implementation of Spectrum Aggregation in LTE-Advanced Over 800 MHz and 2 GHz Bands

An innovative formulation to compute the average SINR in the context of SA in LTE systems is proposed, which comprises an iCRRM entity from Sect. 3.3.2 and schedules the users between the two LTE systems operating at 800 MHz and 2 GHz, whilst considering the integer programming optimization and considering RR schedulers within each LTE system. Firstly, considering a topology with a frequency reuse pattern $K = 3$ and the COST-231 Hata model [26] for the path loss (L) for two carrier frequencies, 800 MHz and 2 GHz, we obtain the following L model:

$$\begin{aligned} L_{800\ \mathrm{MHz[dB]}} &= 119.6 + 37.2 \cdot \log_{10}(R_{[\mathrm{km}]}) \\ L_{2\ \mathrm{GHz[dB]}} &= 128.1 + 37.6 \cdot \log_{10}(R_{[\mathrm{km}]}) \end{aligned} \tag{3.3.17}$$

Following a similar approach as in Sect. 3.3.2, now working at 800 MHz instead of 5 GHz, the values for the transmitter power are found for different cell radii so that a constant average SINR could be kept. According to [25] (Chap. 12), for a topology with a MS in a given position (y, 0) and for $K = 3$, the inter-cell BSs distance is $3R$, as it is shown in Fig. 3.20.

According to [27], for HSPA, $(1 - \alpha) \cdot P_{ow}(P_{Tx}, x, y)$ with $\alpha \in [0,1]$, denotes the average channel multipath orthogonality factor, i.e., the fraction of the total output power that is experienced as intra-cell interference as the channels from the same cell are no longer perfectly orthogonal. Furthermore, for HSDPA, $\alpha > 0.9$ is assumed most of the time, while for WCDMA values in the range [0.5, 0.9] are recommended. The LTE design from [27] proposes to assume unitary α for simplicity. Thus, for LTE, the own cell interference is not considered in Eq. (3.3.5) from section, due to the lack of codes, as in WCDMA/HSDPA systems.

Figure 3.21 shows the average SINR in both frequency bands with different BS transmitter powers and $\alpha = 1$, achieving the maximum values of 19.5489 dB and 19.8416 dB, for the 800 MHz and 2 GHz frequency bands, respectively. Finally,

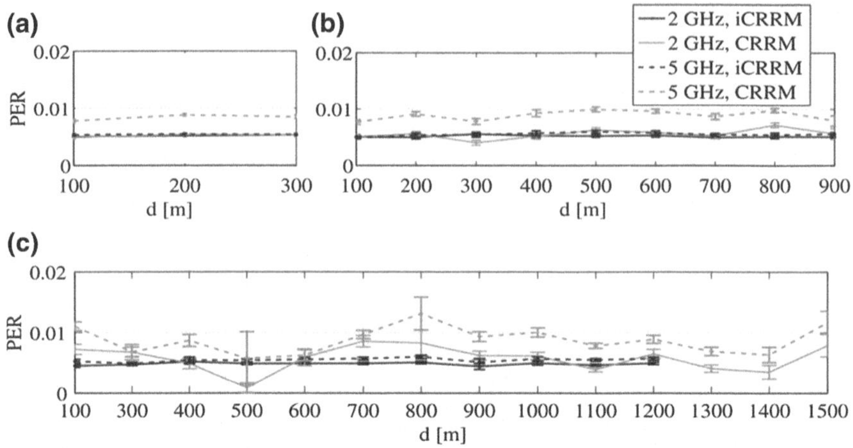

Fig. 3.19 *PER* variation for cell radii. **a** $R = 300$, **b** $R = 900$ and **c** $R = 1{,}500$ m

to calculate the normalized P_{Tx}, required in both bands (to maintain such an average SINR constant), we took $\overline{\text{SINR}}_{\text{max800MHz}} - \text{V}\ \%$, $-V$, where V is the percentage of variation. Table 3.3 shows the results of the normalized transmitter powers, P_{Tx}, required, with variation V equal to 1, 5 and 10 %.

Figure 3.22 shows that, as the cell radius increases, the transmitter power required to keep the envisaged average SINR constant increases as well.

3.3.4 Opportunistic Load and Spectrum Management for Mobile Communications Energy Efficiency

The power consumption of mobile and wireless communications systems can be quite significant. Even an environmentally-aware operator, such as Vodafone, consumes approximately 40 MW in running its business in the UK, the majority of which can be attributed to base stations (BSs) [28]. Lowering the operational power consumptions of BSs is particularly important. Enabling sleep modes for such equipment, and more generally reducing necessary transmission power, can have a very significant effect on the overall power consumption.

3.3.4.1 Power Saving Spectrum Management Concepts

Power Saving by Dynamically Powering Down Radio Equipment. The concept, illustrated in Fig. 3.23a, is the switching off of radio equipment through reallocating load to other bands at times of low load. This is extremely promising as it

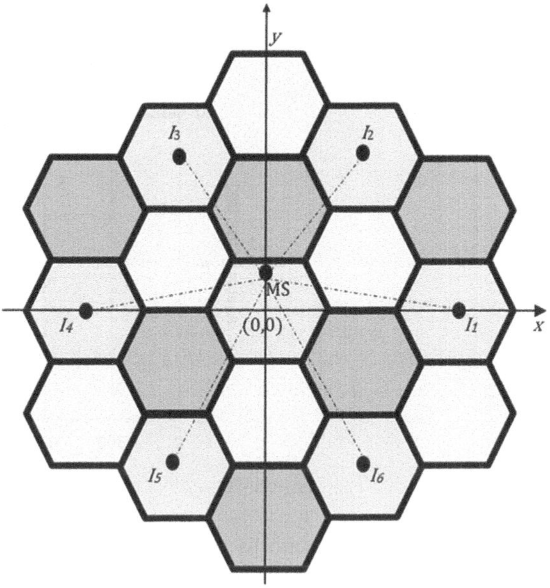

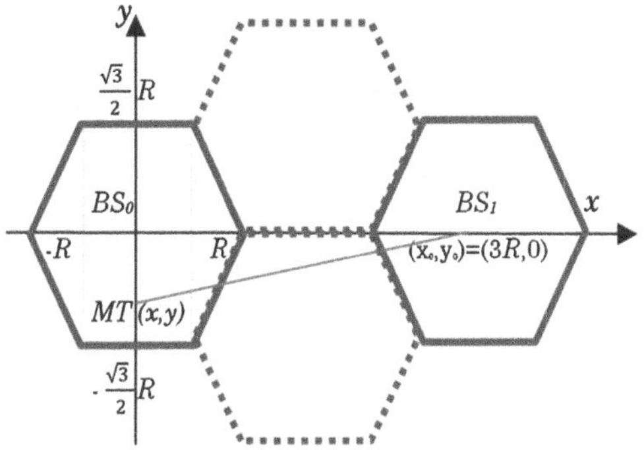

Fig. 3.20 Topology considered in our formulation with $K = 3$, the inter and intra cell interference for an LTE network

implies a guaranteed power saving through radio equipment being virtually "switched off at the socket".

There are two possibilities: (i) turning off cells entirely in one network or spectrum band at that time/location, through traffic being sufficiently carried by a single network or spectrum band, and (ii) using spare capacity of one network/band to cover the required drop in load of another network/band in order to enable that other network/band operate in omnidirectional mode instead of tri-sectored mode. The power saving assessments of this concept reallocates users between bands whenever possible to achieve one or both of these objectives.

Power Saving by Propagation Improvement.
The concept, illustrated in Fig. 3.23b, is the opportunistic reallocation of links or users to more appropriate propagation bands at times when that spectrum becomes

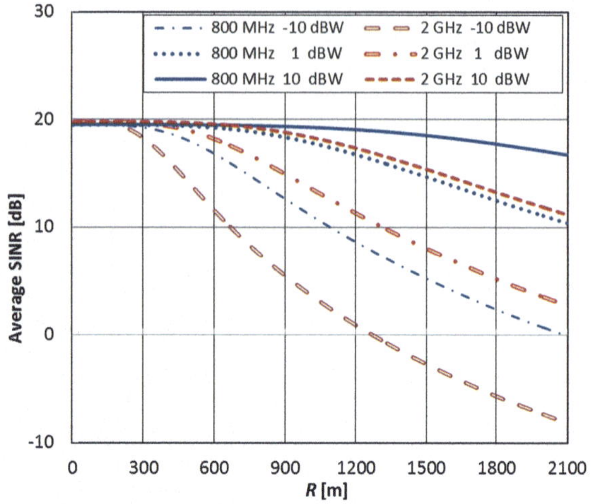

Fig. 3.21 Average SINR (dB) as a function of the cell radius (m) with three values of P_{Tx} (−10, 1 and 10 dBW)

Table 3.3 Values for the normalized transmitter power P_{Tx} (dBW) for the 800 MHz and 2 GHZ bands

V(%)	Freq. Band (MHz)	Radius (m)					
		300	600	900	1200	1500	1800
1	800	−1.971	9.226	15.777	20.424	24.029	26.975
	2000	5.951	17.263	23.883	28.580	32.224	35.201
5	800	−9.113	2.085	8.635	13.283	16.888	19.834
	2000	−1.064	10.253	16.874	21.572	25.215	28.193
10	800	−12.772	−1.575	4.976	9.624	13.229	16.174
	2000	−4.706	6.612	13.233	17.931	21.574	24.552

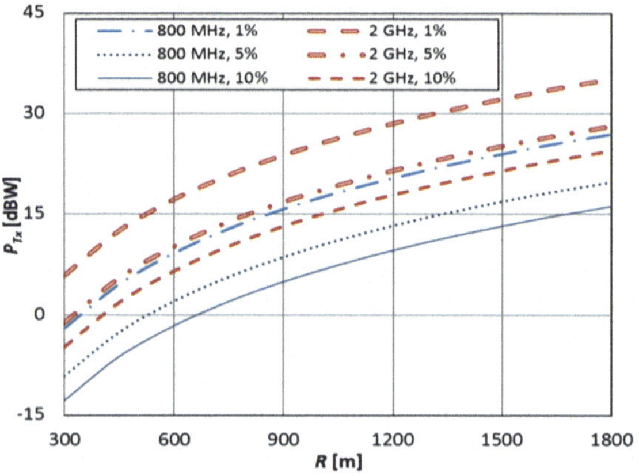

Fig. 3.22 Normalized P_{Tx} required to achieve a constant average SINR (dB), near the maximum, as a function of the cell radius at 800 MHz and 2 GHz

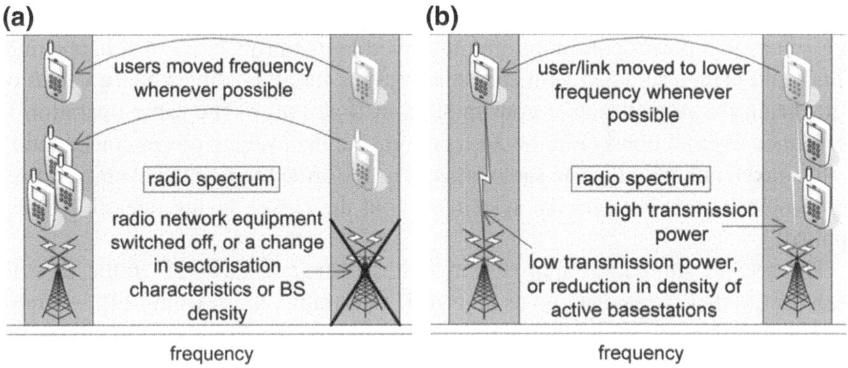

Fig. 3.23 Power saving concepts: **a** Reallocating traffic to enable radio network equipment to be switched off. **b** Reallocating users/links to improve propagation

available. This decreases necessary transmission power due to improved propagation, or alternatively in a frequency reuse scenario, reallocation based on the necessary deployed cell density/radius and the given local propagation environment can be used to reduce inter-cell interference through minimizing power "leaking" into co-channel cells. Both power saving concepts might be employed together, yielding further improvement in power efficiency.

3.3.4.2 Assessment of Power Saving Potential

Power saving potential of the aforementioned concepts is assessed through simulation of cellular systems carrying either video, FTP, or HTTP (web browsing) traffic. Concerning the traffic models used in simulations, it is assumed that the average cell load at time of day *t*, *L(t)*, varies according to a set of statistics on traffic load as a percentage of busy hour load (*BusyLoad*) over a 24 h period, pertaining to traffic in a 3G network in London, UK. Two simulation approaches are taken. In the first, it is assumed that the statistical number of active users in the cell receiving a video flow is Poisson distributed, the mean of which at any one time of day can be taken from the average load at that *L(t)*. The probability of there being *k* active users in the cell at time of day *t* is expressed as:

$$P(k,t) = \frac{L(t)^k e^{-L(t)}}{k!} \tag{3.3.18}$$

Under this first approach, our numerical assessment cycles in outer loops through a 24 h period in steps of *t* of one hour, and uses the value of *L(t)* (given *BusyLoad*) at each hourly time unit to parameterize Eq. (3.3.1). In inner loops, for each time unit, it then cycles through each possible value of *k* representing each possible number of users in the cell, for all participating frequency bands, and for each set of *k*'s ascertains the power consumption that would be required given power saving solution being applied. The actual power consumption for each case is given as this power consumption, multiplied by the probability of it happening. This result is then summed with equivalent results for all possible chosen values of k to obtain the overall power consumption at time unit t. The same operation is performed over all hourly t in the 24 h period, and the average power consumption is then taken among all t. The same process is performed to find the average power consumption of a conventional system without the power saving operation being applied.

The second simulation approach assumes a separate ON/OFF traffic flow to each user, either parameterized as FTP or HTTP traffic, taken from [29], whereby the number of users receiving flows varies throughout the 24 h period according to $L(t)$.

3.3.4.3 Power Saving by Dynamically Powering Down Radio Equipment

Simulating the LTE system with video traffic sources to users, Fig. 3.24 gives the percentage of from-the-socket power ("Energy Reduction Gain—ERG") that is saved through moving users between bands to dynamically power down radio equipment, for 2, 3, and 4 spectrum bands participating in the process, where the 2-band case considers the network powering down solution with and without sectorization switching in tandem. For all results in Fig. 3.24 *BusyLoad* is the same for all channels. It is clear that power savings in the range of 20–50 % can be achieved if there are lesser bands participating in the process or there is a greater network loading. For the 2-band case, if the networks are heavily loaded the sectorization switching solution considerably improves performance compared with the network powering down solution operating alone, but offers no improvement if networks are lightly loaded.

Figure 3.25 plots power saving results for the FTP and HTTP (web browsing) ON/OFF traffic models over the LTE configuration, where 2 bands are participating in the process and the assumption is that the network powering down solution only is employed.

Results again show a significant power saving potential of up to 50 % for low network loads. In the FTP case, power saving begins to reduce at a *BusyLoad* of $\sim$20 users, reaching as low as 10 % at a *BusyLoad* of $\sim$50 users. In the HTTP case, power saving begins to reduce at a *BusyLoad* of $\sim$150 users, and hits 10 % at a *BusyLoad* of $\sim$500 users.

3.3.4.4 Power Saving by Propagation Improvement

In assessing the power saving of reallocating users/links to improve propagation, we have opted to study an HSDPA system. This is because an HSDPA BS was the most modern specification BS for which detailed data was available on from-the-

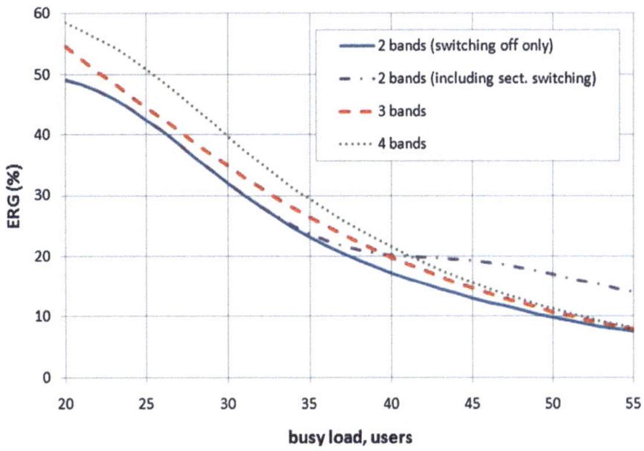

Fig. 3.24 Power saving against busy hour load for powering down solutions (streaming video traffic)

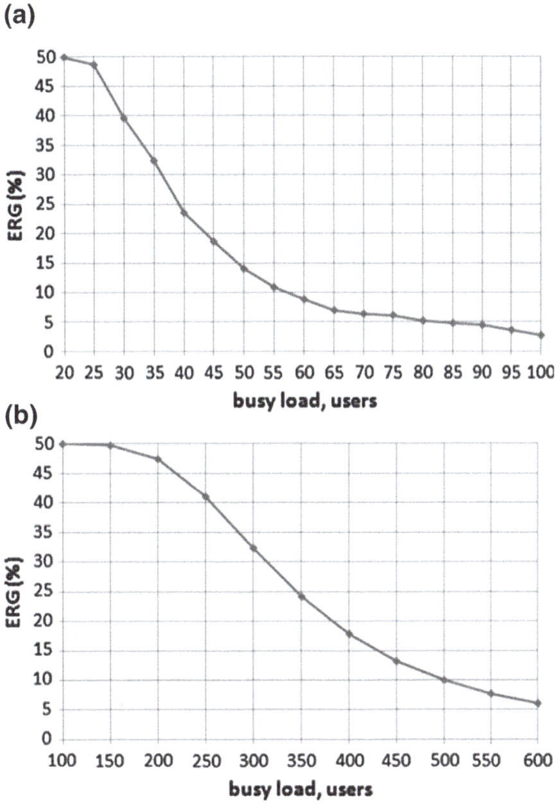

Fig. 3.25 Power saving against busy hour load for network powering down solution: **a** FTP ON/OFF traffic. **b** HTTP ON/OFF traffic

socket power consumption against transmission power. Internal documentation indicates power consumption for an HSDPA BS at 100 % transmission power to be 857 W, and at 20 % transmission power to be 561 W. It is widely observed that from-the-socket power consumption against transmission power broadly varies

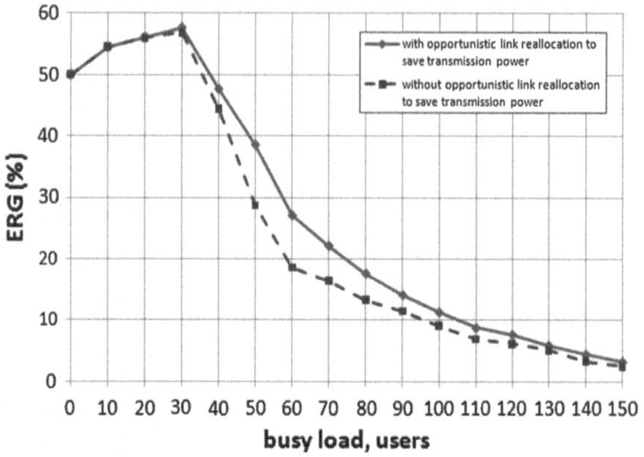

Fig. 3.26 Power saving against busy hour load through opportunistic link reallocation to use better propagation bands (FTP ON/OFF traffic)

with an $m{\cdot}p + c$ relationship, comprising a fixed term c that is independent of transmission power p, and a term that varies with transmission power, $m{\cdot}p$. The above figures regress to give 487 W, as the fixed part from-the-socket power consumption c, and the gradient of variation with transmission power m as 9.25 W per transmission Watt.

In ascertaining necessary transmission power, we use values in Table 3.3 of Ref. [30], with 80 % of the power budget scaled by the number of users present in the system and 20 % allocated to pilot transmission. The comparison in [30] is between full HSDPA networks operating at 2 GHz, and at 5 GHz. A 600 m cell radius is chosen, where again we assume the FTP ON/OFF traffic model. Figure 3.26 show that there is significant transmission power saving potential through the opportunistic reallocation scheme.

Power saving initially increases to some 58 % as the busy hour load is increased to 30; this is because it is always possible to reallocate users to power down radio equipment. However, as the traffic load is increased further power saving decreases and a difference begins to emerge in performance for the solutions with and without opportunistic reallocation to save transmission power.

3.4 Opportunistic Unsynchronized Cognitive Radio Networks Using Filter Bank Multicarrier (FBMC)

Maurice Bellanger

The Conservatoire National des Arts et Métiers (CNAM), Paris, France

The radio spectrum has two essential characteristics for communication, it is a limited resource which can be accessed from everywhere. The concept of cognitive radio has been created with the objective to make the best of this situation, by providing the highest spectral efficiency and offering the maximal access flexibility. The ultimate implication of the concept is that a communication system, for example a base station and its users, which detect an unoccupied frequency band in the spectrum should have the freedom to exploit it, without having to go through a lengthy clearance procedure and coordinate with other systems in the same geographical area. Hence, the denomination of opportunistic unsynchronized networks. From the operational perspective, the approach is particularly appealing due to its potential agility and the possibility to use a light infrastructure.

In practice, however, the introduction of such new systems amid existing communication networks or broadcasting devices requires a minimal level of regulation and standardization, in order to ensure coexistence and gain acceptance by all the players in the field. In fact, it appears that some global supervision will be required, at least at the beginning of the spectrum sharing deployment process. To that purpose, the most promising technique is the geolocation data base concept, according to which the users receive relevant information about the spectral occupancy in their local environment and instructions for their operation conditions. Depending on the level of authority the data base will have on the users and its ability to impose connection parameters, the cognitive radio networks will be semi-opportunistic or not opportunistic at all.

In any case, truly opportunistic or not, the cognitive radio networks must rely on the capabilities of their terminals to face the challenge of maximizing spectral usage under the constraint of coexistence. The physical layer is particularly critical in that respect and it must have the following features:

- capability to handle unsynchronized users with minimal loss in spectral use. For example, primary and secondary users cannot be synchronized.
- guaranteed protection of other users, for coexistence.
- capability to exploit fragmented spectrum in case of broadband transmission.
- capability to establish a link without preliminary distant alignment, for multi-user efficiency.
- high performance for real time spectrum sensing/monitoring in terms of resolution and latency, for autonomous access decision or to complement the information provided by the geolocation data base.

Multicarrier transmission techniques have the potential to provide these features. However, the technique employed by existing systems, OFDM, does not fully satisfy the above requirements. An improved physical layer is needed and the FBMC technique is proposed for cognitive radio systems [31, 32], as an enhancement of OFDM. The approach consists of complementing the FFT of the OFDM scheme by a set digital devices, called a polyphase network, so that a filter bank is obtained, in which a protype filter is shifted in frequency around the sub-carriers. Then, the transmission channel is split into a set of sub-channels and the attenuation characteristics of the prototype filter controls the spectral separation of the sub-channels. In order to minimize system latency and computational complexity, a sub-channel overlaps with its immediate neighbours, so that it is sufficient to introduce an idle sub-channel between two blocks of sub-channels to separate these blocks and make

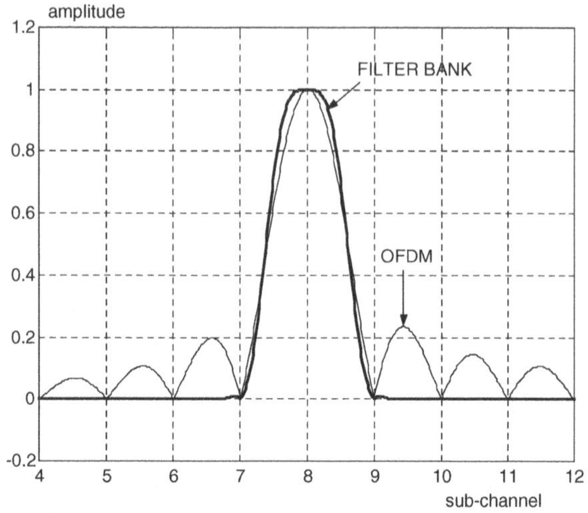

Fig. 3.27 Frequency responses of OFDM and FBMC

them spectrally independent. The comparison with OFDM is illustrated in Fig. 3.27. The OFDM frequency response exhibits large sidelobes so that full synchronization of the sub-carriers is necessary. The filter bank has no sidelobes, and only neighbouring sub-channels overlap.

As shown in Fig. 3.28, users with different transmission parameters may occupy the spectrum, even with different modulation schemes. Regarding spectral efficiency, there is no cyclic prefix, which means an increased bit rate and easy streaming. If necessary, for example in broadband communications, the distortions of the transmission channel can be compensated by equalizers at the sub-channel level.

Radio networks generally transmit by bursts. With FBMC, the transmission of a burst requires some extra time to accommodate the initial and final transitions associated with the filter impulse response. An illustration is given in Fig. 3.29. In fact, these transitions ensure that the adjacent users are not disturbed when a user accesses or quits the spectrum. If a frequency gap is present between the users or if a temporary reduction in performance is tolerable, the transitions can be significantly shortened, for example to $Ns + 1$ symbols if Ns symbols have to be transmitted. The practical consequence of the increase of the burst length is that FBMC favours longer bursts.

In any case, when compared with the OFDM burst, the FBMC burst carries more data symbols in a given time interval, because there is no guard time, or cyclic prefix, between the multicarrier symbols.

The filter bank in the receiver can also be employed for spectrum sensing and the conventional techniques can be applied at the sub-channel level, namely energy detection, matched filter, cyclostationary feature detection or self-correlation. The specificity of FBMC lies in the spectral separation provided by the filters and the absence of spectral leakage which improves the performance. In particular, significant gains can be obtained with respect to OFDM concerning

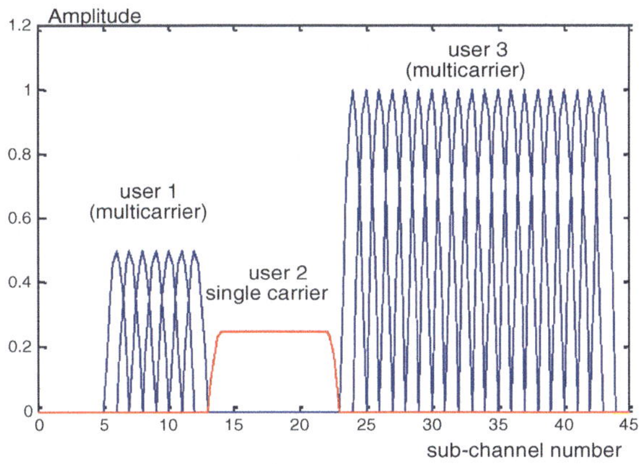

Fig. 3.28 Multi-user transmission with FBMC

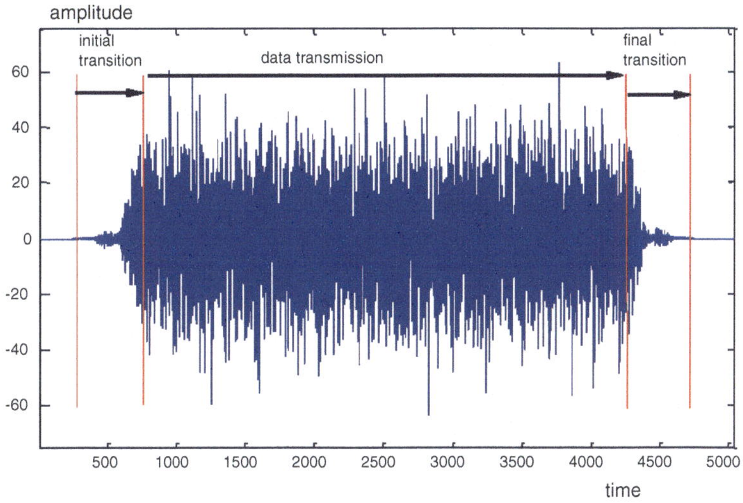

Fig. 3.29 Structure of the transmitted burst with FBMC

noise power estimations. The counterpart is the latency, in which the filter impulse response must be included.

An important aspect is that continuous spectrum monitoring is readily achieved if sub-channels are reserved for this function. In particular, if a group of 3 sub-channels is left idle, the center sub-channel does not overlap with the active sub-channels and it can sense during transmission.

Finally, it can be stated that a system based on the FBMC physical layer can operate efficiently in an unsynchronized environment. The next step towards building the opportunistic or semi-opportunistic network is the definition of a suitable protocol.

3.5 Detection of Malicious Users in Cognitive Wireless Ad-Hoc Networks: A Statistical Approach

Ferran Adelantado[1] **and Christos Verikoukis**[2]

[1]Universitat Oberta de Catalunya, Barcelona, Spain

[2]Telecommunications Technological Centre of Catalonia (CTTC), Barcelona, Spain

The Opportunistic Spectrum Sharing (OSS) concept [33] is based on two main premises: the maximization of the spectrum usage and the protection of the incumbent primary systems. Both premises are highly coupled and depend on the accurate knowledge of the nearby wireless environment, i.e. the detection of primary users (PU) transmitting/receiving in the vicinity of the secondary users (SU).

The research community is currently debating between two cognition approaches: the local sensing [34] and the usage of location databases [35]. The former is based on the acquisition of spectrum usage knowledge by sensing the air interface whether in a cooperative or non-cooperative manner. On the contrary, the latter assumes the existence of a spectrum usage database, which is updated periodically by the PUs, where all SUs address queries before leasing any primary channel. If, for a given SU location, time, and spectrum resource, the database response certifies the idleness of the PUs, the SUs lease the spectrum. Each approach is deemed appropriate for different applications.

Focusing on the local sensing approach, it relies on the sensing capabilities of the so-called Secondary Users (SU). Usually, SUs are assumed to be able to detect and identify the energy transmitted by the primary Users (PU), process the information and proceed accordingly [33]. However, such capabilities are limited in nature and the SUs present non-null false alarm and missed detection probabilities [34]. Both inadequacies, i.e. false alarm and missed detection, affect the two premises on which OSS is based. Whereas false alarm diminishes the spectrum usage efficiency, missed detection endangers the protection of the incumbent PUs. In the aforementioned context, the cooperation among SUs comes up as a proper manner to improve the environment knowledge obtained during the sensing process [36]. However, although cooperation is effective to smooth the effect of sensing errors, it turns out to be inefficient to cope with the attacks of malicious SUs [37, 38]. Thus, cooperation poses new vulnerabilities, since false information might be easily propagated across the secondary network. In order to limit the impact of inaccurate/false sensing information, malicious SUs detection mechanisms are required.

In this section, a new algorithm based on statistical non-parametric methods is proposed to detect the malicious SUs. First, it is worth noting that any detection mechanism should consist of two stages:

Detection of the existence of malicious SUs: In the first stage, given a set of users, it must be decided whether there are malicious SUs or not.

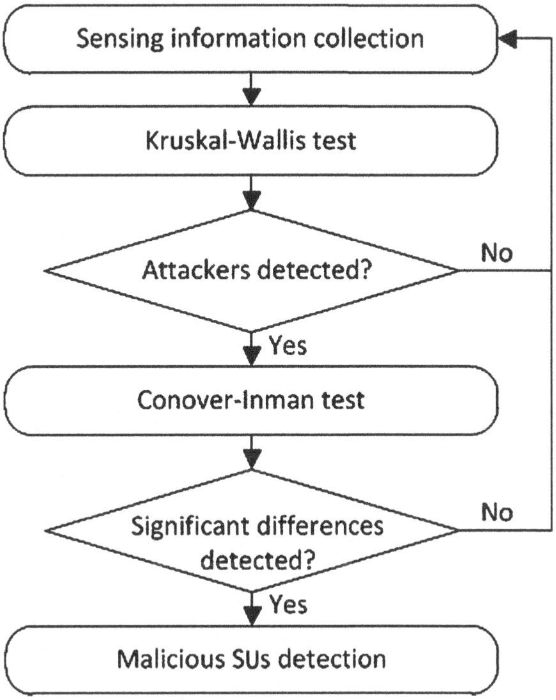

Fig. 3.30 Flowchart of the detection mechanism

Identification of the malicious SUs: Given a set of users where at least one of them is malicious, in the second detection stage the malicious SUs must be identified.

In the context of cognitive wireless ad-hoc networks, the assumptions regarding the knowledge of the primary channel activity have a huge impact on the feasibility of the proposals. Hence, there are two hypotheses that limit the real application of the detection mechanisms: the knowledge on the existence of malicious SUs and the knowledge of the primary channel activity.

In the system under study, the detection of malicious users can be done by comparing the sensing information provided by the cooperative SUs. As little a priori information is known about the behavior of the SUs, non-parametric statistical methods are appropriate to effectively perform such a task. In particular, the Kruskal–Wallis [39] analysis is used to detect the existence of attackers, whereas, after detecting it, the post hoc Conover-Inman [40] analysis is aimed to identify the malicious users. The flowchart of the process is shown in Fig. 3.30.

The described mechanism has proven to detect the malicious users even when there is not a priori information regarding the activity of the primary channel and the characteristics of users. Yet, it is also true that the optimum performance of the proposed mechanism depends a great deal on its parameters, e.g. the number of employed samples or the significance level, which should be tuned according to the scenario characteristics.

3.6 Spectral Efficiency for the Benefit of CR and Coexistence with Iterative Water-Filling

Maurice Bellanger[1] **and Aurimas Anskaitis**[2]

[1]The Conservatoire National des Arts et Métiers (CNAM), Paris, France
[2]Vilnius Gediminas Technical University, Vilnius, Lithuania

3.6.1 Efficiency and Coexistence

Waterfilling (WF) improves the spectral efficiency of communication systems whenever the channel signal-to-interference-plus-noise-ratio (SINR) exhibits large variations in the exploited frequency band [41, 42]. In radio communications, it can be applied to broadband transmission, provided the channel remains nearly static or varies slowly [43].

The WF technique, associated with multicarrier transmission, is implemented in several steps. The first step is the measurement of the channel SINR as a function of frequency, with the granularity of the multicarrier scheme. Then comes the bit-loading assignment, whereby the measurement results are used to distribute the number of bits to be transmitted in every symbol across the sub-carriers. The rule is that the sub-carriers carry a number of bits related to their received SINR. Specifically, the curve of inverted SINR is filled with energy, namely the sum of the received signal power and noise power, up to a constant level, hence the term "waterfilling". With this procedure, sub-carriers with high SINR receive more energy and carry several bits, while sub-carriers with low SINR receive less energy and carry a single bit or are just discarded. If the total number of bits per multi-carrier symbol is imposed, then the emitted signal power is adjusted, along with the "water level". Conversely, if the emitted signal power is specified, then the bit rate is adjusted. Next, once the bit loading has been decided, the information is forwarded to the distant terminal for implementation. Of course, for the procedure to be successful, the channel characteristics should not change significantly between successive symbols and, in any case, the roundtrip transmission delay of the system must be kept minimal. Also, it must be pointed out that the accuracy of the SINR measurements is critical for the benefit of the approach.

In fact, the WF algorithm is optimal for the radiated power distribution, in case of single user communication. Therefore, WF contributes to the efficiency of the spectrum usage.

Now, in the context of cognitive radio and multiple users, WF helps coexistence because the signal of a primary user, or that of another secondary user, will be incorporated in the global SINR measurement of a potential new secondary user and the overlapping sub-carriers are likely to be either not loaded at all, or weakly loaded. In other words, the WF procedure automatically leads to separating successive users in the spectral domain. Moreover, the fact that WF leads to the minimum of radiated power for a specified bit rate is an important contribution to coexistence. Finally, it can be stated that modems, when they are equipped with WF algorithms and devices, autonomously adapt their power and spectra for coexistence.

Now, after the single modem case, the global behaviour of a system consisting of several modems implementing simultaneously and independently the WF technique is illustrated [44].

3.6.2 Iterative Waterfilling (IWF)

Several communicating pairs in some area use the same (or overlapping) frequency band and they cause interference to each other. IWF basically reduces this interference and separates the transmitted spectra of the users. Every transmitter takes its turn sequentially and performs WF power allocation (assuming perfect channel knowledge to its corresponding receiver). Transmitted signals of other users are considered as interference, they in effect decrease the measured SINR level. In the real world, this process often converges very quickly to some stable state called Nash equilibrium [45], but in general the convergence is not proven. Also, the resulting sum channel capacity is often good but not necessarily optimal.

As an illustration, a situation when two communicating pairs share the same channel bandwidth is simulated. Transmitters and receivers are placed in the corners of a rectangle. All four channels used in the simulation are generated independently from model ITU-R vehicular A [46] and path loss between all nodes is calculated using standard Hata model. The same power is allowed for both transmitters and omnidirectional antennas are assumed.

The power spectral density (PSD) of allocated power for both transmitters at every allocation step is shown in Fig. 3.31. It can be seen that, at first, only the first transmitter allocates power, while the second transmitter is silent. Then second transmitter allocates power into more or less unallocated regions. The process continues for two more iterations. Usually excellent convergence is achieved after three iterations. In fact, even after two iterations the results are good.

In order to estimate the gain of IWF, some channel rates are calculated for both constant power allocation and IWF allocation. In this particular experiment data rate for constant power allocation is 16 Mbit/s and for IWF allocation it is 21 Mbit/s. Such gains are typical and they increase even more when signal-to-noise ratios decrease.

Overall, the IWF approach is attractive as communicating pairs do not require central governing organization, though, for best results some exchange of information is valuable.

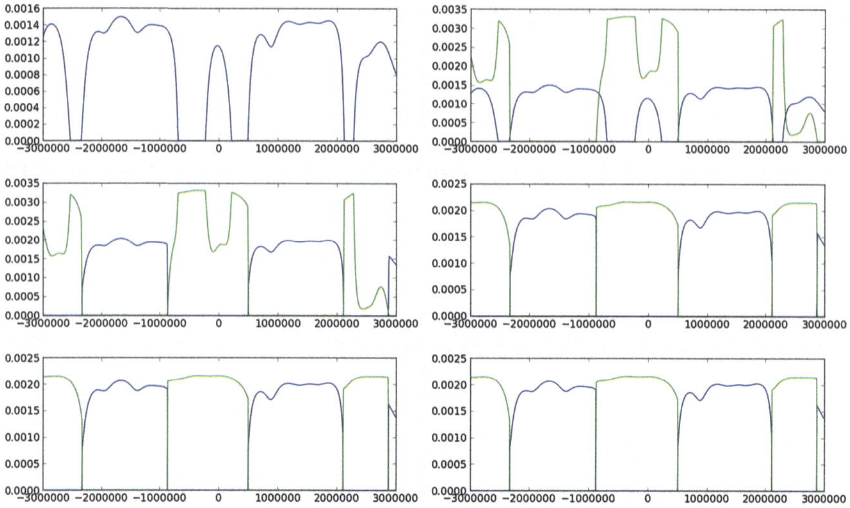

Fig. 3.31 Iterative waterfilling for 2 communicating pairs

3.6.3 Conclusions

Waterfilling is a promising technique for the deployment of cognitive radio networks and for spectral efficiency, particularly in the presence of quasi-static channels. It has the potential to bring noticeable gains in global performance and facilitate coexistence of opportunistic secondary users. However, a number of issues remain open, regarding, among others, the impact of the modulation schemes, the feedback information, the fairness of access, and the global optimality of the system.

3.7 Assessing the Amount of Spectrum that may be Available for DSA

Liljana Gavrilovska and Vladimir Atanasovski

University Ss. Cyril and MethodiusSkopje, FYR of Macedonia

3.7.1 Technology Enablers for Spectrum Assessment: Radio Environmental Maps (REM)

The problem of re-usage of available spectrum opportunities faces difficult practical challenges because of the spatio-temporal and contextual dependence of the available spectrum. The theoretically envisioned Dynamic Spectrum Access (DSA) and Cognitive Radio (CR) solutions require novel, environment-aware and self-organizing capable, approaches towards practical implementations in order to tackle the aforementioned challenges. The Radio Environmental Maps (REMs) [47–50] represent powerful technical enablers of DSA and CR. They were originally envisioned as a two-dimensional representation of the radio field strength [47], but today are foreseen as a rich hierarchical database or knowledge base that stores various kinds of radio environmental information. This information can be subsequently used for a plethora of optimization procedures in different DSA and CR scenarios [51].

REMs can store various type of information that can be easily fetched and used for a particular use-case. Generally, the stored information can be classified as either being *directly measured* (i.e. empirical) or being *indirectly derived* (i.e. derived through modelling). In any case, the information can be viewed (from a DSA and CR perspective) as:

- *Static*, e.g. locations of transmitters and/or receivers, terrain model etc. and
- *Dynamic*, e.g. propagation environment, up-to-date spectrum measurements, users' activity patterns etc.

The REM information can be used by many entities requiring reliable spectrum assessment. For instance, regulators and dedicated public bodies can use REMs for large-scale estimation of spectrum usage in order to track compliance to regulations, estimate frequency-planning effectiveness etc. Cellular network operators can interpret measured results through drive tests or mobile subscribers in order to perform Minimization of Drive Tests (MDTs), network planning and fault detection etc. Finally, consumers can perform self-optimization and/or learning of patterns and habits leading to higher QoS, lower prices etc.

The focal point of every REM is the capability to accurately assess the spectrum occupancy, i.e. accurate decision upon possible spectrum opportunities. As a result, the available spectrum opportunity detection techniques strongly influence the design and practical implementation of a REM. Currently, there are two distinct spectrum opportunity detection techniques [53], i.e.:

- Sensing-based spectrum opportunity detection and
- Database-based spectrum opportunity detection,

each with its own advantages and disadvantages [11]. The optimal way towards practical REM implementation should encompass features of both techniques [51, 53].

Figure 3.32 depicts the FARAMIR REM architecture [53] as the world's first complete practically deployed and fully operational REM creation architecture.

It comprises four distinguished architectural elements:

- **Measurement Capable Devices (MCDs)**—responsible for performing spectrum measurements (continuous and on-demand) in different band of interest.
- **REM data Storage and Acquisition unit (REM SA)**—responsible for storing the spectrum measurement data coming from the different types of MCDs, as

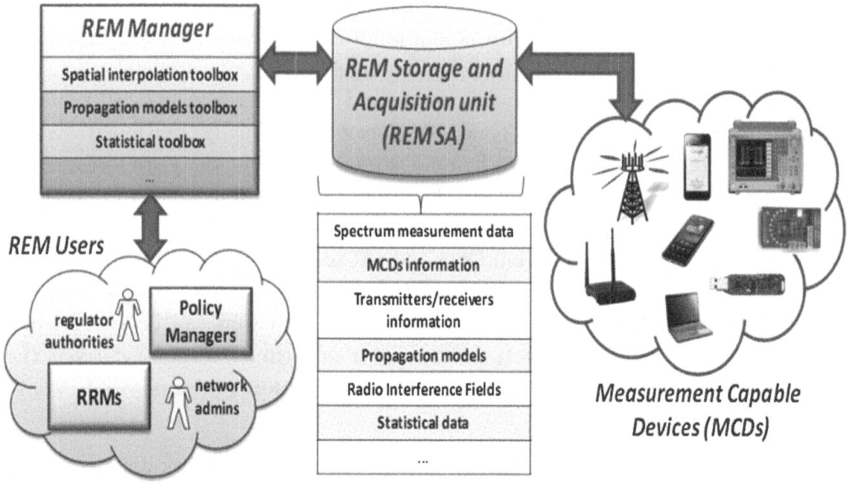

Fig. 3.32 The developed functional REM architecture [51, 52, 54]

well as information about the positions and configuration of radio transmitters and receivers, environment characteristics and REM processed data.

- **REM Manager**—responsible for processing regarding the REM data creation and evaluation (modularly constituted comprising various toolboxes that serve for estimation of Radio Interference Fields (RIFs), localization of transmitters, statistical analyses of the spectrum usage, assessment of the environment propagation characteristics etc.).
- **REM Users**—entities that use the REM information, i.e. entities that govern the frequency/power allocation, spectrum access/(re-)usage, network optimization etc.

An example of the FARAMIR REM's architecture capability to act as technology enabler for spectrum assessment is its *spatial interpolation capability of sparse spectrum measurements* [50, 51, 53, 54] and its *transmitter localization capability* [55]. Both capabilities are embedded in the REM Manager as scalable toolboxes able to run real-time spectrum occupancy analysis. The spatial interpolation toolbox relies on local neighborhood based spatial interpolation whereas the transmitter localization toolbox uses an ML-like technique to detect potential transmitter fast and with sufficient accuracy. Figure 3.33 visualizes the obtained REM information for a particular case of using 16 independent heterogeneous spectrum sensors indoor and performing spatial interpolation in 900 points.

Additionally, the figure also shows other possible REM related functionalities such as real-time monitoring of the current signal strength level from every MCD in the field, real-time calculation of the duty cycle of a particular MCD for a given time frame and a given threshold of interest and statistical analysis of the field.

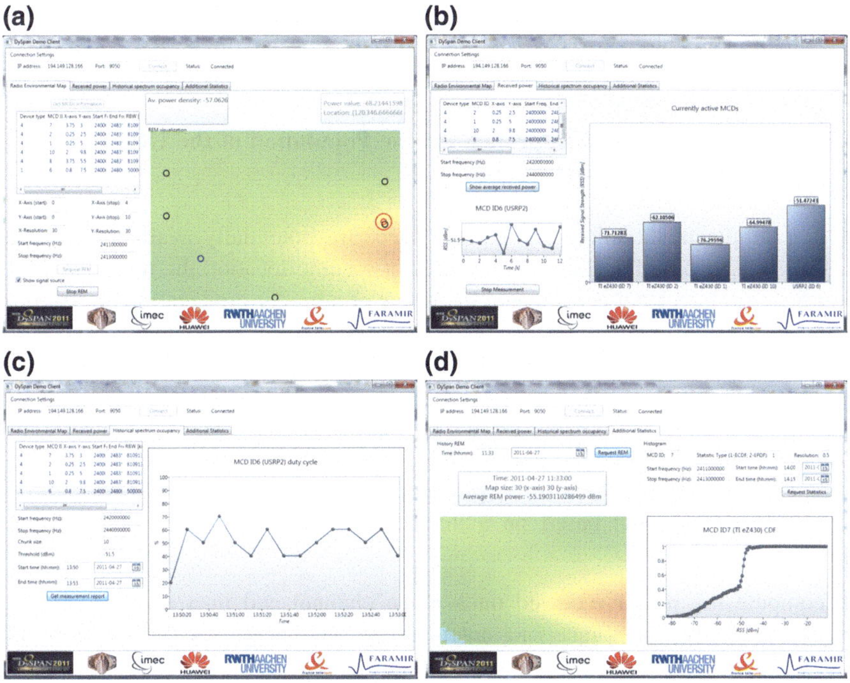

Fig. 3.33 Visualization of the REM information. **a** REM with MCDs and estimated transmitter's position. **b** Current Received Signal Strength (RSS) values for a single and multiple MCDs, **c** Calculated duty cycle for selected MCD, **d** Old REM and statistics graph

3.7.2 White Spaces Potentially Available in Italian Scenarios Based on the Geo-Location Database Approach

Marina Barbiroli[1], **Claudia Carciofi**[2], **Doriana Guiducci**[2] **and Valeria Petrini**[2]

[1]Bologna's University, Bologna, Italy

[2]Fondazione Ugo Bordoni, Rome, Italy

The estimation of the amount of spectrum potentially available as white space depends on several factors, as the White Space Device (WSD) characteristics, the topology of the area, the national rules governing the use of spectrum, the applications or services using the adjacent bands, and many others: these factors have to be fully defined by national or international regulators.

Spectral resources potentially available for White Space Devices (WSDs) in the UHF band (channels from 21 to 60) have been investigated by FUB [56] in the framework of the CEPT SE43 work group [57] based on the geo-location database approach [58].

Three different methodologies are implemented to estimate the amount of potentially available white space in different Italian scenarios [59].

3.7.2.1 Threshold-Based Approach for the Population of the Geo-Location Database

A first simple methodology to identify spectrum potentially available as white space is based on a threshold approach: in a specific location the received signal power on each channel is evaluated by means of a proper propagation model [60], and if the estimated power on a channel is below a selected threshold it is deduced that there are no licensed users for that channel in the proximity of the investigated location. Therefore a WSD is allowed to transmit its signal, provided that the specific emission requirements are met [61, 62]. The threshold can be determined as a function of different parameters such as the incumbent service to be protected (e.g., DTT, PMSE) or the level of protection to be granted.

The WS estimation based on a threshold approach has been performed as follows:

1. compute the power received on a given channel and in a given pixel (e.g. $600 \times 600\ m^2$) by a receiving antenna (e.g. omnidirectional with 0 dBi gain) assumed at a specific height above ground level.
2. compare the received power against a specific threshold;
3. if the received power is below the threshold, the channel is considered as vacant;
4. iterate steps 1–4 for all the channels from 21 to 60 and for all the pixels of the considered area.

3.7.2.2 Location Probability Approach for the Population of the Geo-Location Database

The second approach is particularly focused on the protection of the DTT service which is the one of paramount interest for the Italian scenario. According to this approach, the usage of a specific channel is prevented within the coverage area of each DTT transmitter employing that channel, in order to guarantee a wanted Location Probability related to the proper field strength threshold [63]. In Table 3.4 the minimum median received field strength levels of the GEO-06 RPC at 650 MHz frequency are shown for fixed reception with respect to different location probability values.

DTT field strength levels are evaluated using accurate propagation models, which take into account also diffraction phenomena. Predicted values are then employed to identify the DTT protection area and the paired zone outside the coverage area where the channel is potentially available for WSDs.

Table 3.4 DTT reference planning configurations (fixed reception, frequency 650 MHz)

Location probability (%)	$F_{k,min}$ dBμV/m at 10 m
99	60
95	56
50	48
1	34

The potentially available white spaces are calculated based on the following approach:

1. for each pixel and for each channel, the field strength level E_{rx} (dBμV/m) considering all possible DTT transmitters is evaluated with a suitable propagation model;
2. the calculated field strength level E_{rx} is compared with the selected planning configuration threshold $F_{k,min}$ [63].
3. if E_{rx} is $>$ $F_{k,min}$ the pixel is within the protected service contour, hence the channel is occupied; if E_{rx} is $<$ $F_{k,min}$ the pixel is outside the protected service contour, hence the channel is vacant.

3.7.2.3 Combined Geo-Location Database and Sensing Approach

In this approach geo-location database and sensing methodologies are combined to determine white spaces potentially available.

Vacant and occupied channels are evaluated based on the following algorithm:

1. apply the methodology described in Sect. 3.7.2.2 to identify white spaces (pixels and frequencies) potentially available based on the geo-location approach;
2. implement field strength measurements in different pixels and channels and compare the results with a proper sensing threshold to identify occupied and un-occupied channels;
3. only when "un-occupied channels" are obtained from both step 1 and step 2 the considered pixel and frequency are potentially available for the WSD otherwise the channel is occupied.

3.7.2.4 Threshold-Based Approach Results

We apply the estimation procedure described in Sect. 3.7.2.1 to the case of West Piedmont. Using public information on ERP values and positions of each DTT transmitter, we assessed the power received in each channel and location by means of the Recommendation ITU-R P.1546-3 [60] assuming the receiving antenna height at 1.5 m and 10 m.

We considered two different thresholds set to -114 dBm and -120 dBm.

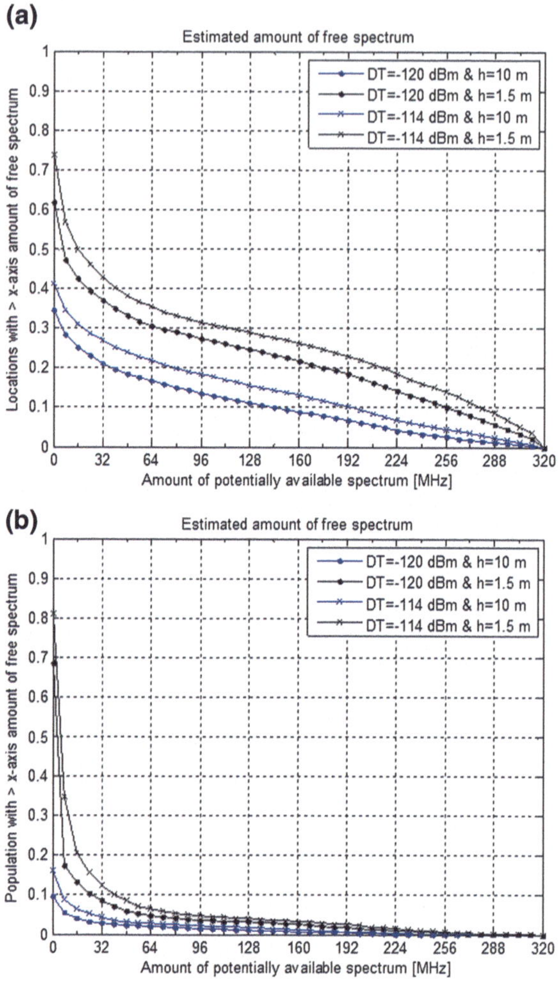

Fig. 3.34 Cumulative distribution function of the amount of White Spaces per pixel **a** and per population **b** (West Piedmont)

Figure 3.34a and b show the Complementary Cumulative Distribution Function (CCDF) of the estimated amount of spectrum available as white space respectively as a function of the territory and population density.

These figures confirm that densely populated areas have little or no spectrum available as white space since the areas where there are more available channels are rural. It can be easily noted that the CCDF referring to the population percentage (Fig. 3.34b) has a steeper descent than the one referring to the location percentage (Fig. 3.34a). For example, while almost 20 % of locations have more than 64 MHz available with DT = −120 dBm and h = 10 m, only 2 % of population actually resides in these areas. The amount of spectrum available as white space strongly depends on the chosen threshold. For instance, assuming a threshold equal to −120 dBm and an antenna height of 1.5 m, the percentage of

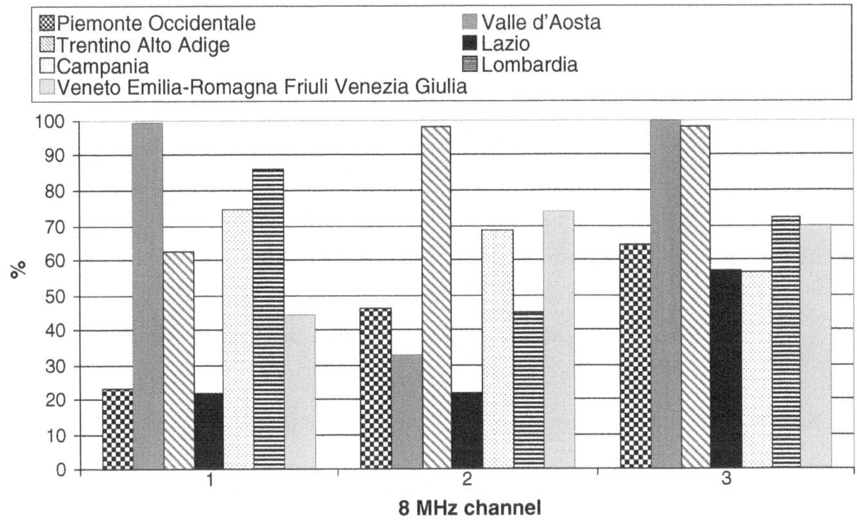

Fig. 3.35 Percentage of population living in potential white space areas for three different channels in several Italian regions

pixels where there is at least 1 available channel is 47.19 %, while raising the detection threshold to −114 dBm this percentage becomes 56.99 %.

Although these results refer to a delimited area of Italy, a similar behaviour is observed in other studies on EU and non-EU countries [64–65].

3.7.2.5 Location Probability-Based Approach Results

DTT coverage simulations have been performed in a real scenario in other different Italian regions. Predictions have been carried out using a proprietary prediction tool, where the propagation model takes also into account the diffraction phenomena according to ITU-R Recommendation P.526 [66]: occupied and vacant DTT channels are identified according to different levels of protection of the incumbent service (i.e., different percentage of Location Probability).

According to the adopted criteria for DTT protection, it is possible to identify those pixels where specific channels could be available for WSDs.

In Fig. 3.35 the usage of three different DTT channels in some Italian regions is considered and the percentage of population living in the white space locations is calculated on a regional basis. It can be noticed that there are regions where the chosen channels are almost completely un-occupied (e.g. mountain regions such as Valle d'Aosta and Trentino Alto Adige), while in more densely populated areas (e.g. Lazio) the considered channels are potentially available to serve a lower percentage of population.

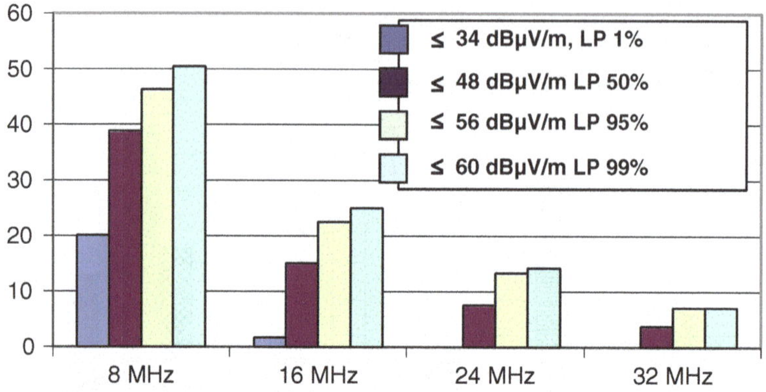

Fig. 3.36 Percentage of locations where 1, 2, 3 or 4 adjacent channels could be available

In Fig. 3.36 the percentage of locations where a least one or two or more adjacent DTT channels could be available in a central-north Italian region is calculated as a function of the protection level of the incumbent service.

It can be noticed that the percentage of locations where more bandwidth is potentially available strongly decreases as the number of the considered adjacent channels increases. Moreover the selected protection level for the DTT service has a strong impact on the percentage of locations where one or more channels could be available.

3.7.2.6 Combined Sensing and Geo-Location Results

The combined sensing and geo-location approach has been applied in a real scenario in Italy, in the province of Bologna. In the considered scenario field strength levels for DTT channels from 21 to 60 are calculated over square pixels (400×400 m^2). In Fig. 3.37 simulated field strength levels (dBμV/m) are shown for a DTT channel and it can be noticed that most of the pixels of the province of Bologna are occupied, as only some hilly and mountainous areas quite far from Bologna are not reached by the DTT signal.

Further information on channel occupancy can be achieved by means of sensing techniques. To this aim measurements have been implemented in six different locations using the "channel power" mode, in order to evaluate the total amount of power received in the DTT channel bandwidth.

In Table 3.5 the percentage of channels for which both geo-location database and sensing identify unoccupied channels is reported.

It can be noted that the percentage of potentially available WS is strongly dependent on the DTT protection threshold used to populate the geo-location database.

A comparison between results obtained with the combined approach and with the geo-location database alone highlights that the percentage of WS is similar

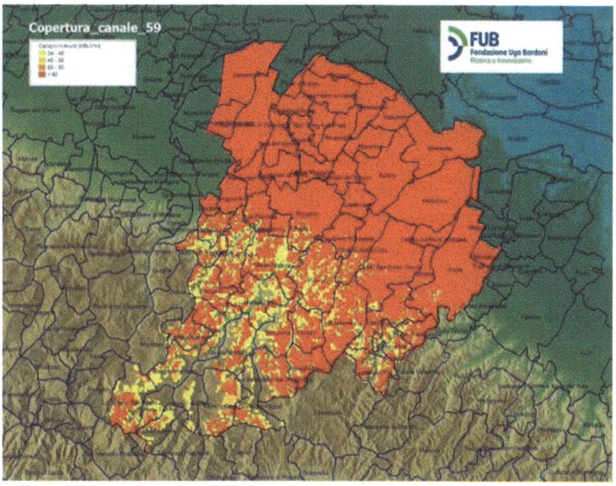

Fig. 3.37 Example of received DTT signal strength in the province of Bologna

Table 3.5 Combined geo-location and sensing approach

Different field strength levels for DTT reference planning configurations (fixed reception) dBμV/m	Percentage of available WS
<=34	20.08
<=48	38.89
<=56	46.15
<=60	50.43

considering the geo-location database alone and the combined and sensing approach. This is mainly due to the scarce sensitivity of the probe equipment (−80 dBm) with respect to the sensing detection threshold specified in [67] (e.g. −120 dBm). Therefore to improve the accuracy of the combined approach we will provide other measurements with a probe which is characterized by a more accurate sensitivity level (<−80 dBm).

3.7.2.7 Conclusions

Some general considerations and remarks can be derived from the obtained results: we have applied different approaches to identify potentially available white spaces in real Italian scenarios. From simulation results, it is evident that in Italy the 470–790 MHz band is densely utilized and most of the potentially available WS are located in low populated zones such as mountain or hilly areas. Higher degree of the protection of the incumbent service strongly reduces the amount of un-occupied channels especially if contiguous bandwidth is needed to achieve broadband WSD transmission. Moreover, the above analysis confirms that a combined geo-location and sensing approach may give higher protection to incumbent services, provided that a proper detection threshold is applied.

References

1. Report ITU-R SM.2152. Definitions of Software Defined Radio (SDR) and Cognitive Radio System (CRS). International Telecommunication Union (2009)
2. CEPT Report 24. Technical considerations regarding harmonisation options for the Digital Dividend. A preliminary assessment of the feasibility of fitting new/future applications/ services into non-harmonised spectrum of the digital dividend (namely the so-called "white spaces" between allotments). Electronic Communications Committee (2008)
3. ECC Report 159. Technical and operational requirements for the possible operation of cognitive radio systems in the 'white spaces' of the frequency band 470–790 MHz. Electronic Communications Committee (2011)
4. ECC Report 186. Technical and operational requirements for the operation of white space devices under geo-location approach. Electronic Communications Committee (2013)
5. Draft ECC Report 205. Licensed shared access. Electronic Communications Committee (2013)
6. ECC Report 185. Complementary Report to ECC Report 159. Further definition of technical and operational requirements for the operation of white space devices in the band 470–790 MHz. Electronic Communications Committee (2013)
7. Mesodiakaki, A., Adelantado, F., Alonso, L., Verikoukis, C.: Energy-efficient contention-aware channel selection in cognitive radio ad-hoc networks. In: IEEE CAMAD 2012 (2012)
8. Yucek, T., Arslan, H.: A survey of spectrum sensing algorithms for cognitive radio applications. IEEE Commun. Surv. Tutor. **11**(1), 116–130 (2009)
9. Wang, B., Liu, K.J.R.: Advances in cognitive radio networks: a survey. IEEE J. Sel. Topics Signal Process. **5**(1):5–23 (2011)
10. Lunden, J., Koivunen, V., Huttunen, A., Poor, H.V.: Collaborative cyclostationary spectrum sensing for cognitive radio systems. IEEE Trans. Signal Process. **57**(11), 4182–4195 (2009)
11. Zhao, Y., Song, M., Xin, C.: FMAC: A fair MAC protocol for coexisting cognitive radio networks. In: IEEE INFOCOM 2013 (2013)
12. Zhao, Y., Song, M., Xin, C., Wadhwa, M.: Spectrum sensing based on three-state model to accomplish all-level fairness for co-existing multiple cognitive radio networks. In: IEEE INFOCOM 2012 (2013)
13. Mesodiakaki, A., Adelantado, F., Alonso, L., Verikoukis, C.: Fairness Evaluation of a Secondary Network Coexistence Scheme. In: IEEE CAMAD 2013 (2013)
14. Jain, R., et al: A quantitative measure of fairness and discrimination for resource allocation in shared computer systems. Technical Report TR-301, Digital Equipment Corp (1984)
15. Cesana, M., Cuomo, F., Ekici, E.: Routing in cognitive radio networks: challenges and solutions. Ad Hoc Netw. **9**(3), 228–248 (2011)
16. Akyildiz, I.F., Lee, W.Y., Chowdhury, K.R.: Crahns: cognitive radio ad hoc networks. Ad Hoc Netw. **7**(5), 810–836 (2009)
17. A. Akhtar et al., "Multi-hop Cognitive Radio Networking through Beamformed Underlay Secondary Access," IEEE ICC 2012, Ottawa, Ontario, Canada, June 2012
18. Akhtar, M., Holland, O., Le, T., Nakhai, M.R., Aghvami, A.H.: Cooperative cognitive radio beamforming in the presence of location errors. In: IEEE 73rd Vehicular Technology Conference (VTC 2011-Spring), May 2011
19. Mihovska, A., Meucci, F., Prasad, N.R., Cabral, O., Velez, F.J.: Multi-operator resource sharing scenario in the context of imt-advanced systems (invited paper). In: Proceedind of Second International Workshop on Cognitive Radio and Advanced Spectrum Management (2009 -CogART 2009), Aalborg, Denmark, May 2009
20. Kellerer, H., Pferschy, U., Pisinger, D.: Knapsack Problems. Springer, Berlin (2005) ISBN 3-540-40286-1
21. Karlof, J.K.: Integer Programming: Theory and Practice. 1st edn. CRC, New York (2005)

22. Monteiro, V., Rodriguez, J., Gameiro, A., Cabral, O., Velez, F.: RAT and cell selection based on load suitability. Invited paper in Proceeding of VTC2008 The 68th IEEE Vehicular Technology Conference, Calgary, Canada (2008)
23. Holma, H., Toskala, A.: WCDMA for UMTS—HSPA evolution and LTE, 1st edn. Wiley, West Sussex, England (2007)
24. TR25.211. Physical channels and mapping of transport channels onto physical channels (FDD), 5th. edn. 3GPP. (2005)
25. Chen, K.-C., Roberto, J., de Marca, B.: Mobile WiMAX, 1st edn. Wiley, West Sussex (2008)
26. 3GPP TR 25.892 v6.0.0, Feasibility Study for Orthogonal Frequency Division Multiplexing (OFDM) for UTRAN enhancement: The 3rd. Generation Partnership Project. Technical Specification Group Radio Access Network (2004)
27. Salo, J., Nur-Alam, M., Chang, K.: Practical Introduction to LTE Radio Planning [Online]. Available: http://www.eceltd.com/lte_rf_wp_02Nov2010.pdf (2010)
28. Vodafone UK Corporate Responsibility Report 2009-10: http://www.vodafone.com/content/dam/vodafone/UK%20CR/downloads/online.pdf (2010). Accessed April 2010
29. 3GPP TR 25825: Technical Specification Group Radio Access Network; Dual-Cell HSDPA Operation. V1.0.0 (2008)
30. Cabral, O., Meucci, F., Mihovska, A., Velez, F.J., Prasad, N.R., "Integrated Common Radio Resource Management with Spectrum Aggregation over Non-Contiguous Frequency Bands," Wireless Pers. Commun. **59**(3), 499–523 (2011) (available online DOI:10.1007/s11277-011-0242-6)
31. Bellanger, M., et al.: FBMC physical layer: a primer, EU-FP7 project PHYDYAS (2010). http://www.ict-phydyas.org
32. Noguet, D., Gautier, M., Berg, V.: Advances in opportunistic radio technologies for TVWS. EURASIP J. Wirel. Commun. Netw. **2011**(1), 1–12 (2011)
33. Akyildiz, I., Lee, W., Vuran, M., Mohanty, S.: Next generation/dynamic spectrum access/cognitive radio wireless networks: a survey. Int. J. Comput. Telecommun. Networking **50**(13), 2127–2159 (2006)
34. Yucek, T., Arslan, H.: A survey of spectrum sensing algorithms for cognitive radio applications. IEEE Commun Surv. Tutorials **11**(1), 116–130 (2009)
35. Zhang, J., Zhao, Q., Zou, J.: The IEEE 802.22 WRAN system based on radio environment map (REM). In: Proceedings of First International Workshop on Education, vol. 1, pp. 98–101 (2009)
36. Hossain, E., Niyato, D., Han, Z.: Dynamic Spectrum Access and Management in Cognitive Radio Networks. Cambridge University Press, Cambridge (2009)
37. Arkoulis, S., Kazatzopoulos, L., Delakouridis, C., Marias, G.: Cognitive spectrum and its security issues. In: Proceedings of Next Generation Mobile Applications, Services and Technologies (NGMAST), pp. 565–570 (2008)
38. Parvin, S., Hussain, F.K., Hussain, O.K., Han, S., Tian, B., Chang, E.: Cognitive radio network security: a survey. J. Network Comput. Appl. **35**(6), 1691–1708 (2012)
39. Gibbons, J.D., Chakraborti, S.: Nonparametric Statistical Inference. CRC Press, Boca Raton (2003)
40. Conover, W.J.: Practical nonparametric statistics. Wiley, New York (1998)
41. Baccarelli, E., Fasano, A., Biaggi, M.: Novel efficient bit-loading algorithms peak-energy limited ADSL-type multicarrier systems. IEEE T. Signal Proces. **50**(5), 1237–1243 (2002)
42. Yu, W., Cioffi, J.: Constant-Power waterfilling: performance bound and low-complexity implementation. IEEE T. Commun. **54**(1), 23–28 (2006)
43. Dickmese, S., Srinivasan, S., Renfors, M.: FFT and filter bank based spectrum sensing and spectrum utilization for cognitive radios. Proc. of ISCCSP'12 conference, Rome, 2–5 May 2012
44. Yu, W., Ginis, G., Cioffi, J.M.: Distributed multiuser power control for digital subscriber lines. IEEE J. Select. Areas Commun. **20**, 1105–1115 (2002)

45. Scutari, G., Palomar, D.P., Barbarossa, S.: Simultaneous iterative water-filling for gaussian frequency-selective interference channels. IEEE Inter. Sym Inform. Theor. Seattle, 600–604 (2006)
46. Recommendation ITU-R M. 1225: Guidelines for evaluation of radio transmission technologies for IMT-2000 (1997)
47. Zhao, Y. et al.: Performance evaluation of radio environment map enabled cognitive spectrum-sharing networks. IEEE MILCOM 2007, Orlando, Florida (2007)
48. Atanasovski, V., van de Beek, J., Dejonghe, A., Denkovski, D., Gavrilovska, L., Grimoud, S., Mahonen, P., Pavloski, M., Rakovic, V. Riihijarvi, J., Sayrac, B.: Constructing Radio Environment Maps with Heterogeneous Spectrum Sensors. IEEE DySPAN 2011, Aachen, Germany (2011)
49. Denkovski, D., Rakovic, V., Pavloski, M., Chomu, K., Atanasovski, V., Gavrilovska, L.: Integration of Heterogeneous Spectrum Sensing Devices Towards Accurate REM Construction. IEEE WCNC, France (2012)
50. Denkovski, D., Atanasovski, V., Gavrilovska, L., Riihijarvi, J., Mahonen, P.: Reliability of a Radio Environment Map: Case of Spatial Interpolation Techniques. CrownCom, Stockholm (2012)
51. EC FP7-248351 project FARAMIR. Deliverable D6.2: Prototype Description and Field Trial Results. Information. http://www.ict-faramir.eu/fileadmin/user_upload/deliverables/FARAMIR-D6.2-Final.pdf
52. EC FP7-248303 project QUASAR. Deliverable 2.2: Methodology for assessing secondary spectrum usage opportunities – final report. Information. http://quasarspectrum.eu/images/stories/Documents/deliverables/QUASAR_D2.2.pdf
53. EC FP7- 248351 project FARAMIR. http://www.ict-faramir.eu/
54. Gavrilovska, L., Atanasovski, V.: Dynamic REM towards flexible spectrum management. 11th International Conference IEEE TELSIKS, Nis, Serbia (2013). Accessed 16–19 Oct 2013
55. Gavrilovska, L., Atanasovski, V., Rakovic, V., Denkovski, D., Angjelicinoski, M.: REM-enabled transmitter localization for ad-hoc scenarios. IEEE MILCOM, San Diego (2013). 17–20 Nov 2013
56. www.fub.it
57. ECC Spectrum Engineering Working Group SE43: Cognitive Radio Systems—White Spaces. http://www.cept.org/ecc/groups/ecc/wg-se/se-43
58. ECC Report 159: Technical and operational requirements for the possible operation of Cognitive Radio Systems in the 'White Space' of the frequency band 470–790 MHz
59. Barbiroli, M., Carciofi, C., Guiducci, D., Petrini, V.: White Spaces potentially available in Italian scenarios based on the geo-location database approach. DySPAN (2012)
60. ITU, Recommendation ITU-R P.1546-2: Method for point-to-area predictions for terrestrial services in the frequency range 30 MHz to 3000 MHz. Ottobre (2009)
61. Karimi H.R.: Geo-location databases for white space devices in the UHF TV bands: Specification of maximum permitted emission levels. In: IEEE Symposium on New Frontiers in Dynamic Spectrum Access Networks (DySPAN 2011)
62. Petrini, V., Karimi, H.R.; TV white space database: Algorithms for the calculation of maximum permitted radiated power levels. DySPAN (2012)
63. ITU: Final Acts of Geneva (2006)
64. Mishra, M., Sahai, A.: How Much White Space Is There?" Technical Report UCB/EECS-2009-3, Electrical Eng. and Computer Sciences Department Univercity of California, Berkeley. http://www.eecs.berkeley.edu/Pubs/TechRpts/2009/EECS-2009-3.html (2009)
65. Van de Beek, J., Riihijarvi, J., Achtzehn, A., Mahonen, P.: TV White Space in Europe. IEEE Trans. Mobile Comput. **11**(2):178–188
66. ITU- R Recommendation P.526: Propagation by diffraction. Approved in 2012-02
67. Harrison, K., Mishra, S., Sahai, A.: How much white-space capacity is there? In: Proceedings IEEE Symposium New Frontiers in Dynamic Spectrum Access Networks (DySPAN'10) (2010)

Chapter 4
Economic Aspects of CR Policy and Regulation

Keith Nolan and Vânia Gonçalves

Abstract This chapter turns to evaluation of techno-economic aspects of CR development and regulation, considering both the attractiveness of existing regulatory frameworks and the benefits of creating the new ones. This is important since it may be shown that the regulatory framework may have significant impact on economic benefits and viability of CR market adoption. Section 4.1 offers discussion of the potential for new business cases centred on the use of white space spectrum in the context of cellular networks. Section 4.2 is focusing on business scenarios and models for use of GDBs in TV white spaces. The following Sect. 4.3 provides a primer regarding the dynamics of the wireless communication market and how these can strongly influence the success or failure of a new technology. Section 4.4 considers potential business scenarios for spectrum sensing based on a set of parameters—ownership, exclusivity, tradability and neutrality. Section 4.5 looks at the prospects of business case for CR against the uncertainties of the spectrum market and opportunistic spectrum access circumstances. The chapter is concluded with the case study of techno-economic analysis in Sect. 4.6 that contemplates economic value of CR and secondary access. This builds a solid basis for answering the ultimate questions about business viability of CR, including considerations of cost versus capacity, investments, uncertainty and risk.

K. Nolan (✉)
CTVR/The Telecommunications Research Centre, Trinity College Dublin, Dublin, Ireland
e-mail: knolan2@tcd.ie

V. Gonçalves
iMinds-SMIT, Vrije Universiteit Brussel, Brussels, Belgium
e-mail: vania.goncalves@vub.ac.be

A. Medeisis and O. Holland (eds.), *Cognitive Radio Policy and Regulation*, Signals and Communication Technology, DOI: 10.1007/978-3-319-04022-6_4,

4.1 The Emergence of Whitespace Network-Based Business Cases

Keith Nolan and Jacek Kibilda

CTVR/The Telecommunications Research Centre, Trinity College Dublin, Dublin, Ireland

4.1.1 Introduction

The emergence of whitespace networks, and whitespace communications in general, provides an opportunity to, at least partially, meet the ever-growing demand for mobile data communication and to support new business cases. Many whitespace network solutions proposed so far realise coordination and rendezvous over licensed or unlicensed spectrum. In this section we explore a protocol for networks that rely solely on whitespace spectrum.

This work builds on [1], to where the reader is guided for further information beyond this section.

The proposed protocol allows both communication to the broader network (via the access point) and direct device-to-device links over whitespaces. To showcase the capabilities of the proposed solution we investigate a proof-of-concept software defined radio experiment. Using the experimental platform, we have evaluated the overheads of whitespace operation, which come in the form of an extra delay in association and a throughput loss of approximately 15 % of that achievable with licensed spectrum. The goal is to provide the groundwork for new business cases based on the use of wireless communications systems operating in whitespace spectrum.

Studies show that 100 and 58.6 % of Internet traffic generated by smartphones and PCs, respectively, is carried over wireless interfaces. 69 and 57 % accounts for Wi-Fi, which operates in unlicensed spectrum, and 31 and 1.6 %, respectively, accounts for cellular interfaces operating in licensed spectrum [2]. These numbers show that licence-exempt (or unlicensed) spectrum already plays a vital role in meeting the capacity challenge related to the mobile data crunch. The amount of transmitted mobile data will continue to grow, at an estimated compound annual growth rate (CAGR) of 78 % from 2011 to 2016 [3]. To support this demand we need even more pervasive Wi-Fi deployments, which are, however, limited by interference stemming from unlicensed operation.

Another possible solution to meet this demand is to increase the cellular network's density, which comes in the form of small cells (e.g. femtocells) that operate in a licensed spectrum underlay to macro cells. However, there is an alternative at hand—whitespace spectrum and CR technologies.

Whitespaces are defined by the Internet Engineering Task Force (IETF), as portions of the frequency spectrum that are assigned to a particular use but are unoccupied at specific locations and times [4]. This definition implies the existence of incumbent services, which have prioritised access to the spectrum and whose signals should be protected from harmful interference stemming from other whitespace-operating services. An example of whitespaces are TV whitespaces (TVWS), which are portions of the frequency spectrum made available after the digital TV switchover in the UHF/VHF spectrum in certain geographical locations. To protect incumbent services of the UHF/VHF spectrum, such as Digital Video Broadcasting—Terrestrial (DVB-T), communications regulators in the US and Europe selected a GDB technique as the most feasible, and, thus, the only mandatory solution [5, 6]. Hence, devices that desire to operate in the TVWS will have to interact with GDBs to obtain complete information about spectrum availability.

One of the objectives of the TVWS regulation in Europe was to allow high efficiency and flexibility in spectrum usage at the widest possible ranges of uses and technologies [5]. CRs are ideally crafted for this purpose, as they are wireless communication systems aware of their environment, which learn from this environment and adapt to any statistical variations in it, to achieve, for example, higher reliability or spectral efficiency [7]. A number of scenarios are envisaged for CRs operating in TVWS, for example, remote sensing and machine to machine communications, indoor/outdoor local area networks or ad-hoc (direct) communication between portable devices [5]. Realisation of these scenarios will require a certain level of control and coordination between CR devices; in other words, the formation of a network. In [8], networks over TVWS are formed based on an enriched Wi-Fi protocol and spectrum availability information determined based on local spectrum sensing. The latter, however, does not conform to the subsequent decisions made by the regulators to mandate GDBs as a mean for protection of incumbent services. A more conservative approach to formation of networks operating over TVWS is to rely on out-of-band control messages using existing radio access technologies in licensed or unlicensed spectrum, e.g. [9, 10].

Drawbacks of this approach include the need for additional channels in some licensed band, or reliance on the congested ISM band. Having in mind these problems and the recent decisions of the major communications regulators, our goal is to design a network that relies solely on whitespace spectrum.

This section focuses on the design, development and evaluation of a spontaneously created whitespace network, i.e. a network which relies solely on whitespace spectrum and an outline of potential business cases.

In Fig. 4.1, we depict an example instance of a whitespace network where control channels are deployed dynamically whenever and wherever possible to enable coordination and rendezvous between devices operating in whitespaces. Some of these devices, which have the capability to directly query the GDBs, may self-select to become whitespace access points, to arbitrate and control whitespace communications of other devices (subordinate devices). The subordinate devices, which could be, for example, sensors that belong to a home automation system, would typically have no means of communication with the GDBs. Moreover, these subordinate devices would use whitespace spectrum intermittently to connect to

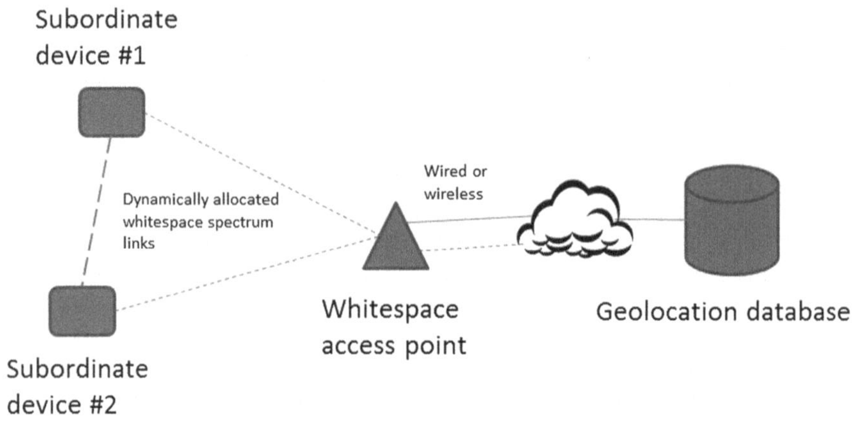

Fig. 4.1 Example instance of a whitespace network

the internet (via the access point's backhaul), or to perform direct device-to-device communications.

Specifically, we examined a protocol that enables operation of whitespace networks with the use of GDBs, dynamic control channels deployed depending on the whitespace spectrum availability, cyclostationary signatures used for control channel identification, and performance monitoring to improve the whitespace allocation [1]. The proposed protocol allows both communication to the broader network (via the access point) and direct device-to-device links. As part of our work, we have implemented a proof-of-concept software defined radio experiment that showcases the capabilities of the proposed solution. Using the experimental platform we have evaluated the trade-offs related to operating exclusively in whitespaces, without relying on licensed spectrum for control channels. These trade-offs come in the form of an extra delay in the order of hundreds of milliseconds and a throughput of up to 85 % of that achievable with licensed spectrum links.

4.1.2 Key Enablers of Dynamically Created Whitespace Networks

In order to build a network that solely operates in whitespace spectrum one needs to overcome a challenge related to the protection of incumbent services and to ensure coordination and rendezvous among the whitespace devices. In our work we overcome these challenges by relying on: GDBs, dynamic control channels, and cyclostationary signatures. In the following we give a brief introduction to each of the above mentioned concepts.

4.1.2.1 GDBs

In principle, a GDB is a database that contains up-to-date information on the spectrum available at any given location and time instance, enriched with other types of related information, such as the duration of availability, maximum effective radiated power permitted, or adjacent channel leakage ratio [4]. GDBs are populated with information created by modelling the propagation of known incumbent transmitters (for example as in [11]), where the model's parameters and algorithms are selected by the authority operating the database. Such whitespace information is provided to the devices on a temporal basis, and whitespace devices need to periodically request the information, where the period is set according to the requirements of the local regulator. Whitespace devices are not allowed to transmit until they have successfully received up to date information on the available channels. When a device has no possibility to directly (without the use of whitespaces) connect to the database, another whitespace device may act as a proxy for the device's queries [4]. In recent years, communication regulators world-wide have mandated GDBs as the only required solution to protect the incumbent services in the TV whitespaces, e.g. [5, 6]. Hence, in our work, we rely solely on GDBs to protect incumbent services and to provide information on the whitespace spectrum opportunities.

4.1.2.2 Dynamic Control Channels

In general, control channels are deployed to organise mobile devices and convey network control information, for example, identification, synchronisation, channel allocations (restrictions) or network policies. In order to facilitate the distribution of control channels for CRs the European Telecommunications Standards Institute Reconfigurable Radio Systems Technical Committee (ETSI TC RRS) has recommended two ways forward: (1) out-of-band, where the control channels are distributed over a globally dedicated physical channel, (2) in-band, where the control channels are transported over a specific radio access technology using separate or an existing control channel. The former has the disadvantage of requiring additional spectral resources and global harmonisation.

The latter is a viable solution for systems operating in licensed bands with fixed operational frequency and high level of coordination, which use whitespaces only temporally to extend network capacity. However, for systems that intend to rely solely on whitespaces it poses some difficulties, as the allocated operational frequency may change depending on the incumbent user behaviour. Herein, we propose a reliable solution for an in-band control channel in whitespaces, which is dynamically deployed depending on the whitespace availability. The centre frequency of this control channel is allocated based on the GDB information. Subordinate whitespace devices will acquire this centre frequency through the detection of a physical layer signature inserted in the transmitted waveform of the control channel, as described in the following subsection.

4.1.2.3 Cyclostationary Signatures

Communication signals of contemporary radio systems have many inherent periodicities which come as a consequence of coupling stationary signals with, for example, periodical waveforms or training sequences. These periodicities may also arise as a consequence of typical communications procedures, such as sampling or multiplexing. One way to observe them is to perform first order and second order cyclostationary analysis to discover specific correlation patterns in time or in the spectral domain of the signal, respectively. However, these periodicities may also be intentionally embedded into the physical signal as so called cyclostationary signatures. A cyclostationary signature can be inserted in an OFDM signal by mirroring one or more selected subcarriers. The arising periodicities can be observed through the spectral correlation function (SCF), at a cyclic frequency that corresponds to the ratio of the spectral distance between the mirrored subcarrier set and the useful symbol duration. Cyclostationary signatures can be detected by sweeping across the bands of interest and performing circular correlations on the received signal samples. When the signature is present in the received signal, a spike in the SCF is observed and the receiver can start decoding the received signal. In case the signature can no longer be detected, the receiver will start sweeping the band until the signature is found again and a new centre frequency is determined. The cyclostationary signatures can be used to identify specific radio systems, specific access networks in coalitions of access networks or to enable rendezvous in dynamic spectrum access networks.

4.1.3 Addressing the Technical and Business Challenges

In order to meet the challenges discussed above, we have designed a dynamic spectrum access and allocation protocol (DSAAP). This allows for the coordinated operation of a dynamic spectrum access network deployed in whitespaces for OFDM-based systems. In general, the DSAAP operations are performed as follows (the subsequent steps are also depicted in Fig. 4.2):

(1) When a new whitespace device that is able to connect to the Internet is switched on, it checks with the GDB for any available frequency channels;
(2) If a channel is found, the device locally reserves this particular channel for secondary spectrum operation and becomes a whitespace access point. A whitespace access point periodically transmits a broadcast signal, which announces the availability of the whitespace access point in the specific frequency channel to any other whitespace devices. The transmitted broadcast signal has an embedded unique cyclostationary signature, which can be assigned as in [12] and detected with a cyclostationary feature detector described earlier. The broadcast signal carries information required to

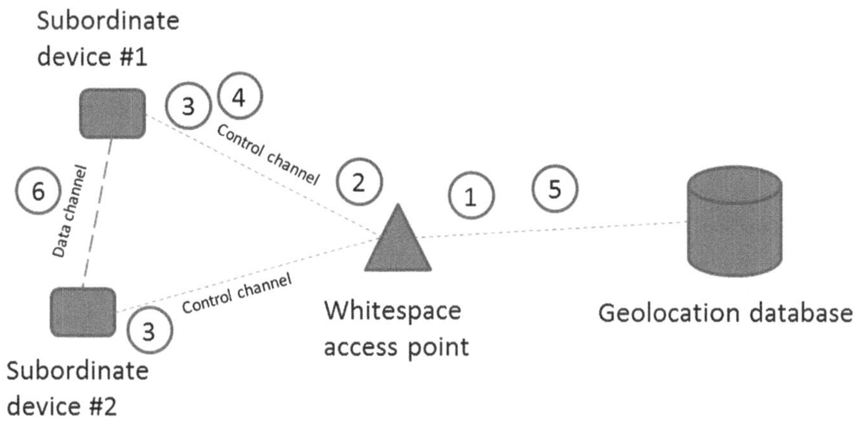

Fig. 4.2 Steps involved in the DSAAP operation

coordinate the cell's operation, such as the rendezvous channel or temporal spectrum allocations for whitespace devices;

(3) When another whitespace device, which has no Internet connectivity, arrives in the coverage region of the whitespace access point, it sweeps the whitespace bands to detect the broadcast signal. If the broadcast signal is detected, the device decodes it and reads the cell's information. Then, using the rendezvous channel it associates with the access point and stays on the detected channel listening to the broadcast signal, becoming a subordinate device. If another whitespace device arrives, a similar procedure follows;

(4) Whenever one of the subordinate devices requires transmission to another local device (or to the Internet), it requests (using the rendezvous channel) whitespace operation;

(5) The access point queries the database and allocates a whitespace channel that meets the demands of the requested transmission, indicating to both the whitespace devices the centre frequency, assigned bandwidth, spectrum availability determination period, and the peer device's MAC address for direct device-to-device transmission;

(6) The information is embedded to the control channel and both devices receiving the information reconfigure their radio front-ends to operate on the specific centre frequency and start the data transmission. During the data transmission, the subordinate devices constantly monitor the connection quality. If the connection quality is sufficient and the spectrum availability determination time elapses, both devices leave the transmission channel and repeat the whole procedure. However, if during the transmission one of the devices observes a significant drop in the connection quality (by means of, for example, an increase in the frame error rate), then that device reconfigures to the rendezvous channel and sends a report to the access point. The access point

will use the measurement information conveyed in the report to improve any subsequent data channel allocations.

The use of licenced spectrum on a nationwide basis can introduce significant cost overheads due to licencing fees in exchange for exclusivity. The other extreme is licence-exempt usage for type-certified devices and non-exclusivity however the trade-offs include low transmission power restrictions, narrow spectrum segments, and uncoordinated usage potentially resulting in interference. A rules-based approach based on TVWS relaxes the requirement for type certification. Coupled with dynamic control channels and database coordination, the viability of new business cases relying on a flexible and scalable wireless communications architecture can be increased.

If pitched as complementary technologies to cellular network deployments, TVWS-based network deployments can support long range, latency-tolerant applications and short range/high building penetration applications. Examples include machine to machine communications, remote sensing, and telematics, wireless data storage and backups where periodic high-bandwidth data transfers can be performed over short ranges, security in remote areas, mobile healthcare e.g. conveying in-ambulance image and patient monitoring information to the emergency ward.

4.1.4 Conclusion

In this section we presented a conceptual wireless network, where all the control and data communications occur using whitespace spectrum. We have outlined a number of potentially viable market opportunities for TVWS networks and have described an access method and spectrum allocation protocol to help enables reliable control channel deployment and efficient data communication over whitespaces. Moreover, our discussion focuses on the use of direct device-to-device links over whitespaces. The goal of our work was to examine how networks relying solely on whitespaces could be used to build the groundwork for future radio systems thus helping to increase the attractiveness of this approach to the market.

Acknowledgments The material is based upon works supported by the Science Foundation of Ireland under grants no. 10/CE/I1853 and 10/IN.1/3007, Framework Programme 7 (FP7) CREW project under grant no. ICT-258301, FP7 CogEU project under grant no. ICT-248560. This work was also supported in part by COST Action IC0905 TERRA through a Short Term Scientific Mission.

4.2 Business Scenarios and Models for Use of GDB in TV White Spaces

Gintarė Sukarevičienė[1, 2], Vladislav V. Fomin[1] and Mike Katsigiannis[3]

[1]Vytautas Magnus University, Kaunas, Lithuania
[2]Vilnius Gediminas Technical University, Vilnius, Lithuania
[3]Aalto University, Espoo, Finland

Geolocation Databases (GDB) as enabler of CR operation is one of the major elements of Dynamic Spectrum Access Information Infrastructure (DSA II). Regulators across the globe have been showing a preference towards a GDB approach for the TV White Spaces (TVWS), as it becomes essential to ensure overall efficiency of radio spectrum for the existing and emerging wireless communication services. Unfortunately, despite the recent advancements in TVWS GDB business scenarios, uncertainties exist with regard to the future technologies and value network configurations for GDB use and access in TVWS spectrum range and elsewhere, for e.g. in some specific spectrum bands such as bands that are allocated for public Digital Audio Broadcasting (DAB) services (e.g. VHF T-DAB band), 1452–1492 MHz (e.g. L band), radar bands and fixed service bands.

Thus, while future business models are a common concern of private operators and regulatory frameworks are under discussion in e.g. European Commission, CEPT and the major Standards Development Organizations (SDOs), such as ETSI, CEN, it becomes important to analyze the economic feasibility of GDB use and access from different points of view within the future DSA II architectures and services. It remains to be seen how collaboration among different stakeholders can be established around the use of GDB, for example:

- For the development of which (novel) services GDB will play a crucial/enabling role?
- How the standardization of GDB access protocols for different wireless services will be unfolding?
- How acceptable to different stakeholders business models for GDB operation and access can be found/developed?

All these questions require both technology-oriented and business-oriented analysis and modeling.

4.2.1 The Concept of DSA II

The concept of Dynamic Spectrum Access (DSA) stands for the opposite of the current static spectrum management policy for particular users in particular geographic areas, and has a large potential to become the crucial enabler of the

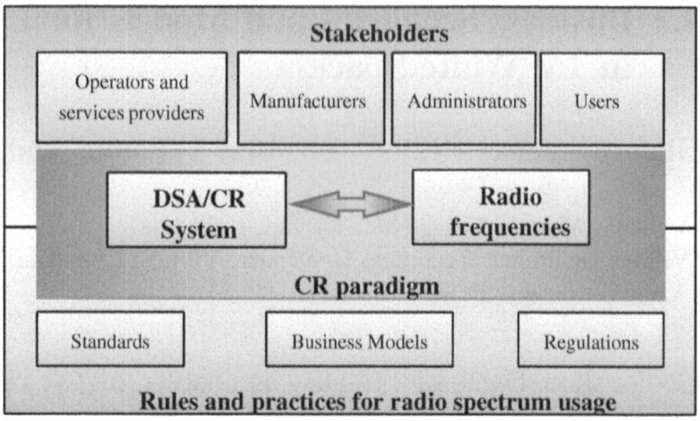

Fig. 4.3 DSA II infrastructure [14]

spectrum reform. Although DSA has broad connotations that encompass various approaches, there are only few ways to get more spectrum: to reallocate it or to allocate unused spectrum for more efficient use, as spectrum is of fixed nature and cannot be grown, manufactured or imported. In this context, DSA could be considered as enabler of CR capability to access and transmit in unoccupied spectrum (white spaces) while minimizing interference with other signals in the spectral vicinity [13].

Development of DSA II requires creating a functional techno-economic model, which describes and analyzes the different stakeholders' interrelationships as well as the technologies, policies, and services, as depicted in Fig. 4.3 [14]:

- At the top of Fig. 4.3, possible stakeholders (directly and indirectly impacted by DSA II) and their roles within the wireless telecommunication services, such as: operators and services providers, manufacturers, administrators and users;
- At the bottom of Fig. 4.3, elements which define the rules and practices (principles) for radio spectrum usage, such as: standards, business models and regulations;
- In the centre of Fig. 4.3, the main elements of the CR paradigm, such as: DSA/CR system and its opportunistic access to radio frequencies.

The Fig. 4.4 depicts the potential relationships among different elements of DSA II, for e.g. one of the ways in which relationships between different elements within DSA II could be established is described below:

- direct relationships (wide double arrows) between CR paradigm and radio frequencies, regulations and standards, administrators and operators, service providers. For instance, incorporation of CR technology into existing frequencies, such as TV White Space (TVWS), radar bands, etc.;
- co-dependence (double arrows) between business models and telecommunication operators (e.g. DNA, Elisa, Tele2), service providers (e.g. Wireless Internet Service Providers, WISP), and users (e.g. primary and secondary users), and wireless device manufacturers;

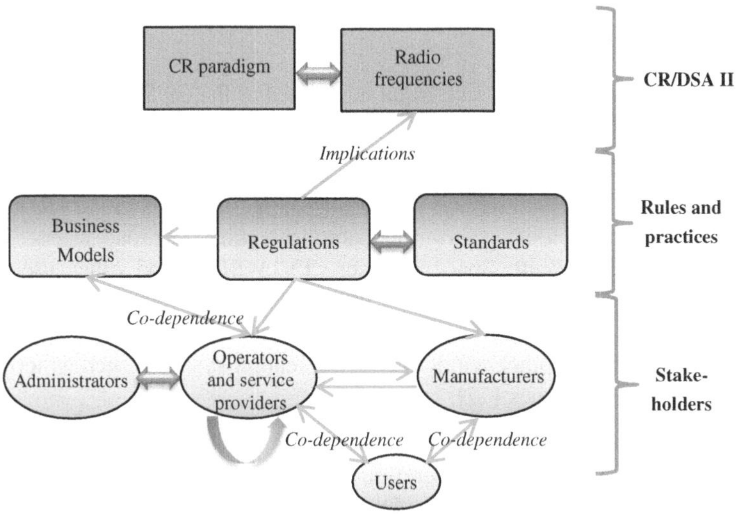

Fig. 4.4 Relationships among different elements of DSA II

- implications of regulations to radio frequencies: the need for the use and assessment of frequencies to be guided by various rules in order not to interfere with services operating on adjacent frequencies. For example, large portions of VHF/UHF TV bands become available on a geographical basis and regulatory entities are already moving towards allowing CR to operate in TVWS bands but it must not interfere with primary users.

4.2.1.1 Standards and Regulations Within DSA II

As described above, standards and regulations can be identified as one of the major element of DSA II, since somehow relate with all other elements of DSA II as shown in Fig. 4.4. This also can be seen in the standardization domain, where three major groups have emerged to work on relevant technologies and architectures [13]:

- IEEE 802.22 and related research that aim to provide DA to vacant TV spectrum;
- DySPAN-SC (formally P1900) working groups;
- ETSI's Reconfigurable Radio Systems Technical Committee on CRs and SDRs.

In general, there are many regulatory bodies that show interest in developing standards or defining norms and regulation for one or another aspect of CR-related telecommunications [15], as reviewed in detail in Sect. 1.4.

4.2.2 The Concept of GDB Within DSA II

Geolocation Database (GDB) access can be defined as the capability of a device to know its geographical position and transmit this information to a database which identifies the suitable channels and transmit powers that the device can use in its current location—other essential element of DSA II [16]. In this way GDB:

- administers principles of spectrum use among regulators, broadcasters, TVWS industry (e.g. TV White Space Devices, TV WSDs), and other users (e.g. Program Making and Special Events, PMSE) in practice.
- controls the frequencies used by TV WSDs and their transmission power so that they do not interfere other wireless communication systems, such as terrestrial TV or radio microphones [17].

Recognizing the importance of the GDB within DSA II (see Table 4.1), business scenarios for GDB for the operation of CR are proposed (Fig. 4.5), as they stand as a basis for further research on DSA and business model related issues (e.g. GDB that is not a big component of the DSA II itself but it is in the center of the market structure), taking into account both technical and business-oriented parameters:

- *Restricted market scenario* (on the top-left corner of Fig. 4.5): it refers to outsource-based business model configuration [18]. The main role here is played by a third party, which is aided by administrator/operator who develops and operates GDBs. It is a solution of generalized GDB that supports all databases.
- *Flexible market scenario* (on the top-right corner of Fig. 4.5): it refers to the user-based and operator-based business model configurations [18]. The main roles here are played by the user and operator, although the available channels are managed by GDB: flexible bands (flexible operators), flexible services (flexible user).
- *Competitive market scenario* (on the bottom-left corner of Fig. 4.5): it refers to the user-based business model configuration [18]. The main role here is played by the users' devices (TV WSDs) in handling available channels. Although it could benefit while introducing the concept of GDB for the operation of CR in TVWS, it could create problem to the existing communication patterns.
- *Hybrid market scenario* (on the bottom-right corner of Fig. 4.5): it refers to the broker-based business model configuration [18]. The main role here is played by TVWS broker by distributing available channels to various service providers.

The business scenarios matrix (Fig. 4.5) is based on two dimensions: technical architecture and industry architecture, (also is relates with previous work [18]):

- *technical architecture* refers to the technology which determines the differences between existing and future architectures: what is the role of GDB in enabling various wireless communication services, how clients/devices of different wireless services will be accessing the same/shared GDB. Based on this parameter GDB scenarios could be split between: centralized technical architecture scenario, and decentralized technical architecture scenario.

Table 4.1 SWOT analysis of GDB

Strengths	Weaknesses
• Interference control	• The changes of the primary spectrum usage have to be updated
• Global view on a radio environment	• Spectrum allocation and radio resource management must be balanced
• Followed primary spectrum usage activities	• Band identification, management, control and cost allocation must be standardized to support successful development of CR
• Sufficient computing power to make complex computations	• Reduced barriers to entry for smaller operators
• Identification of secondary user's location and available frequency on that location	• Higher costs of devices, to validate hardware to meet specific regulatory requirements
• Lower cost-per-bit	
Opportunities	**Threats**
• Improved access to wireless services and applications	• Additional information security issues in traditional wireless communication
• New market opportunities	• Lower communication QoS because of possible interference
• Start of commercial utilization of WS	• Reduction in battery life for the new technologies
• Realization of the CR paradigm for WS in other bands	• Market shift from hardware to software manufacturers
• Greater competition that could lead to value-added services and lower costs	• More complex regulatory regime
• Introduce realization of the CR paradigm for WS in other bands	

		Industry architecture	
		Vertical industry structure	*Horizontal industry structure*
Technical architecture	*Centralized*	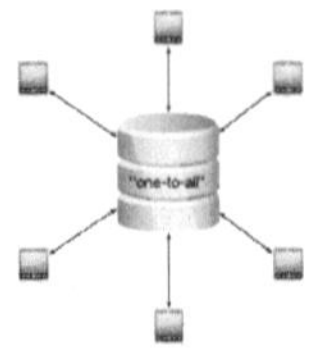 **1. Restricted market scenario** - central GDB; - unnecessary interoperability between third party GDB	**2. Flexible market scenario** - no central GDB; - all third party GDBs are interoperable 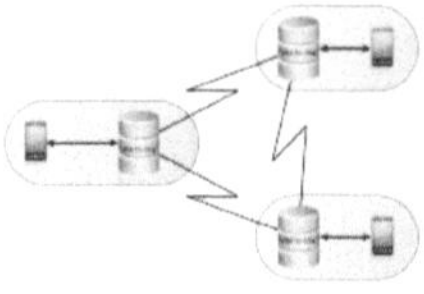
	Decentralized	 **3. Competitive market scenario (internal)** - no central GDB; - not all third party GDBs are interoperable	**4. Hybrid market scenario** - central GDB; - not all third party GDB are interoperable 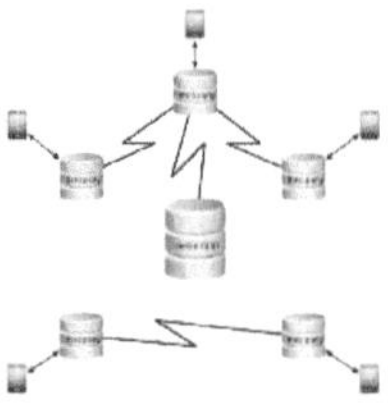

Fig. 4.5 Business scenarios matrix

- *industry architecture* relates to the scope of the GDB in terms of markets and industries in which it competes as well as to the ways in which their roles are combined. The main question here is how GDB business model that would be acceptable/make sense to different stakeholders can be found. Based on this parameter all GDB scenarios could be split between: vertical industry structure scenario, and horizontal industry structure scenario.

Therefore, the *scenarios of centralized technical architecture* refer to the standardized technologies which offer good performance and can scale different use cases and environments there large access network operators are preferred who integrates local area networks into their existing network infrastructure.

On the contrary to scenarios of centralized technical architecture, the *scenarios of decentralized technical architecture* refer to the situation where the access providers may be small and even local. This may lead to the more complex deployments of DSA II.

The scenarios of vertical industry structure refer to the situation when there is one entity (or a group of entities that serve specialized needs to each other) which supports GDB business activity that meets specialized needs of one specific industry, and also is involved in other parts of communication process. This could help eliminating some of the complexity related to the linking of two technologies and business issues related to that technologies, as each generation of technology is as an outcome of tinkering with and meshing together previously unrelated and untried technological combination [19].

Finally, the *scenarios of horizontal industry structure*, by comparison with scenarios of vertical industry structure, are focused on a wider range of GDB business activities of broader range of services and applications grouped according to common requirements to the larger group of customers.

Specific scenarios also encounter specific issues which require specific regulation and standardization as development of DSA II requires creating a functional techno-economic system.

4.2.3 Techno-Economic Studies of Business Scenarios

Techno-economic studies can be conducted using a "bottom-up" approach. This approach firstly focuses on analyzing either new technology architectures (e.g. CR) or industry architectures, from a small group of established stakeholders' point of view, and then the analysis is expanded to create the whole environment of the new evolving technologies and industry architectures. The purpose of the "bottom-up" paradigm is to express more complex variety of detailed business concepts and problems areas from a set of strictly defined concepts of technology, thus for e.g. firstly technology is chosen and after that deciding how to create the whole environment of new evolving technologies and industry architectures.

In general, techno-economic modelling case studies could be classified into two types [20]:

- *Technology-oriented case studies:* analysis and comparison of emerging technologies (focus on network investments and network related OPEX, and less focus on business models, competition, services);
- *Business model-oriented case studies:* analysis and comparison of alternative business models (focus on value network configurations and revenue sharing models, and less focus on technology as proper business models are essential).

In this light, since DSA II is a highly complex phenomenon, as the starting point in the following paragraphs the techno-economic modelling method is used only for evaluating restricted market scenario from the operator point of view (business model-oriented case studies), where GDB is provided. In this model, the responsibility of protecting incumbent's users from interference is taken by third parties, while the central GDB operation is kept under surveillance by the administrator/operator who is aided by the regulatory body, as described in ECC Report 185 [21]. In this case, each geographical area can be controlled by its own database enabling a distributed operation, as well as allowing specific bands to be tradable under the operating and communications on different CR protocols and standards.

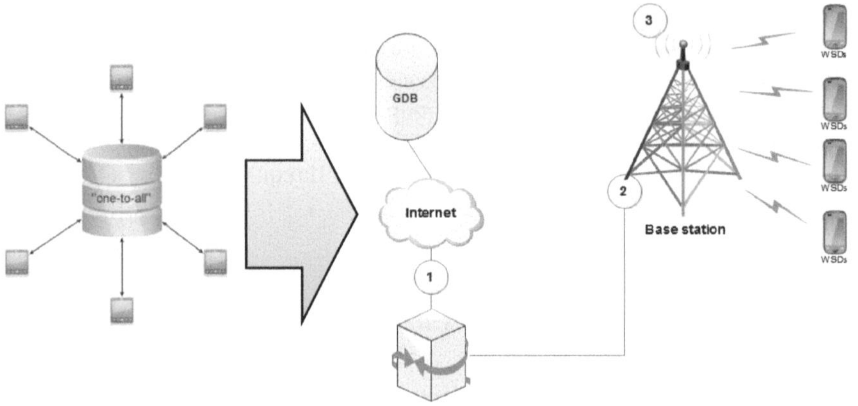

Fig. 4.6 Restricted Market Scenario from the operator point of view

4.2.3.1 The Case of Operator and Service Provider

The main research question is whether operators have direct relationship with the users or not. There is the threat for operators that they may miss their strong position in the market because virtual operators may become real network operators [22]. Due to this, it becomes important to analyze which role has a GDB operator and service provider in the value network. Nevertheless, one thing is clear: the stake of the existing and new operators, as well as service providers, is if they can reach the most positive value through DSA II enabled services that promise a wide range of new opportunities to consumers [14].

To start with, within restricted market scenario (from the operator point of view), there can be two ways (business models) in which GDB exchanges spectrum information with WSDs (see Fig. 4.6):

1. Mobile Network Operators (MNO) offer GDB-based mobile services.
2. GDB operator offers services to end-users utilizing MNO's network.

In the first case, the GDB operator (or licensed spectrum owner) is not a threat to MNO, but instead they are cooperators, because spectrum owner "gives" frequencies to MNO who provides access to WSDs (end-users) in order to get GDB information. In this way, the role of GDB operator is to connect to the MNO's network for offering GDB services to potential users who are also the current mobile subscribers using services such as voice, messaging, internet based video streaming, voice over IP, value added services and other.

Regarding to the cost of the service, there is no need for end-users to pay directly the GDB operator, because the MNO could charge the required amount of money through the customer's bill without knowing about buying access to GDB. However, the GDB operator needs to find the way how to charge the service from MNO and how to offer the new services using mobile network.

In the second case the GDB operator is a threat to the MNO, because when the GDB operator asks to use MNO's infrastructure in order to offer services, the MNO loses some resources. But it becomes too difficult to define the access price to the MNO network because in this way the big fixed provider should charge new players.

In both cases there will be the need for extra investments for upgrading the network (see numbers in Fig. 4.6):

- extra gateway in MNO's network (1);
- base station software upgrade (2);
- base station and antenna system hardware upgrade (3).

All these three points encompass CAPEX for GDB implementation:

- connection to the Internet;
- DB costs (server, database).

In addition, there are also OPEX for deploying GDB, and all these costs need to be covered from MNOs by charging them in some way:

- Operation and Maintenance (O&M);
- electricity;
- personnel costs;
- transmission line of GDB to Internet, etc.

4.2.4 Conclusion

The purpose of this section is to contribute to the on-going discussion on techno-economic analysis of business scenarios, and on classification of business scenarios and identification for business models for the use of GDB in TVWS by narrowing down from general (different stakeholders) to concrete (one stakeholder) point of view.

The final results of this work will allow applying the GDB business scenarios analysis in the future studies, taking into account the same method for modelling new business scenarios for others groups of established stakeholders' of DSA II and then seeing if there can be any common business model derived from the multitude of different models. The results will lead to a model of cost and revenues in which different GDB architectures (centralized/decentralized, horizontal/vertical) are compared from different point of views in order to get very specific view of impact of CR paradigm to the wireless telecommunication services.

Acknowledgments The authors would like to thank their colleagues from COST Action IC0905 (COST-TERRA) for sharing their ideas and support in developing this work.

The first author extends her special gratitude to prof. Heikki Hämmäinen (Aalto University, Department of Electrical Engineering; Espoo, Finland) for hosting her visit to the department in March 2013, and to his team of researchers, especially to Thomas Casey, and Arturo Basaure.

4.3 Underlying Market Dynamics in a Cognitive Radio Era

Thomas Casey and Timo Ali-Vehmas

VTT Technical Research Centre and Nokia Corporation, Tampere, Finland

4.3.1 Introduction

This section explores the possibilities of how the value system around wireless networks could be organised in the future and what would be the underlying market dynamics given the introduction of CR and dynamic spectrum access technologies. Using a combination of systems thinking tools and platform theory, four value system configurations around the future radio platform are introduced and the corresponding underlying dynamics are characterised. Based on this a feedback model using system dynamics and agent based modelling is built, configured with historical market data and used to evaluate future evolution possibilities both for GSM based mobile cellular and Wi-Fi based wireless local area radio platform paths. We explore how the value system could continue on established evolution paths but also deal with the transition to a so called complex adaptive system. Furthermore, for policy makers, we discuss threats associated with winner-takes-all and fragmentation type of scenarios, and highlight the possible importance of aligning the underlying market dynamics with the natural allocation and assignment cycle of spectrum frequency bands. This material is based on works published in [23], to where the reader is guided for further information beyond this chapter.

CR and DSA technologies have the potential to disrupt the current value system and usher in a new era in wireless communications. Under the new paradigm the management of radio resources would be decentralised to the edges of wireless networks where devices would together collaborate and provide wireless services [24]. The paradigm shift could potentially direct the market towards a horizontal and open structure enabling many new service applications and entrants [25] and could thus fundamentally change the underlying dynamics of the market as illustrated in Fig. 4.7. However, established path dependencies on current spectrum management models are strong and it is uncertain whether they can, or even should, be broken. Therefore, as it relates to the deployment of CR and DSA, there is a need to understand the underlying dynamics of the market in addition to the technology itself.

Regarding how actors in the current value system around the radio spectrum resource are organised, one can distinguish different models. Historically, spectrum licenses were given to one actor who was in charge of service provisioning and network deployment and controlled the whole value system from

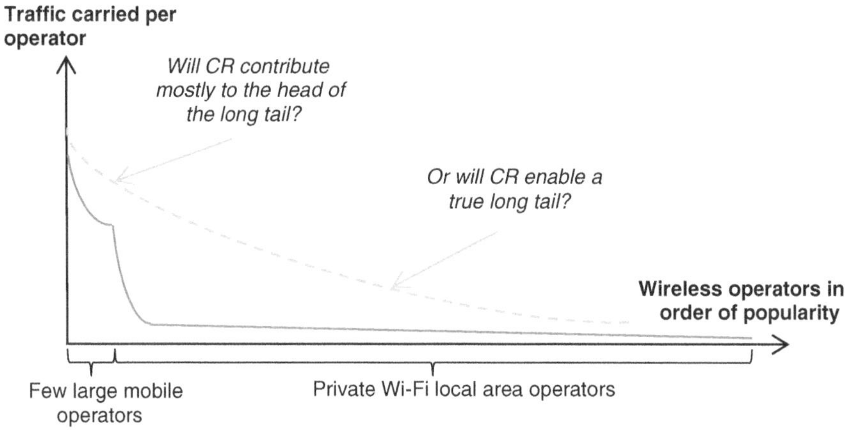

Fig. 4.7 An illustration of the head and long tail of potential application areas and market opportunities enabled by CR/SDR (A long tail results when the tools of production and distribution are democratized and supply and demand are connected [48])

infrastructure to devices (e.g. government monopoly operators) which in turn led to inefficient legacy allocations [25]. Improvements have been made e.g. after telecommunications liberalisation with the introduction of digital cellular mobile communications where licenses have been assigned to a group of operators and where ownership of devices and selection of network (i.e. with the help of SIM-cards) have been given to the end-users [26, 27]. This in turn has fuelled competition between operators and has forced them to use the spectrum resources more efficiently and improve the availability of their networks (both in terms of coverage and capacity). On the other hand, the usage of harmonised technology standards, as was done in Europe following the GSM Memorandum of Understanding of 1987 [28], has enabled large international economies of scale, device circulation and roaming which in turn has been a key ingredient that has enabled the more than six billion mobile subscriptions we currently have in the world. As mobile operators around the world are converging to LTE and LTE-A, CR and DSA technologies could be naturally embedded to this technology path.

As it relates to wireless computer networking, the unlicensed model has diffused widely where access points and base stations can be deployed and services can be provisioned by anybody, provided they follow a simple spectrum etiquette. Wi-Fi certified IEEE 802.11 has become the de-facto standard whose origins can be traced back to FCC's 1985 decision to allow the unlicensed use of spread spectrum techniques on ISM bands [29, 30]. Subsequently, many private enterprises and households have become wireless service providers where the cumulative number of Wi-Fi chipsets sold has surpassed the one billion mark and the installed base of Wi-Fi access points is already in the order of hundreds of millions.

On the other hand public Wi-Fi has remained somewhat limited where e.g. roaming solutions are still rather fragmented and typically proprietary.

Furthermore, given the limitations of the scalability of the IEEE 802.11 MAC protocol the unlicensed model is able to scale and grow in a bottom-up manner only up until a point. Since most of the demand arises from indoor locations [31], more co-ordination and spectrum is needed to enable bottom-up type of growth for which CR and DSA in turn could provide a solution. An example of bottom-up type of infrastructure growth can be observed e.g. with the wide spread diffusion of the Internet Protocol (IP) which has become the generic protocol to interconnect all computers [32]. In a similar manner CR and DSA could enable roaming and mobility between all devices on all possible frequencies which in turn could lead to an open and global network of wirelessly connected devices through which everyone could provide and receive public wireless services on any access point (AP) or device. As it relates to the future of CR and DSA various scenario studies have been conducted [31, 33–36]. In many of these the core question is to what degree the future system (e.g. CR spectrum database structure) is a centralised or decentralised one and to what degree an open (i.e. horizontal) or closed (i.e. vertical) one, a typical pattern that has been identified also on a more generic level [37–39]. However, while static descriptions have been made, the underlying dynamics of these scenarios have not been described. Given the introduction of CR and DSA technologies, the purpose of this paper is to explore the possibilities of how the value system around wireless access provisioning could be organised in the future and what would be the underlying dynamics. Due to the interdependent nature of the problem we take a holistic approach by using a combination of systems thinking tools and platform theory to understand the underlying structures. Based on historical evolution and prior scenario analysis work we introduce four value system configurations around radio platforms and characterise the underlying dynamics for each. Based on this we build a feedback model using qualitative system dynamics and quantitative agent based modelling (ABM), configure it with historical data and use it to evaluate future evolution possibilities both for GSM based mobile cellular and Wi-Fi based wireless local area radio platform paths.

4.3.2 Framework for Underlying Structure of Value Systems

4.3.2.1 Value System Configurations

"Systems thinking", studies how things influence one another within a whole, where a core principle is that underlying structure gives rise to observed trends, patterns and events [40]. The structure between actors and their business (and technical) interfaces can be described as a value system [41]. A value system in turn can be characterised as being organised around a mediating technical platform [42–44] operated by a platform manager [45, 46]. Here we define a radio platform (e.g. a mobile network) as being managed by an operator that provides a wireless service and mediates interactions (facilitated e.g. by a database) between two user groups: end-users using devices and entities hosting base stations (BS) (or access points) who both can create affiliations to the platform. The service itself is delivered through technical interfaces and components (devices and access points) and therefore the other side of the platform (e.g. BS host) might not be directly visible to the other (e.g. end-user).

Based on historical evolution and prior scenario analysis work we define four value system configurations around radio platforms. The platform typology follows the closed or open and centralised or decentralised categorisation.

First, in the centralised and closed value system configuration the radio platform is centred around one actor that controls the spectrum resource and the interactions (and signalling) between end-user devices and base station or access point sites, which would e.g. correspond to old government monopoly operators. In such a system there is only one platform manager with whom everyone has to collaborate since there is no other platform to switch to.

Second, in the centralised and open value system configuration the value system consists of a small set of connected radio platforms managed by a small group of platform managers that both collaborate and compete. The platform managers control the spectrum resource and the interactions between end-user devices and BSs or APs (typically operators operate the BSs and site owners only provide horizontal and value system independent resources for site space and electricity etc.). Since a standardised technology is used the platform users can rather easily switch between platforms. This would e.g. correspond to the competition and collaboration model of mobile operators using GSM based technologies where the end-users can use the same device and switch between mobile networks.

Third, in the decentralised and open value system configuration the value system consists of a large set of small connected radio platforms. Anybody can become a radio platform manager and start providing wireless services for other users. There exists a great heterogeneity of technologies and services with plenty of local innovation and competition. However, actors also collaborate, technologies are made interoperable and radio resources are quickly reassigned between platforms so that valuable services that have high demand are able to flexibly scale bottom-up. End-users can freely switch and roam between platforms and can easily become wireless service providers themselves. Such radio systems do not currently exist, although some open Wi-Fi roaming solutions bear some resemblance (e.g. *Eduroam* and *openWTS3*). Still, examples of decentralised and open systems exist in other fields, such as e.g. IP networks in computer networking.

Fourth, in the decentralised and closed value system configuration the value system consists of a large set of small radio platforms that are isolated from each other where all compete over the radio resources and no (or very limited) co-ordination exists. Isolation and intense competition can lead to the erosion of radio resources where nobody is able to scale their services bottom-up. Anybody can start providing wireless services, but typically only for a closed user group. This would e.g. correspond to private Wi-Fi deployments and fragmented roaming and authentication solutions.

4.3.2.2 Underlying Dynamics of Value Systems

Next we will describe the underlying dynamics of each value system configuration using basic concepts from dynamical systems theory [47]. A dynamical system can be characterised with an attractor, whose type can roughly be divided into four groups: fixed point, limit cycle, strange and no attractor.

First, centralised and closed value system can be seen as being directed by a fixed point attractor which evolves towards a static state (like a damped pendulum).

Second, centralised and open value system can be seen as following the dynamics of a limit cycle attractor which produces periodic and somewhat regular change (like a continuously swinging pendulum).

Third, decentralised and open value system can be seen as following the dynamics of a strange attractor which produces deterministic irregular change and functions on the edge of chaos.

Fourth, decentralised and closed value system can be seen as being characterised as a system that does not have an attractor that would give it structure and thus exhibits complete disorder and random behaviour.

The market share of each operator, i.e. radio platform manager, in each value system configuration is depicted in Fig. 4.8. The dynamics are influenced by the adaptation speed of the actors and the system overall, i.e. how often decisions about platform switches are made, how often resources are re-allocated and re-assigned, and how quickly competitors respond to market changes. In a centralised and closed value system configuration following the fixed attractor dynamics, one actor carries all traffic, as was the case with government monopoly operators. The system is very slow to adapt to changes with long resource allocation and assignment delays where users cannot switch to another provider and can overall be seen as corresponding to the inefficient legacy spectrum assignment model.

In a centralised and open value system configuration following the limit cycle attractor dynamics, few actors carry the traffic, as is typically the case with mobile operator competition today. Here the system adapts to changes cyclically where end-users are able to switch to more valuable networks thus inducing competition and more efficient use of resources. Overall the system allocates and assigns resources in a cyclical manner. In a decentralised and open value system configuration following the strange attractor dynamics, traffic is carried by many actors. The value system is quick to adapt to changes with short delays for resource allocation and assignment and low switching costs for end-users. Here actors form a long tail distribution where actors from the tail can quickly grow and reach the top and vice versa. Such a value system corresponds to the observations of Anderson [48] who states that a long tail distribution results when the tools of service production and distribution are democratised and supply and demand are connected. Overall, the value system would correspond to a so called complex adaptive system [49] where large number of agents interact using simple rules and which is characterised by self-organisation, emergence, and scale-free network

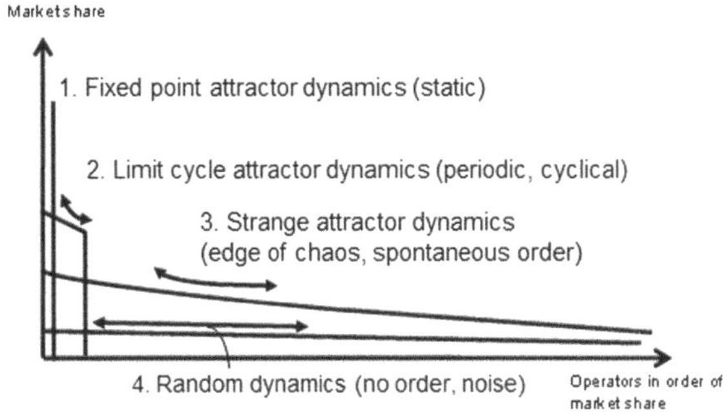

Fig. 4.8 The market share of each operator, i.e. radio platform manager, in each value system configuration

structures with long tail distributions [50]. This has been observed e.g. in the Internet in terms of routers [51] and web pages [50].

Finally, in a decentralised and closed value system configuration following the no attractor dynamics, traffic is carried by many actors but no actor is able to get ahead of others, get more resources and scale up. There is no delay for resource allocation and assignment (as is the case with the unlicensed spectrum licensing), resources do not accumulate and no structure is formed. Overall the system adapts randomly and seems like noise to an outside observer.

4.3.3 Feedback Model of the Underlying Dynamics of the Value System

The above described underlying dynamics are generated by a large set of actors and encompass a large number of feedback connections. Our next goal is to build a model of these underlying dynamics using two feedback modelling tools: qualitative system dynamic modelling [52] and quantitative agent based modelling [53]. As background for the modelling work eight expert interviews were conducted including representatives of device and network equipment vendors, mobile operators, regulators and academia.

As it relates to the modelling approach, it is important to make a distinction between detailed and dynamic complexity. Simply put, dynamic complexity is modelled with feedback structure, whereas detailed complexity is modelled by increasing the number of variables [40]. System dynamics focuses more on dynamic complexity and can easily encompass a wide range of feedback effects, but typically aggregates agents into a relatively small number of states [53]. Agent based modelling, on the other hand, puts more focus on detailed complexity where individuals and their interactions are explicitly represented, which in turn makes it more difficult to link model behaviour to its structure. Therefore, modellers must trade off disaggregate detail and breadth of boundary [53]. Our goal here is to use a combination of detailed and dynamic complexity, i.e. leverage the strength of both system dynamics and agent-based modelling. We start out by characterising the underlying dynamics of the value system configurations with simple system archetype feedback structures [54] and after that use ABM to assimilate the large number of feedback relationships between individual agents simultaneously, i.e. integrate detailed and dynamic complexity together [40].

4.3.3.1 GSM Evolution Path

We now envisage future mobile cellular networking scenarios. CR spectrum licenses to operate mobile networks will be given to all agents during the CR and DSA introduction period (year 2020). We assume that competitive reaction speed (SC) will remain low since rather long term investments are still needed. Furthermore, we conduct sensitivity analysis by adjusting the resource accumulation speed (SR) which reflects the overall spectrum licensing model. In the base case it will correspond to regulated exclusive licenses, i.e. the currently dominant licensing model with large spectrum bands and long license times. In the first sensitivity case SR will be considerably slower and correspond to license-exempt, i.e. unlicensed spectrum. In the second sensitivity case SR will be only slightly slower and reflect light or secondary licensing, where small bands are assigned dynamically with shorter cycles while ensuring that competition prevents extensive resource accumulation. In the third sensitivity case SR is considerably faster and corresponds to unregulated exclusive licenses where all resources can cumulate or be assigned to one operator and no spectrum caps are enforced.

Figure 4.9 shows the market shares of agents in the base case. As can be observed, after the introduction of CR and DSA technologies and the entrance of new smaller operators, competition between the large operators intensifies and they lose some market share. However overall, the underlying dynamics of the value system continue to follow the limit cycle dynamics, i.e. although some additional competition is present the majority of resources still accumulates to and circulates between the incumbent operators and the strength of the success to successful mechanism between the agents remains rather strong.

Changes in competitive efforts are shown in Fig. 4.10 where before the introduction of CR and DSA the competitive efforts of the three large operators are quite close to one another and evolve cyclically (in the model competitive effort ranges from a minimum of 30 to a maximum of 100). After the introduction of CR and DSA and the entrance of new operators, competitive activity between the large operators increases but still, the value system continues to evolve in a cyclical manner, i.e. it has some positive feedback but is still dominated by negative feedback. Nevertheless, this new competition leads to more efficient use of resources and more value overall. One can also observe that the increased possibility for end-users to switch between operator networks increases volatility in the system since the system still remains slow to react to changes.

Next Fig. 4.11 shows results from the sensitivity analysis. As can be observed, introducing an unlicensed model dramatically reduces the market shares of large operators and leads to a situation where the market share of all operators remains

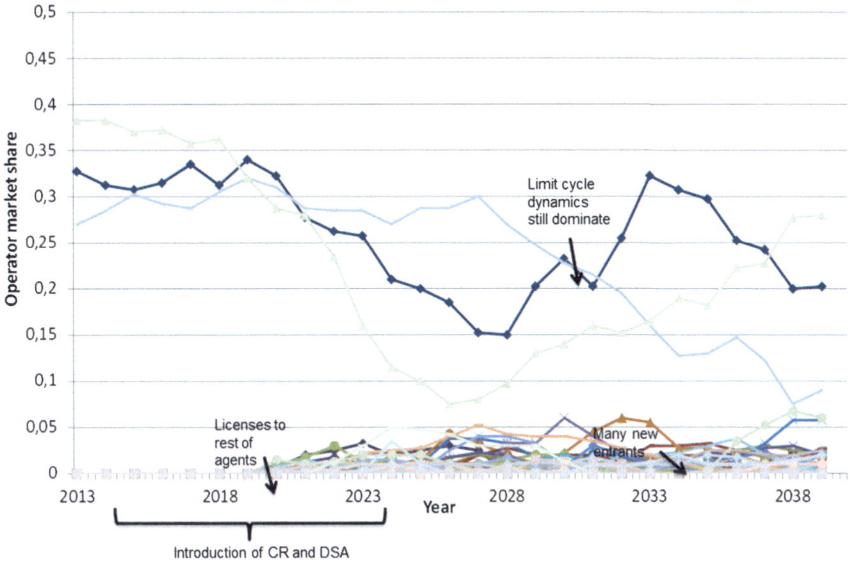

Fig. 4.9 Market share of agents in the base case

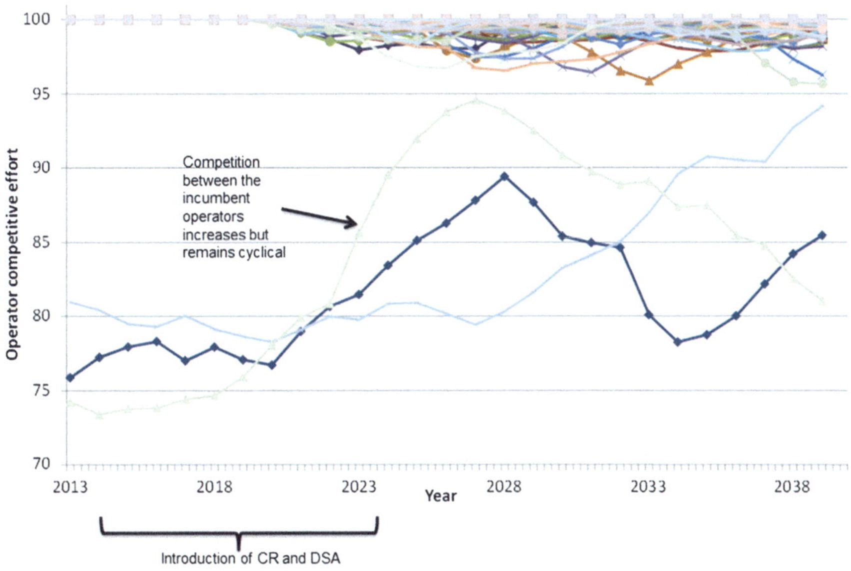

Fig. 4.10 Changes in competitive efforts between agents

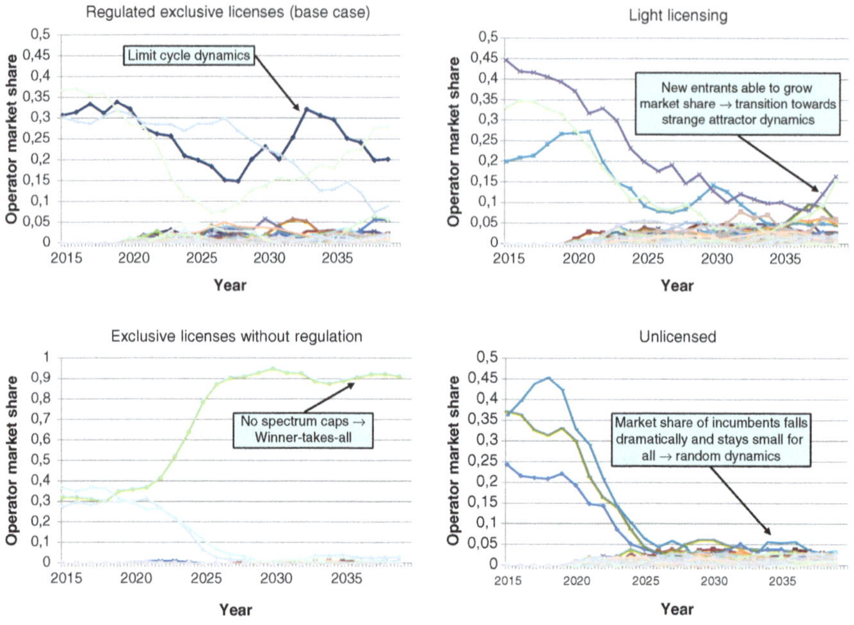

Fig. 4.11 Sensitivity analyses results

small and thus the value system transitions to follow the no attractor dynamics. With a light licensing model incumbent operators are able to sustain some market share but are joined by new entrants who have been able to grow their market share and thus the value system starts transitioning towards strange attractor dynamics. The use of exclusive licenses without regulation leads to a winner-takes-all situation where all resources accumulate to one actor who starts dominating the whole market and thus the value system transitions to follow the fixed attractor dynamics.

In terms of competition, with the unlicensed model all agents compete fiercely, resources do not accumulate and the individual platforms remain limited in value. With the light licensing model competition is less intense and resources are directed to valuable services which in turn are able to grow and scale up but not enough to gain a significant share of the market. With unregulated exclusive licenses competitive effort by the dominating agent drops to a minimum value and therefore, although it controls almost all of the resources, the value of the platform does not increase.

4.3.3.2 Wi-Fi Evolution Path

We now focus on potential future scenarios involving the evolution of Wi-Fi based wireless local area access. We assume that all agents have the existing unlicensed spectrum resources and that competitive reaction speed (SC) will remain the same reflecting local and instantly adaptive behaviour and small scale investments. In

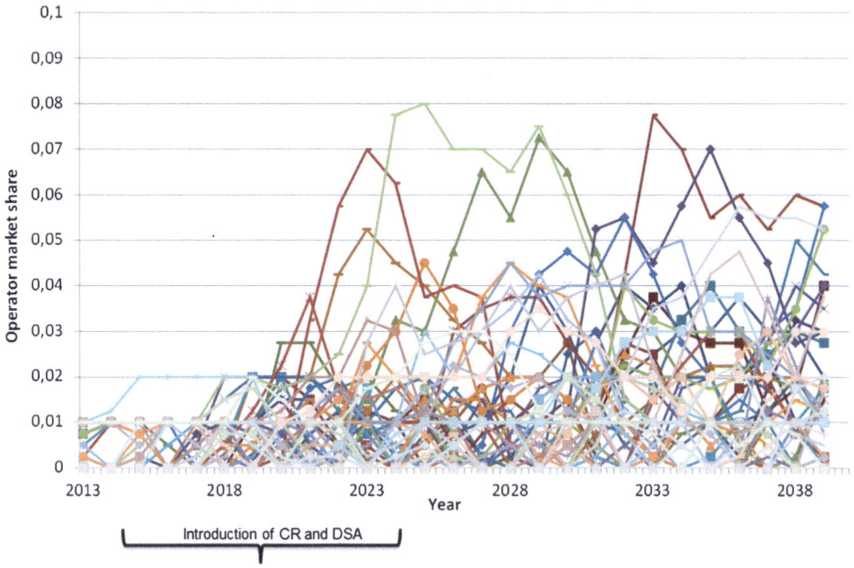

Fig. 4.12 Market shares in the Wi-Fi evolution base case

terms of the sensitivity analysis the resource accumulation speed (SR), corresponding to the spectrum licensing model, will grow to be somewhat faster in the base case (i.e. light and secondary licensing), and in other sensitivity cases will remain the same (i.e. continuation with the unlicensed model), grow to be still somewhat faster (i.e. regulated exclusive licenses), and considerably faster (i.e. unregulated exclusive licenses).

Figure 4.12 shows the market shares of agents in the base case. As can be observed, after the introduction of CR and DSA technologies and light licensing, some operators with valuable services are able to scale up, get more resources and market share. However, the system adapts quickly to changes and resources are re-assigned to wherever new innovations and locally relevant services are created and therefore no single actor or group of actors starts to dominate the value system. Therefore, the value system transitions to follow strange attractor dynamics, where the strength of the success to successful mechanism is low and competition is high. The value system evolves chaotically, i.e. has some negative feedback but is dominated by positive feedback. Overall, the system can be characterised as a complex adaptive system that operates at the edge of chaos.

Changes in competitive effort are illustrated in Fig. 4.13 where one can observe that before CR and DSA, and light licensing are introduced competition between agents is fierce. After the introduction of CR and DSA and light licensing, coordination increases but competition remains still high and fuelling new services and local innovation. However, competition is not so intense that resources erode,

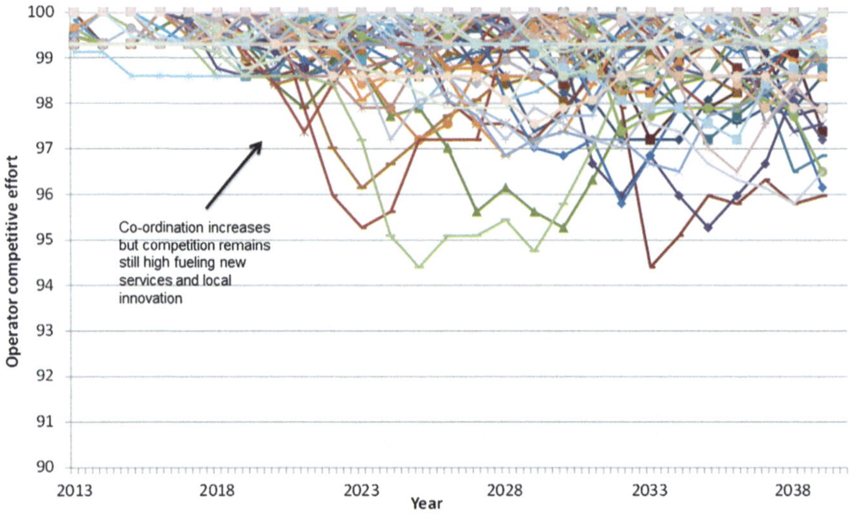

Fig. 4.13 Competitive efforts in the Wi-Fi evolution base case

leading to more efficient use of resources and more value overall as compared to the unlicensed model.

Following Fig. 4.14 shows results from the sensitivity analysis of this case. As can be observed, continuation with an unlicensed model leads to a situation where the market share of all operators remains very small and thus the value system continues to follow the no-attractor dynamics. This would also correspond to the fragmentation of CR technologies and spectrum databases in a similar manner as is the case with Wi-Fi roaming and authentication today.

With a regulated exclusive licensing model, resources accumulate so that two operators start controlling the market and thus the value system transitions to follow the limit cycle dynamics. In the case of unregulated exclusive licenses, resources accumulate to one actor leading to a winner-takes-all situation and fixed attractor dynamics. The dominant actor or actors in both of these cases could come from the group of incumbent mobile operators but could also come from outside the value system e.g. if a large internet player controlled the spectrum database and leveraged network externalities arising from elsewhere.

In terms of competition, with the unlicensed model all agents compete fiercely and the individual platforms remain limited in value, with the regulated exclusive licenses model the two dominant actors that get most of the resources slow down and start competing cyclically and with unregulated exclusive licenses competitive effort by the dominating agent drops to a minimum value. Figure 4.15 shows the top 30 operators in order of market share at the end of the historical simulation (year 2012) and at the end of the simulation in the different sensitivity cases.

When comparing the base case to the historical situation the market shares of wireless service providers especially in the head have increased. With the unlicensed

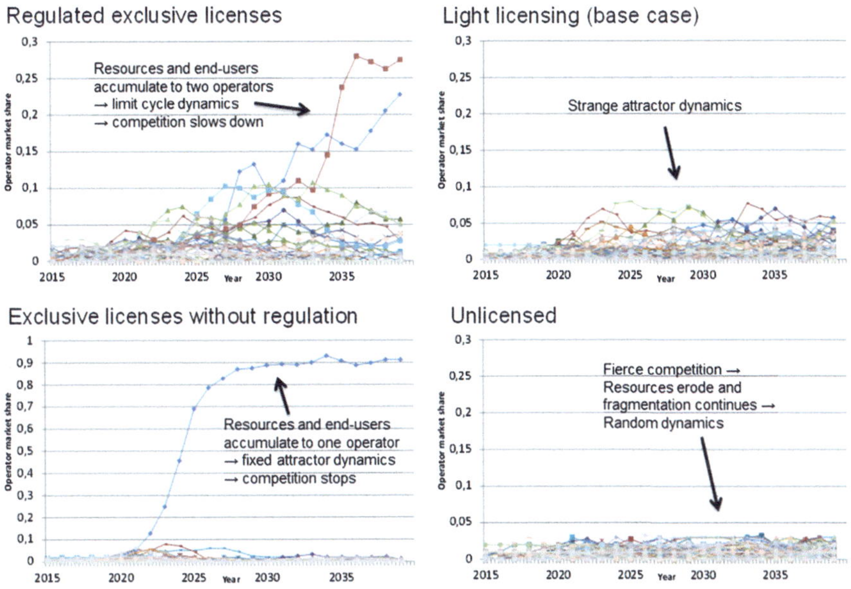

Fig. 4.14 Market shares in the Wi-Fi evolution sensitivity analyses

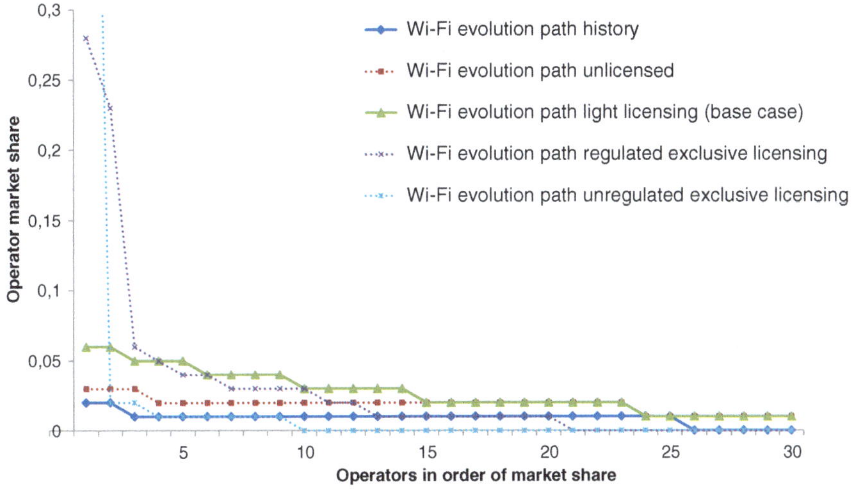

Fig. 4.15 Top 30 operators in order of market share year 2012 and at the end of the simulation in the different Wi-Fi evolution sensitivity cases

model the head has also grown slightly but the tail has become considerably longer than with light licensing and the number of active wireless service providers stabilises to roughly 70 agents. This would correspond e.g. to a situation where most of the agents are operating their smartphones as Wi-Fi access points for themselves. With the regulated exclusive licensing model the two operators in the head have taken most of the market share where the tail in turn has lost market share and most of the operators have become passive. With unregulated exclusive licenses one agent in the head gets all of the traffic and practically no long tail exist.

4.3.4 Discussion

The implications of the underlying dynamics of future CR scenarios and the corresponding spectrum database structure also highlight issues specifically relevant for policy makers. As it relates to the GSM evolutionary path, the value system continues to follow the limit cycle dynamics and to be dominated by few incumbent operators. In such a case CR and DSA technologies are likely to be embedded to the technology standards used by the mobile operators (i.e. LTE-A and its future versions). The possible spectrum databases and indoor sites would also be mostly controlled by mobile operators.

In terms of the Wi-Fi evolutionary path, the value system evolves to a complex adaptive system where the CR and DSA technologies would establish themselves as an independent technology standard enabling roaming and mobility between all devices on many frequency bands. The database infrastructure would follow an open and decentralised architecture (resembling that of IP) and be operated by many entities. Furthermore, as shown in the sensitivity analysis, it is also possible that a collision occurs between the two evolution paths and that the overall value system transitions from a centralised to a decentralised one or vice versa corresponding to the more general level descriptions of [37, 39]. The value system around the mobile cellular network platform could evolve towards strange attractor dynamics (i.e. entrance of many small operators and a diminishing role for incumbent operators) and vice versa the Wi-Fi path could evolve towards limit cycle dynamics (e.g. Wi-Fi access points controlled by incumbent mobile operators or other large actors).

From a policy maker perspective the results also point out future threats. There is a possibility that CR and the corresponding database technologies will become fragmented, much like Wi-Fi roaming and authentication now, and the roles of CR databases will remain very limited, isolated and local. Yet another threat is a winner-takes-all type of situation where one of the existing operators, or another strong player outside the value system, controls the CR database infrastructure and uses closed proprietary technologies which might in turn slow down diffusion overall. The results could also have implications as it relates to different spectrum frequency bands and their characteristics. As discussed by [47], dynamical systems tend to naturally synchronise with one another and transition to follow the same dynamics. For example, roughly put, one can say that low frequency bands propagate far and need more centralised co-ordination and long assignment cycles whereas high bands in turn do not propagate far, remain as a local resource (especially in indoor locations) and thus need less co-ordination. Therefore, one could pose a question whether there is a natural allocation and assignment cycle for the spectrum frequency bands and if so, how would these characteristics relate to the described underlying dynamics. For example in terms of the GSM evolution path, the usage of standardised technologies, cellular network planning and competition following the limit cycle market dynamics has led to rather efficient use of 900 and 1800 MHz bands. Subsequently, one can question, to what degree should CR and DSA technologies even be used to disrupt these underlying dynamics. Still, one can argue that there exists an upper limit for frequencies after which building cellular networks becomes inefficient. Unlicensed private Wi-Fi deployments, on the other hand, have led to rather efficient use of use of the 2.4 GHz ISM band and correspondingly one can question are the unlicensed and light licensing models more naturally aligned with higher spectrum bands and short range sites.

Since the policy maker can influence the underlying dynamics of the market with the spectrum licensing model it could be beneficial if the value system would be orchestrated so that the underlying market dynamics are aligned with the natural allocation and assignment cycle of the radio resources. This would correspond to a few core applications (such as mobile voice, text messages and managed mobile internet connectivity) enabled by mobile cellular technologies and governed by cyclical competition.

Strange attractor dynamics and light and secondary licensing models would be aligned with high spectrum bands and base stations and access points working on sites with short range with instantly adaptive behaviour and small scale investments needed where somewhat unreliable assets, e.g. light or secondary licenses, would be sufficient. This would correspond to many different types of applications, locally relevant public services enabled by CR and DSA technologies and be governed by chaotic competition with just enough co-ordination to ensure system operation. No attractor dynamics and the unlicensed model would be aligned with very high frequency bands and with access points and devices working on very short range sites. This would correspond to private and personal use and applications, enabled by low power levels, simple spectrum etiquette and decentralised medium access protocols with collision avoidance mechanisms (e.g. CSMA/CA) but otherwise isolated governance In reality such alignment is of course difficult (if not impossible) to reach and therefore the dynamics could work on all frequency bands (such as CR devices on TV white spaces) and on all site types. Nevertheless, as a general rule, one can argue that this would be the most natural alignment, which in turn would mean that CR and DSA technologies could reach their highest potential if they were used with short range sites and high spectrum bands.

Furthermore, what is interesting to note is these underlying dynamics might be better aligned with the market characteristics of particular countries. For example the limit cycle dynamics are commonly observed in many European countries with a strong harmonisation legacy, such as e.g. Finland, where only GSM based technologies have been used, three network operators compete using the same technology and SIM-card based post-paid subscriptions are common leading to moderate churn rates (e.g. annualised churn typically above 10 % in Finland [55]). Markets in countries such as e.g. India are already more decentralised and follow strange (or no) attractor type of dynamics where many operators are present and pre-paid subscriptions and multi-SIM phones are common leading to very high churn rates (e.g. annualised churn roughly 40 % in India [34]) which in turn could make the market better compatible with CR and DSA systems as pointed out by [56].

On the other hand, in countries with vertical market structures, such as e.g. Japan, operators have traditionally had tight control of the technologies deployed, each operator having their own application stack, where the operators can internally be seen as following the fixed attractor dynamics with dedicated operator devices and high switching costs leading to low churn rates (e.g. annualised churn well below 5 % in Japan [55]). Although in our simulations it was assumed that CR and DSA increase device flexibility and the probability of switching between operators, this might not be the case if operators are in a position to limit and control the deployment of CR and DSA technologies in the devices.

Overall, these simulations show that only small changes in some parameters might change the market dynamics significantly. Therefore, as it relates to technology standardisation, it is important to preserve the opportunity to manage the market dynamics during the entire lifetime of the system technology and to avoid undesirable deadlocks and market failures. Since it is not possible to define all the parameters precisely right today it would be beneficial to preserve flexibility and configurability in standards and technologies in order to be able to control and adapt to the market dynamics later. The right architectural technology decisions are therefore very important for CR and DSA technologies.

4.3.5 Conclusion

In this section, we have studied value system evolution around future radio platforms given the introduction of CR and DSA technologies. We have used a combination of systems thinking tools and platform theory to characterise four value system configurations around the future radio platform and the corresponding underlying dynamics and have built a feedback model to evaluate future evolution possibilities both for GSM based mobile cellular and Wi-Fi based wireless local area radio platform paths. The results showed how the value system could continue on established evolution paths but also how it could transition to a so called complex adaptive system. For policy makers, the results have pointed out threats of winner-takes-all and fragmentation type of scenarios. The results also highlighted the possible importance of aligning the underlying market dynamics with the natural allocation and assignment cycle of the spectrum frequency bands, a hypothesis that could be explored more in future research. Furthermore, the overall framework introduced here, could in the future also be used to model the evolution of value systems around other technologies and e.g. explore the relationship of CR and DSA to other ICT technologies, e.g. Internet and cloud computing.

Acknowledgments The authors would like to thank Professor Heikki Hämmäinen and Mr. Ankit Taparia for their contributions. This work has received funding from the EECRT project and has also been conducted under the context of COST-TERRA (COST Action IC0905) research framework.

4.4 Business Scenarios for Spectrum Sensing-Based DSA

Vânia Gonçalves, Simon Delaere and Pieter Ballon

iMinds-SMIT, Vrije Universiteit Brussel, Brussels, Belgium

4.4.1 Introduction

The scarcity of spectrum that is experienced and/or anticipated today, is caused by the ever-growing use of wireless applications and by the way in which spectrum is managed. The long-term allocation of spectrum blocks to specific radio access technologies (RATs), specific services and specific operators is often cited as an inflexible spectrum management mechanism leading to suboptimal results. It is well known that most of these blocks of spectrum are not fully utilised.

Therefore, it is widely expected that measures allowing more efficient use of radio spectrum will include a shift from classic "command-and-control" to more dynamic forms of spectrum management and access will be a crucial part of the future telecommunications [57, 58]. In many markets, significant moves towards such dynamic spectrum management have already been made, including the introduction of selling and leasing of frequencies, collective use of spectrum and technologically neutral spectrum licenses. A technological advance that supports this objective is the development of CR and spectrum sensing prototypes. In its Report on Cognitive Technologies [58], the Radio Spectrum Policy Group defines spectrum sensing as follows: "[Spectrum sensing] provides a real-time 'map' of the radio environment. The main focus is on identifying unused areas in the intended frequency range that can be used by [Cognitive Radios]." The intended frequency range of our concept of spectrum sensing is considered to cover the entire spectrum, resulting in a RF tuning range of 100 Hz to 6 GHz. Furthermore, the spectrum sensing concept used for this research can sense very fast (29.5–88.5 ms) and requires low power amounts (7.8 mJ), making it ideal for implementation in terminals.

Spectrum sensing research often takes as point of departure a limited number of use cases, in order to sketch out a number of typified actors and their interactions. However, it is seldom addressed whether the conclusions drawn from such analysis are valid for other implementations of spectrum sensing. The hypothesis put forward here is that many contexts in which spectrum sensing technologies may be applied are so distinct from a business and regulatory point of view, that the characteristics and viability of one use case cannot be determined from analysing other use cases. It is therefore essential to determine which parameters are critical for distinguishing fundamentally different business scenarios. Four of such fundamental variables are identified and discussed below.

In order to test this hypothesis, the following research questions will be discussed: are there important differences in spectrum sensing scenarios that have to be considered in any business or regulatory analysis? If so, what are the business parameters that explain these differences? Is it possible to construct a business classification based on these business parameters? What added value would such a classification have and who would benefit from it? The results presented here could be used as a starting point for future research and decision-making related to spectrum sensing.

4.4.2 Business Parameters

The business parameters proposed below are the main differentiators between distinct classes of spectrum sensing business scenarios. They have been derived from an analysis of the use cases currently outlined in a variety of academic and consultancy research and industry white papers on spectrum sensing (see a.o. the references below). Based on these differences, four fundamental business variables have been derived, namely: ownership, exclusivity, tradability and neutrality.

4.4.2.1 Ownership

The main differentiator between spectrum sensing business scenarios is ownership. The concept of ownership used points out to ownership of a license and thus, the right of use for a given frequency band conferred by a regulatory authority, which still differs from ownership of spectrum. Using this business parameter in the classification, two major groups of business scenarios arise: the *unlicensed* spectrum business scenarios and the *licensed* spectrum business scenarios. The latter ones include every business scenario in which a regulator has issued licenses for a certain band of spectrum, independent of the way it is used and whether or not this license grants exclusivity rights to a dedicated frequency band.

4.4.2.2 Exclusivity

Drilling down within the group of licensed spectrum business scenarios, the exclusivity business parameter addresses the question whether or not frequency bands are exclusively assigned to a licensee. A regulator can decide to assign a specific frequency band for every licensee, thus making the frequency band *exclusive*. If the regulator would decide to group multiple frequency bands in a spectrum pool and make it available for multiple licensees, there would be *no exclusivity*. Note that the concept of exclusivity does not imply that only the licensee can have access to the frequency band. In some cases, users that do not have a license for the specific frequency band can utilise some (or all) of the frequencies in that band. The following business parameter will further discuss this topic.

4.4.2.3 Tradability

A third business parameter that is bound to affect future business models and regulatory consequences is tradability. This business parameter questions whether or not it is permitted for terminals to switch between different operators' frequency bands. If tradability is allowed, an operator can buy or lease a licensee's frequency band. Motivations for an operator to do so could include (but are not limited to) offloading of its own over-utilised bands, better coverage for its clients on the competitor's network, better quality of service, etc. In return, the primary user can be compensated. However, if tradability is either not allowed, or impossible, the use of the frequency band is restricted to the licensee itself.

4.4.2.4 Neutrality

A final differentiator is technology neutrality in licensed spectrum bands. Some frequency bands may be open to a variety of radio access technologies (RATs), while others only allow one specific technology. It is obvious that the latter case limits the efficient use of spectrum, but in terms of regulatory consequences, it can be assumed that a technology neutral frequency band would need to address more issues, such as setting technical conditions to access the band and coordinating the cooperation between multiple technologies.

4.4.3 Business Classification

Based on the aforementioned parameters, it is possible to derive a variety of distinct spectrum sensing business scenarios as shown in Fig. 4.16. Each category of business scenarios entails different regulatory issues and approaches. Furthermore, different roles and main beneficiaries can be identified in different cases.

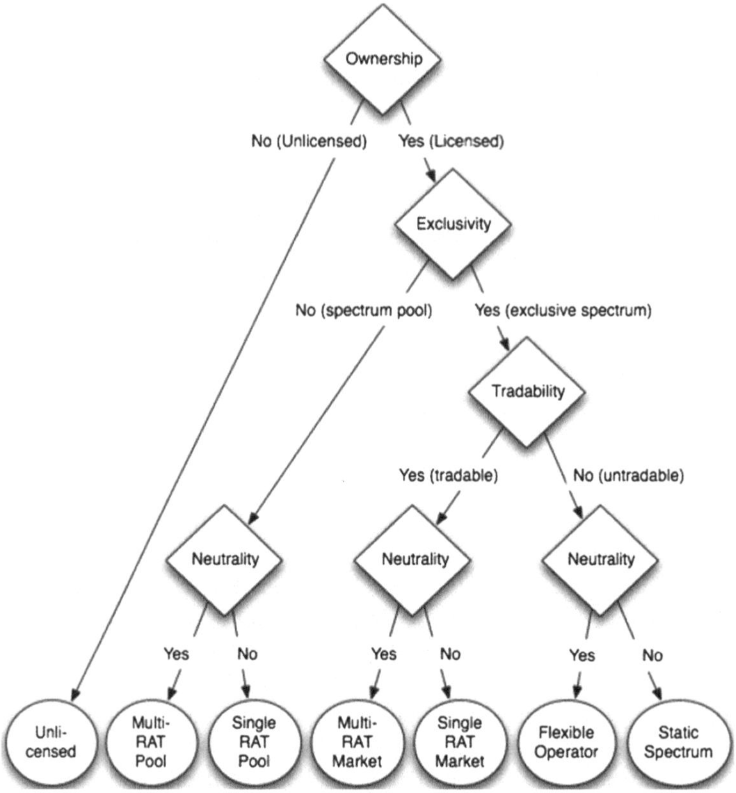

Fig. 4.16 Classification of spectrum sensing business scenarios

This classification differs from other classifications, such as [59] and [60], because (as far as the specific scenarios go) it is focused on spectrum sensing, it is not a technical classification and it uses a very detailed level of scenario groups.

For every class of the proposed classification, an exemplary business scenario has been chosen for discussion. In the following subsections, examples of the Unlicensed business scenario, the Single RAT Pool business scenario, the Multi-RAT Market business scenario, the Single RAT Market business scenario and the Flexible Operator business scenario will be discussed. The Static spectrum business scenario will not be discussed, as there is no use for spectrum sensing in a frequency band with restricted use for the licensee and one specified technology only. Furthermore, the Multi-RAT Pool business scenario will not be reviewed, as it can be argued that there is presently no realistic scenario in which a frequency band would be awarded in the near future with full flexibility as described by the four business parameters.

4.4.3.1 Unlicensed

The unlicensed case is different from most other business scenarios because there is no ownership of a license involved. Examples of business scenarios are mostly found in the unlicensed bands or ISM bands. Like many other technologies, both Zigbee and Wi-Fi (802.11 g/n) operate in the 2.4 GHz ISM band. This may cause problems of interference, resulting in a failure for the radio access technologies to send and receive data. Since Zigbee's data loss is more apparent, it is up to Zigbee to adapt and move to another frequency. In order to choose the optimal frequency or channel, Zigbee can use spectrum sensing. This way, it can dynamically detect the ideal location that provides the least risk of interference. By moving away from the Wi-Fi signals, both technologies are able to coexist.

The main benefit of spectrum sensing for this case is the fact that multiple technologies and users can coexist in the same band. This is achieved by avoiding interference. For unlicensed business scenarios, the most important actors are the unlicensed users and the regulator. The unlicensed users are allowed to share unlicensed spectrum, but they need to comply with certain rules put forward by the regulator. Most importantly, the bands accessible without license are defined by ITU-R and national radio authorities. Additional rules mainly contain technical requirements for the devices, accepted power levels, field strength limits and regulations regarding interference. Every potential unlicensed user should comply with these rules before accessing the ISM band.

The above describes the current workings and regulations for the business scenario. A question that can be asked is whether the implementation of spectrum sensing in unlicensed devices would change this situation. It can be assumed that additional regulations will not be needed. On the contrary, some technological device requirements that have the purpose of limiting and avoiding spectrum could be redeemed by spectrum sensing, since it could by itself solve all interference issues. In order for this to work, however, an additional condition a regulator might set, is that every device that wants to enter the unlicensed band should be equipped with spectrum sensing engines.

Additionally, a regulator might have issues with the fact that spectrum sensing could also lead to frequency hoarding. Since everyone will be able to sense the ISM bands for available frequencies, some users may block all of these frequencies, just in case they might need more bandwidth. If the regulator is aware of this sort of behavior, it is very likely it would act against it.

4.4.3.2 Single RAT Pool

If license ownership is a fact, but no exclusive frequency bands are assigned to every single licensee, then those licensees will have to share spectrum from a spectrum pool. In case different radio access technologies could operate in this spectrum pool, while sensing for appropriate frequencies, spectrum efficiency would theoretically be maximised, although some experts argue that the diversity of technologies and their propagation characteristics would make interference mitigation measures so stringent that parts of the gained spectrum efficiency would again be lost, for example due to excessive 'largest common denominator' guard bands and spectrum masks. This Multi-RAT Pool business scenario is still rather unrealistic at this moment. Therefore, the focus will be on spectrum pool in which all licensees of just one technology share spectrum by sensing the pool and occupying appropriate and available frequencies.

To assess business and regulatory issues, the Open spectrum LTE business scenario has been chosen; in which all LTE licensees share all available LTE bands. If this scenario is compared to the unlicensed one, the huge differences immediately become clear. This is an operator-based scenario, which does not require off the shelf equipment, but expensive industrial scale networks that need to be used more efficiently because of the huge investments. This being said, it is incomparable to most other scenarios. For spectrum sensing, it is important to know that LTE can operate on multiple frequencies, in a variety of frequency bands and even in various slices of bandwidth ranging from 1.4 MHz up to 20 MHz. Considering that this variety of frequency bands is to be found in the spectrum pool, spectrum sensing becomes essential in rapidly finding available frequencies. Furthermore, spectrum sensing could lead to more efficient use of the spectrum pool, by optimally filling it.

One of the apparent downsides of this model comes down to the willingness to fairly share between competitors. Imagine five mobile network operators all utilising the same LTE "spectrum pool". The regulator must guarantee access for all operators and a fair distribution of the spectrum. A first issue to address here is how such a fair distribution could be defined. Among other options, the regulator might take into account the number of mobile subscriptions, and set bandwidth boundaries accordingly.

A second issue the regulator may struggle with is the actual use of frequencies for the right purposes. In other words, how can the regulator control whether or not occupied frequencies are actually used for serving the customers? Furthermore, how can it act if occupied frequencies are not used for serving customers? It is needless to say that these issues still have to be resolved on a regulatory level, before spectrum sensing could ever be implemented in a spectrum pooling business scenario.

4.4.3.3 Multi-RAT Market

In this scenario licenses are issued, specific bands are exclusively assigned to every single licensee and tradability is allowed. In this case secondary users can, under specific conditions, access the licensee's frequency band. Again, this is a very prominent difference with the previously discussed scenarios. Since the context is again entirely different, different conclusions can be drawn from this group of scenarios.

For this scenario, two business cases will be explored below: emergency and public services and TV White Spaces business scenarios.

4.4.3.4 Emergency and Public Services

Every European country has a designated emergency band, for which the emergency operator has an exclusive license. This is the 380–400 MHz band. For routine situations, this band offers more than an adequate amount of frequencies. However, in crisis situations, the need for bandwidth exceeds the available band. Summarised, the emergency operator usually has excess and occasionally experiences a shortage of spectrum. Obviously, the latter could have serious consequences as all radios would have to queue before being able to communicate.

Spectrum sensing could offer two main benefits to solve these problems. First of all, secondary users could sense the emergency band, looking for available frequencies during routine situations. In return, the emergency operator could receive compensation. Second, the emergency operator itself could sense for available frequencies in other bands during times of crisis, when the need for bandwidth is exceeding the emergency band. Again, the emergency operator would also have to compensate the primary user (licensee) for utilising its frequencies.

Since crisis situations are impossible to predict, it is crucial that in these rare cases, the emergency operator can push all secondary users from its frequency band. The emergency operator would thus require guarantees concerning the availability of spectrum before opening up its band. Even though most agreements are bilateral (between the primary and secondary user), the emergency operator would still want the regulator to be involved. An emergency operator for example, would want the regulator to first check whether the sensing technology works. Second, the emergency operator wants regulatory guarantees that the technology would never fail. Third, the technology for 'pushing' secondary users during crisis situation should be examined, and lastly, the emergency operator would want the regulator to do some research on correct pricing and negotiation platforms.

Even if all these conditions are met, the question of which secondary users would put up with the occasional push, still remains. It is rather unlikely that mobile operators would risk offering a bad quality of service, with a bad reputation as a consequence, for a minor spectrum gain in return.

4.4.3.5 TV White Spaces

The TV White Spaces scenario is distinct from the emergency and public services scenario as the latter is using public spectrum, while this scenario will focus on commercial spectrum. As the context differs, it becomes clear that these scenarios cannot be treated equally.

The analog to digital switch over in television broadcasting has had some positive consequences on spectrum use. As digital signals require less bandwidth to provide the same or even better quality of television, previously occupied frequencies become available. Moreover, in many places empty channels exist between channels used for broadcasting, in order to avoid interference. These so-called TV White Spaces could be used for other, licensed or unlicensed, services. Because they are situated in the lower areas of the radio spectrum, these available channels have very good propagation characteristics, making them well suited for long range broadband access technologies (e.g. WiMax), particularly in areas where fixed broadband access is hard to realise. In order to make use of this potential without refarming the frequencies altogether, the TV band can be opened up to secondary users, which would scan the licensed band, looking for available channels. If the secondary user wants to make use of the available channel, it has to adhere to certain conditions (such as avoiding interference) and possibly compensate the primary user for using its spectrum band. Not only will sensing allow the secondary user to identify available channels, it will also allow the secondary user to (dynamically) avoid interference [61]. In case the license holder starts to make use of frequencies previously lying idle, the secondary user can again detect this through sensing—possibly aided by a database, as in the U.S.—and move away to other, available channels. Summarised, spectrum sensing would enable efficient use of abundant spectrum. In return, the licensee could receive compensation [62].

It can be assumed that the licensee would be willing to open up its band if correct compensation is foreseen and if the broadcasting of its content does not experience any interference. Secondary users, from their side, would be very willing to access the TV band. They could lease the excess frequencies to deploy mobile services, such as last mile broadband city coverage using IP-based WiMAX. For many of these secondary users, spectrum sensing is crucial to find available spectrum portions to operate. Without the existence and detection of these available portions, most of these operators would not be able to transmit any data, as they do not have appropriate frequencies or licenses at their disposal.

Given the fact that licensees would trade frequencies with secondary users, this business scenario deals with the secondary market principle. In the RSPG report [58] it is proposed that the conditions for such a secondary market should be set by the regulator. However, in case there is only one licensee, a "marketplace model" would be unlikely to be deployed in the future. On the contrary, bilateral contractual agreements between the licensee and the secondary users would be more likely to occur. This also implies less control of the regulator over the trade process. The licensee will most likely be the actor deciding on the different conditions, such as compensation, technical requirements and interference issues. However, the current general regulatory framework should always be taken into consideration, even in case of bilateral contractual agreements.

4.4.3.6 Single RAT Market

For this group of scenarios, a *Secondary Market LTE* scenario is discussed. In this business scenario, only the use of LTE is allowed. As opposed to the *Open spectrum LTE* business scenario discussed earlier, this business scenario does not deal with a spectrum pool, but with exclusively assigned frequency bands that can be conditionally accessed by secondary users. As a consequence, this Single RAT Market should also be regarded as a separate group of scenarios.

The rights of spectrum use, acquired by the primary user, can be traded or leased. In other words, the licensee of a frequency band for LTE use would be allowed to be remunerated by a secondary user, in return for opening up a certain portion of its frequency band. The secondary user would sense this band, looking for available frequencies. The primary user's motivations would be the compensation it would receive from the secondary user, making up for the high fee paid to acquire the license to operate in the frequency band. The secondary user's objectives could be offloading of its own over-utilised bands, better coverage to its clients on the competitor's network, better quality of service, etc.

On a regulatory note, there has been some discussion about regulatory reform to be able to allow this secondary market. In the RSPG report [58] it is proclaimed that the national regulator can decide on the conditions for such a secondary market. It is even expected that in the future, a real-time marketplace and negotiation platform could come into place.

Another question that arises in this secondary market model is whether or not this will create new actors in the telecommunications industry. If a marketplace would come into place, who would be in control of this market? Would this be the regulator? Would this be an LTE operator? Or would this even be a third party, acting as a broker? Would there even be a marketplace accessible for all operators, or would the secondary market just exist in bilateral relations, when one operator privately contacts another operator to buy or sell?

In any case, it is believed that the need for a regulator would be less stringent than in the *Open spectrum LTE* business scenario. Contracts not only set the technical conditions for entering the primary user's spectrum between operators, but they also decide on other conditions (such as compensation, duration, interference limits, Quality of Service guarantees, etc.).

4.4.3.7 Flexible Operator

A last business scenario can be situated in the licensed and exclusively assigned bands that are not tradable. In other words, the assigned frequency band can never be accessed by other users. If only one access technology can be used in such a band, spectrum is used in a static way, similar to the situation today. Since spectrum sensing would make no sense in such a business scenario, it is of no use to elaborate on it. However, if multiple technologies can be used, spectrum sensing could do its part. For this case, an LTE—femtocell handover business scenario will be analysed.

Femtocells are smart cellular access point base stations that use the Internet as backhaul. The femtocells are designed to solve the problem of reduced coverage and data rates, when using cellular technology indoors. Multiple femtocell 'heads' connect to a base station controller, which performs the handover (between macrocell and femtocell) and radio resource management. Besides better quality of service (higher data rates and increased coverage), the use of femtocells can be advantageous because they are cheap, in terms of CAPEX and OPEX, and require less power.

Spectrum sensing in mobile phones could be used to connect to better performing networks (femtocells). The network operator would encourage this, because it can offload its macrocell networks. Furthermore, the operator can save on OPEX and enjoys customer lock-in, as the bond between end-user and network operator has tightened, considering the purchase of an operator's femtocell. On the other hand, the end-user will enjoy better quality of service, guaranteed coverage and higher data rates, enabling innovative services. Additionally, it could be possible that he has to pay less for service through femtocells.

The most important actors in this business scenario are, without any doubt, the mobile network operator and the end-user. Presumably, the end-user already has a mobile subscription with the mobile network operator.

From a regulatory perspective, not much has to change vis-à-vis the current regulatory framework. The end-user and network operator play by the rules that were agreed in their contract. Still, one consideration can be made: a mobile network operator would want the femtocell to only operate in its own frequency bands. As a consequence, the operator enjoys customer lock-in. This may be in conflict with the general regulatory preference of interoperability. A few years ago, number portability came into place to ensure end-users the freedom to switch between operators. Therefore, it can be assumed that the regulator would want a femtocell to serve not for one operator only, but for all operators in the market.

4.4.4 Conclusion

The idea that a set of spectrum sensing business classes can be distinguished which refer to strongly divergent actors and interactions, and subsequently also to different consequences and conclusions, has been tested in this section. Four business parameters have been proposed, which are the basis of a business classification of distinct spectrum sensing classes and scenarios. The purpose of such a classification is providing a starting point for future research of spectrum sensing and CR implementation. Furthermore, such a classification could be of value to business actors and regulators, as they could use this classification for further analysis and decision-making.

It is clear that spectrum sensing cannot be managed and regulated as a whole. Because different business scenarios have different actors, roles and consequences, this paper indicates that the proposed scenario groups are fundamentally distinct and incomparable. As a result, conclusions for one set of scenarios should be assumed to be potentially widely different from other spectrum sensing business scenarios. In other words, every scenario should be analysed separately to evaluate its viability and the way spectrum sensing can contribute to this. Future research will further detail and analyse the fundamentally different business and regulatory logics behind the proposed classes and scenarios in real-life cases.

Acknowledgments The findings presented in this paper are based on research performed in the IWT ESSENCES project (Flanders, Belgium). Contributions from colleagues and project partners are here acknowledged, with special contribution from Matthias Barrie.

4.5 Possible Business Opportunities for CR

Peter Anker

Ministry of Economic Affairs and Delft University of Technology, Delft, The Netherlands

It may be assumed that the regulatory regime and the fundamental choices that will have to be made on the use of CR technology will create certain business opportunities and at the same time will pose limitations on other business opportunities for CR and dynamic access to spectrum (see Sect. 5.2 for in-depth discussion on this). There needs to be a fit between the regulatory regime, the fundamental choices on technology and a perceived business opportunity.

Opportunistic spectrum access based on sensing will always have a likelihood of interference and there are no guarantees that an OSA-device can find an opportunity to communicate. This will depend on the amount of OSA-devices and their communication needs in relation to the amount of capacity available. This sets limitations to the use and on the types of applications that can be supported. Since there is no need to build infrastructure there is a match with a device oriented open access regime of a commons. OSA based on sensing is expected to be restricted to low-end applications involving low power devices.

Opportunistic spectrum access can be used to share bands between licensed users and unlicensed short-range devices in bands that were difficult in the classic scenario. A good example of this is the use of the 5 GHz band. RLANs use sensing to detect and avoid incumbent radar systems.

OSA is also of interest to military users but for a completely different reason. A true OSA-device acts solitary without the need for coordination with the outside world. This makes it possible to communicate without making the whereabouts and communication needs of the military radios known to others. This will make their communications less vulnerable.

Since sensing is in its present form is not reliable enough, regulators around the world have turned their focus from sensing towards a GDB. This will require investments in a database and related infrastructure that need to be recouped. Entrepreneurs will only invest in this infrastructure if there is long-term assurance for access to spectrum and willingness to pay from customers. This shifts the orientation from a device centric approach to a service centric approach. Such a business case is better supported by a regulatory regime based on property rights.

A possibility to ease the problem of the (un)reliability of sensing is to focus sensing in a band that is not too-wide in a completely unlicensed environment to create a true commons for short range devices. The regulator should pinpoint a band for dynamic spectrum access in cooperation with industry. To reach economies of scale this band could be designated on a regional level, for example on a European level.

A very promising application for a true commons whereby unlicensed devices pool their spectrum is in-house networking. An in-house network is an ad-hoc network by its very nature. No two in-house networks are exactly alike and devices are turned on and off during the day, new devices are brought in, devices leave the house and the neighbouring houses have the same ad-hoc way of working. The number of wireless devices in a household is rising while the users want to have new equipment that is "plug and play". A new device that is put into service should be able to find its own possibilities to communicate within the in-house network. OSA can be used to realise this goal. A new OSA device senses its environment and coordinates its use within the local in-house network. A possible band to start is e.g. the 60 GHz band.

A second example of ad-hoc networking is the radio network between vehicles as part of Intelligent Transportation Systems (ITS). Restricting access to the pool for certain applications with a polite cognitive protocol, may alleviate the tragedy of the commons. In that case, the number of devices outnumbers the available spectrum in such amount that the spectrum is of no use to all. However, even if a polite cognitive protocol is used and the band is restricted to a certain type of applications, the amount of spectrum that is made available must be enough to cater for the intended business case.

Another possibility is to use sensing in a more controlled environment between licensed users. This will give more control over the environment, because the users are known. This type of sharing could be used to broaden the amount of accessible spectrum for temporarily users who need a guaranteed Quality of Service. This makes this type of sharing a perfect fit for e.g. Electronic News Gathering (ENG) and other Programme Making and Special Events services. ENG only requires spectrum for short periods of time and for a restricted local area but it requires guaranteed access during the operation.

Another service that needs guaranteed access to spectrum but only in a very local area and for a short period of time is public safety. Public safety organisations have their own network for day-to-day operations. However during an emergency situation they have a huge demand for communications on the spot [63]. A public safety organisation might make an agreement to alleviate their urgent local needs with other frequency users. In the agreement sharing arrangements are covered but the actual spectrum usage can be based on the local conditions and spectrum sensing of the local use of the primary user.

A good opportunity to start this form of sharing is in bands of the military. The military already have a longstanding practice of sharing with both the ENG community and public safety organisations. This may raise the level of trust to a level that is high enough to start an experiment.

In a true property rights regime dynamic access to spectrum is obtained through buying, leasing or renting access rights from the owners of the spectrum. This regime provides the possibility for active coordination between the incumbent user and the cognitive user about the likelihood of interference, and on guarantees about access to spectrum. If the barriers to instant trading are removed, the opportunity to buy and sell rights to access spectrum can be based on the actual demand for spectrum. This creates the opportunity to use DSA systems for higher valued services, such as mobile telephony, and for a spot market to be introduced. A spot market is a perfect means to acquire or sell rights to spectrum access based on the actual demand at any given moment in time.

This property rights regime can be used among operators to pool the spectrum in such a way that the rights to spectrum access are based on the actual demand for spectrum by their respective users. One of the suggested implementation scenarios is that mobile operators use a part of their spectrum to provide the basic services to their respective customers and pool the rest of their spectrum to facilitate temporarily high demands for spectrum. However, cooperation between mobile operators that are in direct competition to each other is not likely to happen [64].

This kind of sharing spectrum might be a more viable option for implementation in border areas to ease the problem of border coordination. Nowadays the use of spectrum in border areas is based on an equal split of the use of spectrum between neighbouring countries through the definition of preferential rights. However, there is no relationship with the actual demand for spectrum at either side of the border. A prerequisite is that the spectrum market is introduced at both sides of the border or in a region, e.g. the European Union.

Pooling spectrum between different services that are not in direct competition to each other might be a more promising approach. A property rights regime can help to make licensed spectrum that is not fully used available to others users. In this case access to spectrum is based on an negotiable acceptable level of interference, instead of the worst case scenarios based on harmful interference that are used by regulators to introduce a new service in an already used band. This may open bands for alternative use which might otherwise be kept closed. The incumbent licensee may now have an incentive to open its spectrum for other, secondary, users. The incumbent licensee is in full control because it can earn money with unused spectrum, whilst the access to its spectrum of the secondary user is on the incumbents own conditions.

Licensed owners of spectrum can also grant access to parts of their spectrum that they do not need in a certain geographic area and/or for a certain period of time to secondary devices. These devices can get access to this spectrum after an explicit request for permission to the owner of the spectrum. The owner will need a mechanism to facilitate requests from secondary devices for permission to use spectrum. Cellular operators can use their existing infrastructure to handle these requests. E.g. a mobile operator can set aside a mobile channel for this purpose. The owner of the spectrum and the secondary user can negotiate their own terms under which the secondary user may have access to spectrum. This provides possibilities for active coordination between the incumbent and the secondary user about the acceptable level of interference and guarantees to access spectrum.

A spectrum market can only function if information about the actual ownership of the spectrum property rights is readily available to facilitate trading. The regulator is ideally positioned to perform the task to keep a record of the ownership of these rights. Inclusion of monitoring information about actual usage of spectrum can further facilitate trading by giving more insights in the possibilities for secondary usage.

A second incentive might be to introduce easements in spectrum property rights. In other words, if a spectrum owner is in possession of spectrum that (s)he actually does not use, everybody is entitled to use this spectrum in an opportunistic way as long as the transmissions of the rightful owner are not subject to interference from this opportunistic spectrum access. This is an incentive which might prevent market players from hoarding spectrum [65].

A special case of licensed spectrum pooling is pooling whereby a single operator who is the exclusive owner of the spectrum uses CR technology to perform a flexible redistribution of resources among different radio access technologies within its own licensed frequency bands to maximize the overall traffic by an optimum use of spatial and temporal variations of the demand. This could be used by mobile operators to realise a flexible spectrum allocation to the various radio access technologies in use or to have an optimal distribution of spectrum between the different hierarchical layers of the network. For example to realise an optimal allocation of spectrum to femto-cells that takes account of the actual user demand without affecting the macro network. The prime requisite for such a scenario is that the license from the operator is flexible enough and is technology neutral.

4.5.1 Conclusion

CR holds an interesting promise for improved utilisation of the radio spectrum. However, there is a considerable degree of uncertainty regarding the potential application of CR. In addressing these uncertainties the business case for the CR is to be considered as centre point.

Both the regulatory regime under which the CR will operate and the specific characteristics of the CR technology will pose limitations to the business opportunities for the CR. Successful introduction of CR will require alignment between the characteristics of the CR and the regulatory regime under which the CR will operate. This is further discussed in Chap. 5.

4.6 Value of TVWS Spectrum and Analysis of Business Feasibility of CR for Mobile Broadband Services

Jan Markendahl[1] and Bengt G. Mölleryd[2]

[1]Royal Institute of Technology (KTH), Stockholm, Sweden
[2]Swedish Post and Telecom Authority (PTS), Stockholm, Sweden

In this section we present an overview of approaches for valuation of spectrum and describe characteristics and differences between valuation of licensed and non-licensed spectrum. Cost and cost structure for CR are introduced. The impact of deployment costs and spectrum prices on total costs are illustrated for a number of business scenarios where deployment using CR is compared to conventional mobile broadband. Finally we look into uncertainty and risk in terms of control of spectrum and availability of CR equipment.

4.6.1 Value of Licensed Spectrum and Approaches for Valuation of Spectrum

4.6.1.1 Introduction: Industry Transition Push Up Demand for Spectrum

The on-going transition from a voice to a data centric business is challenging for mobile operators as it undermines the established business model. This could be illustrated by the fact that mobile voice generates the equivalent of EUR 240 per GB while mobile data generates around EUR 5 per GB. This forces operators' to launch efficiency programs, cut operational expenditures, like network operational cost. However, in order to cope with the steep traffic growth and capacity constrains operators are forced to continue investing despite declining revenues (Fig. 4.17).

In order to increase capacity spectrum is essential, as spectrum could be seen as a substitute to additional sites, and secondary spectrum access using CR could potentially provide mobile operators with a cost efficient addition of capacity.

4.6.1.2 Valuation of Spectrum and Network Deployment

The necessity to release more spectrum is at the heart of most countries digital agendas. However, Plum Consulting[1] underscores that the majority of spectrum

[1] Plum Insight, August 2013, available at: http://www.plumconsulting.co.uk/pdfs/Plum_Insight_August_2013_The_role_of_spectrum_valuation.pdf

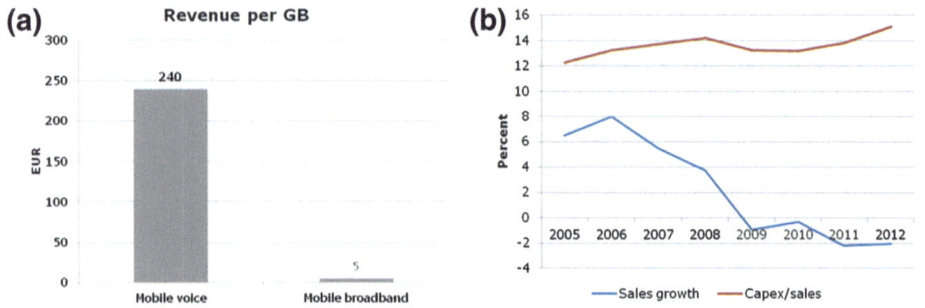

Fig. 4.17 **a** Revenue per GB for mobile in Sweden 2012; **b** Average CAPEX-to-sales and sales growth for European operators 2005–2012, based on an average on company ratios for operators: BT, DT, FT, KPN, Swisscom, Telefonica, and TeliaSonera. **a** *Source* PTS statistics and authors calculation **b** *Source* Bloomberg

suitable for mobile communications have been allocated which implies that it is required to transfer it from other applications in order to make spectrum available for mobile communications. In order to make these decisions valuation of spectrum is essential. Consequently, the area of valuation of spectrum generates a growing interest from industry, operators, consultants, academia, regulators and governments.

Plum presents a review of the value of spectrum licenses, model values based on expected revenues and costs for a hypothetical operator [66]. The Australian government (ACMA) applies an opportunity cost modeling, which it defines as the highest value alternative forgone, but underscores that the opportunity cost pricing differs according to circumstances [67]. Doyle state that it is necessary to take account of the opportunity cost values associated with alternative uses and across different frequency bands used by different users [68]. Yeo estimates spectrum values based on calculations from auction data and with an analysis of observed bidding behavior through an econometric model [69].

Ard-Paru captures the value of spectrum commons in Thailand through a cost and benefit analysis, in combination with an engineering valuation which could be used as an indicator for the regulator to decide to license spectrum or not [70]. ITU presents an approach to valuation of spectrum in order to facilitate for spectrum regulators to determine reasonable expectations on market-based revenues for the spectrum in beauty contest or administrative distribution processes, and for spectrum auctions to determine reserve prices [71].

Altogether, the valuation of spectrum could be based on the opportunity cost approach as it builds on the fundamental idea to capture the value of the alternative use, or expressed as what have to be forgone when one alternative is chosen rather than another one [72]. Moreover, Doyle [68] underscores that it is challenging to calculate opportunity cost values of spectrum and that it will generate a wide range of estimates.

Given that the value of spectrum is a function of network capacity as spectrum and base stations sites could be regarded as substitutes it is motivated to highlight the fundamentals for network deployment, which is followed in the next section.

4.6.1.3 Coverage, Capacity and Cost

Capacity in mobile networks can be increased by replacing existing radio equipment with more efficient technology, by deploying new base stations or by adding more radio equipment to existing base station sites using additional spectrum. The relation between network costs, capacity, bandwidth and service area has been established by Zander [73], which stipulates that for a specific amount of spectrum and for a specific radio access technology the following relation holds for capacity limited systems: *"the deployment of N times more capacity requires N times more base stations"*.

Operators that are unable to obtain additional spectrum are forced to deploy more base stations which require more investments compared to competitors who can add more spectrum and re-use existing base stations sites. Zander describes basic relationships that can be used for comparing different network deployment options [73]. For example, if a mobile operator with a 3G network at 2.1 GHz wants to expand the capacity one option is to build a denser network using the 2.1 GHz band. Another option is to acquire new frequencies in the 1.8 or 2.6 GHz band and reuse existing sites. This is feasible since these bands have almost similar propagation characteristics. Analysis of network deployment and sharing strategies for operators with different amount of spectrum and existing number of base stations are presented in [74, 75] (Fig. 4.18).

4.6.1.4 Cost Structure Modeling and Analysis

For macro cellular network deployment the main components in the cost structure of the Radio Access Network (RAN) are the base station sites, the radio equipment and transmission. It is, however, not the cost of radio equipment that is the dominating component in the cost structure. The largest costs are associated with the base station sites, including costs for towers, masts, non-telecom equipment, power, installations and site leases [76].

When 3G and HSPA system was deployed the costs for the radio equipment (and the capacity) were comparable with the site costs. The fierce competition among equipment manufactures in combination with technology advancement has pressed down prices on network equipment during the last decade, improving the cost-capacity ratio significantly. This enables operators to replace existing radio equipment with new equipment (LTE) for approximately EUR 10K per base station. This can be compared to typical costs of EUR 100K in Europe for

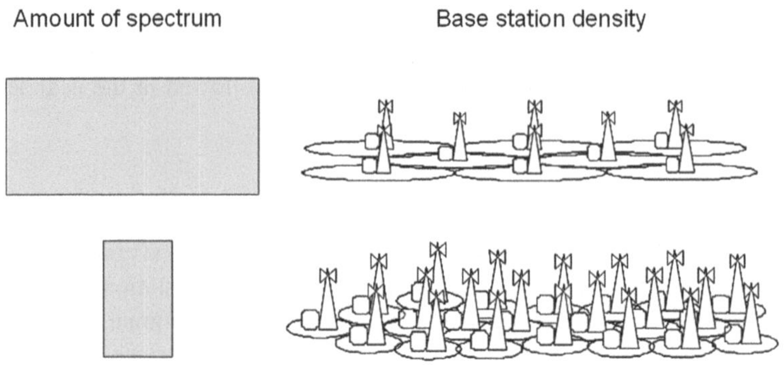

Fig. 4.18 Higher capacity can be provided by more sites or with a larger amount of spectrum

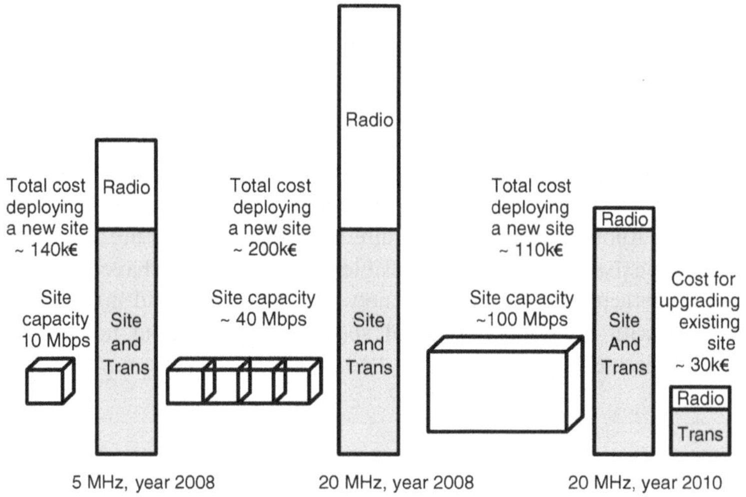

Fig. 4.19 Site capacity and costs illustrating cost reduction by re-use of existing sites [75]

deployment of a new site and EUR 20–30K for upgrading an existing site with fiber connection [75], see Fig. 4.19 for a comparison of site capacity and costs. The most recent base station equipment supports three sectors, bandwidths up to 20 MHz and multi-standard solutions, e.g. GSM, WCDMA and LTE.

The main driver for network costs is the amount of new sites that needs to be deployed. Hence, this is a key aspect when alternative deployment options are investigated. The capacity is related to the amount of radio equipment. Additional spectrum means that operators can re-use existing sites and hence capitalize on existing infrastructure investments.

4.6.1.5 The Overall Approach

The estimation of the opportunity cost of spectrum is based on an analysis of network capacity and cost for different network deployment options which use different amounts of spectrum. The cost comparison is the basis of the opportunity cost of spectrum and represented by the cost savings facilitated by additional spectrum bands compared to building out existing networks that provide the same capacity as the network with additional spectrum. The approach applied below, which is a high system level analysis, builds on [77–79].

The approach has been explored in several papers [80–83] and the applied analysis consists of three steps: (1) Selection of the network deployment and spectrum allocation cases to compare, (2) Analysis of the deployment cases including user demand, capacity and cost structure, and (3) Comparison of network costs for the options resulting in the opportunity cost.

If operators do not obtain additional spectrum they need to deploy a denser network in order to enhance capacity in areas with capacity constraints. The operators' strategies for network deployment and spectrum portfolio management are vital parts of overall business strategies, which varies between operators depending upon regulatory and market conditions and operators' market position.

4.6.1.6 Calculation of Opportunity Cost

The user demand expressed as capacity per area unit (Mbps per km^2) is based on user density and the data usage per subscriber. It is based on monthly user demand (GB/month) and an approximation on how the usage is spread out over the day. For example, a usage of 5 GB per month spread out over 8 h per day is roughly equal to a continuous demand of 0.05 Mbps per user.[2] By calculating the demand of all users in the area an estimate is obtained of the total area demand (Mbps per km^2). This is compared with the capacity per area unit provided by the base stations, calculated as follows:

Site capacity = *bandwidth* (MHz) * *spectral efficiency* (bps/Hz) * *number of cells/sectors per site*.

With a LTE system and a re-use factor of 1, i.e. all the frequencies can be used in all cells (or sectors), translating into that with 20 MHz and an average spectral efficiency of 1.7 bps/Hz the capacity for a three sector site is 100 Mbps.

Total investments to deploy a mobile network are calculated by taking the capital expenditures (CAPEX) for electronics and civil works per site multiplied with the total number of sites. The total cost per site for the active equipment (electronics, radio) is currently around EUR 10K. The cost for civil works is depending upon cost for material and labour implying that the CAPEX is

[2] The estimate of 0.05 Mbps per user is based on a usage of 5 GB per month: 5 * 1024 * 1024 * 8 = 4194304000/30 = (1398101333/24/3600) * 24/8 = 49 kbps.

determined by national cost levels. The opportunity cost of spectrum is estimated by analyzing substitution between spectrum and base station sites, and calculating cost savings provided by additional spectrum bands compared to increase the number of sites. The basis is operators' current spectrum holding, and the geographical coverage of the network. It is followed by an estimation of the number of existing sites, and the range of the cell radius. The spectral efficiency gives the basis to calculate network capacity for the different deployment options providing the similar amount of capacity per km^2.

4.6.2 Aspects and Approaches for Valuation of Non-licensed Spectrum

The objective of this sub-section is to highlight the differences between the valuation of licensed and non-licensed spectrum as the "valuation logic" differs substantially. Basically, it makes no sense to apply the opportunity cost approach if the non-licensed spectrum is the only type of spectrum that an actor has. On the other hand it is relevant if the actor has other types of spectrum.

4.6.2.1 Key Differences in Valuation of Licensed and Non-licensed Spectrum

The valuation of licensed spectrum is based on the opportunity cost approach where the key idea is substitution between spectrum and base station sites. The basic assumption is that the value of the alternative use, or expressed as what have to be forgone when one alternative is chosen rather than another one [72]. The used assumption is that the resulting capacity and availability of spectrum is well defined and stable, this is the case considering licensed spectrum.

If there is just one type of spectrum used, no opportunity cost analysis is possible. If the actor can use more than one type of non-licensed spectrum a modified opportunity cost approach based on substitution between sites and spectrum could be used. For example TV white space spectrum can be used as replacement or complement to a LTE wide area network or to a WiFi network, i.e. instead of deploying a denser LTE or WiFi network. Estimation of the value of non-licensed spectrum applying the opportunity cost approach makes sense if other spectrum resources are available for the operator under study.

With just one single type of non-licensed spectrum, open access (like WiFi), secondary access (like TV WS) and shared access (some type of LSA), the value is that it enables operators to offer services "at all". Hence the value of the non-licensed spectrum depends on potential revenues in relation to the costs for exploiting the non-licensed spectrum bands. The costs are both related to the network deployment and to the overall business. Network costs are e.g. to build base station sites and transmission, to rent space in exiting sites, to buy and install CR equipment and maybe to develop CR solutions. The overall costs are those typical for an operators business, e.g. to build up and maintain a customer base (i.e. marketing & sales, CRM) and to provide service and billing platforms.

For the cases where the approach with opportunity costs and substitution of sites and amount of spectrum can be used another type of aspect needs to be considered—different types of uncertainties. Unlike licensed spectrum the use of non-licensed spectrum is uncertain in many aspects, availability, interference level and resulting quality for end-users. In addition, the complexity and implications for cost of the CR equipment is associated with large uncertainties. For LTE base station equipment both the performance and costs are well known, see Fig. 4.20 for a comparison.

4.6.2.2 How to Estimate Spectrum Value for Non-licensed Spectrum Bands?

In order to estimate value of non-licensed bands different approaches are used depending on what kind of actor we consider and if that actor can make use of other types of spectrum resources. In the rest of this section we will discuss this situation considering mobile broadband (MBB) services using licensed band and conventional LTE systems and compare this to a system using TV white space spectrum and CR equipment.

The analysis approach is outlined in Fig. 4.21. It is applicable to both actors with licensed spectrum, i.e. mobile operators, and actors with making use of non-licensed spectrum only. The first steps are common and include: (i) estimation of spectrum availability, (ii) estimation of capacity that can be provided for a specified type of deployment and inter-site distance, and (iii) a check if the supplied capacity can meet the estimated demand. If not, another (larger) site density needs to be applied. When the demand is satisfied the analysis is split into two branches depending on what actor that is considered.

For an actor with just non-licensed spectrum the next step is to estimate the willingness to pay by end-users. The resulting revenues are then compared with the estimated investments for networks and for other components in the operator overall cost structure.

For a mobile operator with licensed spectrum the opportunity costs approach can be used. The two following build out approaches are compared:

1. Build new sites using existing LTE technology and licensed spectrum
2. Re-use exiting sites and deploy new (CR) radio technology using non-licensed spectrum

Below we will in more detail describe the approach in Fig. 4.21 with focus on the common steps and how a mobile network operator can exploit TV WS. For a new actor the spectrum value depends on the potential revenues and the overall business case, this is beyond the scope of the section.

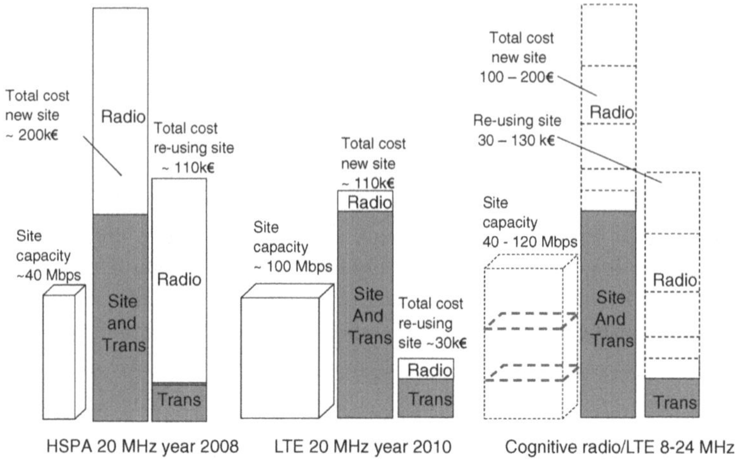

Fig. 4.20 Example of capacity and cost structure for different types of radio access technologies (For the CR solution the indicated variations for capacity and radio costs depend on the amount of available bandwidth and uncertainty about radio complexity and implementation, picture modified from [88])

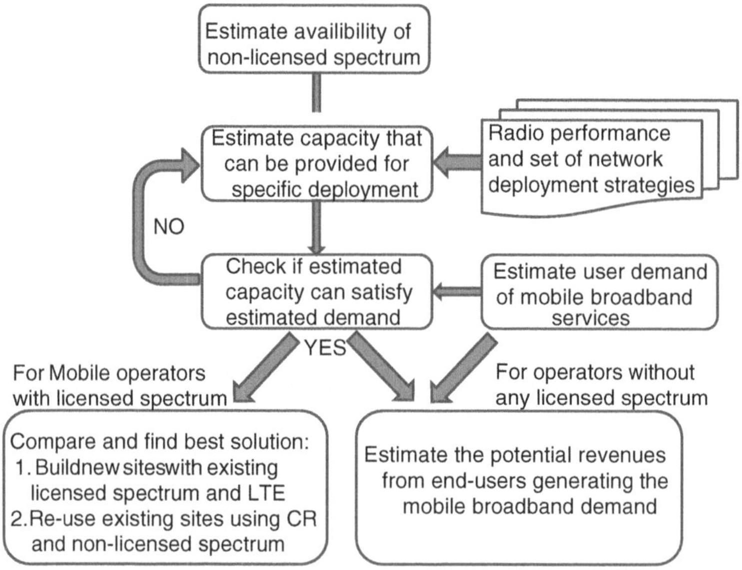

Fig. 4.21 Overall work flow for estimation of value of non-licensed spectrum

4.6.2.3 Network and Capacity Modeling and Analysis

Radio access technology

In order to see if the use of TV WS is feasible we need to do some general modeling of capacity. In this case where we consider "cellular use" of TV white spaces (TV WS) we mean mobile broadband access (MBBA) services. One motivation for this choice is the increasing demand for MBBA services and the relatively low amount of bandwidth that is currently allocated to mobile operators in the 800 or 900 MHz band. For mobile operators TV WS can be used as a complement to licensed spectrum possibly offering improved cost-efficiency. In the 800 and 900 MHz bands TV WS could be used as complement to or as replacement for licensed spectrum.

We assume that the MBBA service will be provided by a radio access technology like LTE with varying system bandwidth up to 20 MHz. We will compare the deployment of networks using the TV WS with deployment of MBBA using LTE in the 800 MHz band. In the analysis we consider cases with a relatively low number of available TV channels, 1–4 TV channels corresponding to a bandwidth of approximately 8–32 MHz.

Availability of TV white space spectrum

In the Quasar project[3] the number of available TV channels has been estimated for a number of countries. The number of "un-used" TV channels is very low in most part the country. "Many" TV channels are available in rural areas in northern Sweden, areas where the population density (and demand) is low [84–86].

Please note that the availability of spectrum for secondary use depends on the type of services and the type of network deployment that is used. By using macro base stations with high towers the mobile broadband will cause interference over large distances, hence the spectrum availability is low. If the spectrum is used for indoor deployment using low power base stations then the secondary usage will cause interference in limited area and hence the number of "available" TV channels will be much larger.

For Sweden less than five channels are available in most parts of the country [85]. Only in some rural areas in northern Sweden more than 20 channels are available, in these areas the demand is very low. One and four TV channels correspond to in total 8 and 32 MHz respectively. This can be compared with the spectrum allocation for the frequency bands intended for LTE in Sweden:

- At 800 MHz the operators have 10 MHz (downlink and uplink);
- At 2.6 GHz the operators have 10–20 MHz (downlink and uplink).

Offered capacity

The offered capacity for the mobile broadband access service depends on the available bandwidth and the spectral efficiency. The offered cell capacity in Mbps equals bandwidth (MHz) * spectral efficiency (bps per Hz). The bandwidth

---

[3] http://www.quasarspectrum.eu/

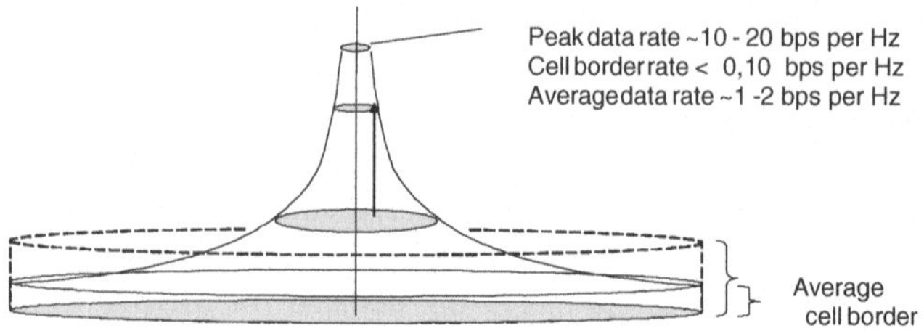

Fig. 4.22 Spectral efficiency target values for LTE

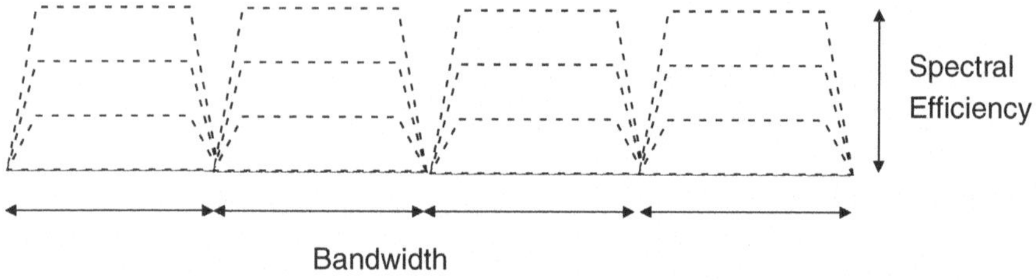

Fig. 4.23 Bandwidth and spectral efficiency

depends on the number of TV channels available for secondary access, the spectral efficiency depends on the network deployment and interference from other secondary users. In our estimates we will use cell average values although we know that the spectral efficiency for MBBA depends on the location of the end-user. In Fig. 4.22 the ITU target data rates are shown for the peak, average and cell border values.

The estimated capacity for a base station site with three sectors is 3 * the spectral efficiency * the bandwidth (3 * SE * BW). Both the spectral efficiency and the bandwidth in terms of number of TV channels can vary according to Fig. 4.23. With this model the key parameter is the product SE * BW with the dimension "bits per second".

The parameter set {SE = 1; BW = 8} gives the same results as {SE = 0.50; BW = 16} and {SE = 0.25; BW = 32}. The impact of interference and different cell sizes can be reflected in the spectral efficiency. For deployment in urban and rural areas we can assume spectral efficiency values in the range 0.50–2.0 and 0.25–0.50 bps per Hz respectively. The lower spectral efficiency for deployment in rural areas combined with a larger bandwidth (more available TV channels) results in values of the product "SE * BW" in the same range as for urban deployment.

Modeling of user demand

For dimensioning of mobile broadband access we define the user demand as the capacity needed per area unit expressed as Mbps per km^2. This equals the average usage per user times the number of users per area unit. Mobile data usage is the amount of data sent and received per user during one month and usually expressed in GB. For Europe the smartphone users typically consume 0.1–1 GB per month

Table 4.2 Examples of required capacity as function of number of users and usage level

Geotype	Users per km^2	Area demand for different usage levels ($Mbps/km^2$)		
		0.1 GB/month	1 GB/month	10 GB/month
Rural	10	0.01	0.1	1.0
Suburban	100	0.1	1.0	10
Urban	1,000	1	10	100
Metro	10,000	10	100	1,000

and laptop users with dongle consume 1–10 GB. The usage needs to be expressed in terms of data rates. Assuming that the data is consumed during 8 h per day all days a monthly demand of 10.8 GB corresponds to an average data rate of 0.1 Mbps. Hence, a monthly usage of 0.1 GB, 1 GB and 10 GB per month roughly corresponds to 1, 10 and 100 kbps respectively.

In order to estimate the demand per area unit we need to consider the population density and the penetration of the service offered by the provider. The orders of magnitude of the area demand are illustrated in Table 4.2.

The demand is shown for different "user" densities and for users with different demand levels. The dimensioning means that these demand numbers need to be matched by the offered capacity.

Analysis of demand and offered capacity

We consider cases where quite few TV channels are available. One and four TV channels correspond to 8 and 32 MHz which can be compared to the deployment of 800 MHz networks with bandwidth in the range of 5–20 MHz.

In Table 4.2 we presented examples of the user demand depending on the number of users per area unit and the usage level per user. The user demand in these scenarios, expressed as Mbps per km^2, is compared to the offered capacity. The assumed bandwidth (BW) is in the range one to four TV channels and the spectral efficiency (SE) is in the range 0.25–1.0. As mentioned elsewhere, the key parameter for the capacity estimates is the product SE * BW, see Table 4.3. We have assumed deployment scenarios where the cell size differ an order of magnitude when it comes to coverage area.

The comparison indicates that for the assumed usage and user densities and coverage areas of sites the demand can reasonably well be met with bandwidth corresponding to a few TV channels. With 32 MHz quite high demand levels can be met. When demand and supply can be matched the deployment strategy needs to be examined in more depth. The cell size and the site density need to be considered from a cost perspective.

The conclusion of this analysis is that since the offered capacity can meet the estimated demand the assumed type of deployment can be used for further assessment. The actor using on TV WS needs to look into revenues and the overall business cases. A mobile network operator needs to investigate what deployment options that is best, to make use of the TV WS spectrum re-using existing sites or to build a denser network using licensed bands. This is to be discussed next.

Table 4.3 Examples of user demand and offered capacity per area unit assuming different coverage areas per site and spectral efficiency * bandwidth (SE * BW)

	Number of users per km^2	Area demand Mbps/km^2)	Coverage area per site (km^2)	Capacity (Mbps/km^2) for varying SE * BW		
				2	8	32
Rural	10	0.1–1.0	100	0.06	0.24	0.96
Suburban	100	1.0–10	10	0.60	2.4	9.60
Urban	1,000	10–100	1.0	6.0	24	96
Metro	10,000	100–1,000	0.1	60	240	960

4.6.2.4 A Trade-Off for Mobile Operators: To Build a Denser Network or to Use More Spectrum

For addition of more capacity mobile operators have two main options. To use more spectrum and upgrade existing base stations sites with new radio equipment or to use existing spectrum bands and to build a more dense network, i.e. to add more base station sites. As an alternative to buying licensed spectrum operators may use secondary spectrum access and hence some type of CR.

To add more sites are more costly since towers etc. dominates the cost structure of base station sites. The value of more spectrum in general is illustrated in Fig. 4.24. The price of licensed spectrum can vary a lot [81]. For cases in Europe en estimated "spectrum cost per site" is equal to or less than the cost of the radio equipment. Hence, operators can make substantial cost savings by using more spectrum, no matter if it licensed bands or bands exploiting secondary access are used.

Also for use of secondary access to spectrum the major cost savings result from the fact that no new base station sites are needed. It is not the zero spectrum costs that it is the main issue even if this has a larger impact for cases where the spectrum prices are very high. The costs for CR equipment are uncertain but anyway costs savings can be substantial as illustrated in Fig. 4.24. The use of secondary access would be interesting for mobile operators for another reason. This type of added capacity is used as complement to licensed spectrum bands. Actors using secondary access to spectrum as the only resources are much more vulnerable. On the other hand mobile operators may hesitate to include yet another type of solution and technology, this will be discussed more below in Sect. 4.6 on investments and risk

The impact of network deployment and spectrum costs will be illustrated in the next section. The total network costs are studied both for fixed used demand and varying amount of spectrum as well as for fixed amount of spectrum and varying user demand.

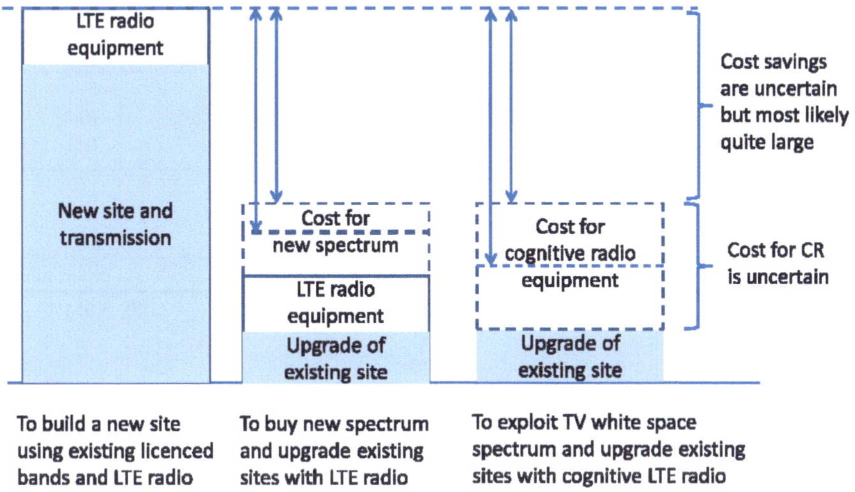

Fig. 4.24 Illustration of cost relations for mobile operators that want to add more capacity using existing licensed spectrum, new licensed bands or TV white space bands

4.6.3 Case: Impact of Deployment Costs and Spectrum Prices on the Business Viability of Mobile Broadband Using TVWS

In this subsection spectrum valuation will be illustrated by looking into the business feasibility of mobile broadband access services using secondary access of spectrum in the TV bands. The capacity-cost analysis considers costs for radio equipment, base station sites and radio spectrum comparing network deployment by a market entrant and an existing mobile operator using either licensed spectrum or TV white spaces. In addition, the impact of high and low spectrum prices is considered. The analysis shows that market entrants will be in a more difficult position than the established actors. No matter the cost-capacity performance of CR equipment, a new operator needs to invest in a new infrastructure with sites and transmission. Only for cases where the spectrum costs are "high" (compared to other cost components) use of TW white spaces turn out to be more cost efficient for both existing operators and new operators [87].

4.6.3.1 Case Description, Models and Assumptions

We consider cases for urban and rural network deployment were we compare the overall network costs for a market entrant and an existing mobile operator using either licensed spectrum or TV white spaces. The impact of spectrum prices is illustrated using examples from Europe and India.

Table 4.4 Example of spectrum prices, data from [84]

Case	Bandwidth (MHz)	Spectrum price (€/MHz/pop)	Cost/Site (k€)
Germany 2.6 GHz	20	~0.05	~1
Sweden 800 MHz	10	~0.50	~10
India metro areas 2.1 GHz	5	~5	~100

Table 4.5 Network assumptions

	Urban environment	Rural environment
(Coverage Area [km^2], Radius [km])	(1; 0.56)	(100; 5.65)
Sectors/base station site	3	3
Bandwidth [MHz]	20	20

Table 4.6 Assumptions of user demand

	Urban area Sweden/India	Rural area Sweden/India
#Users/km^2	2,000/20,000	100/1,000
Usage GB/month/user	10/1	10/1
Demand (Mbps/km^2)	200/200	10/10

Spectrum costs

It is often claimed that one driver for secondary use of spectrum is that the cost of spectrum can be avoided. This is only partly true; it depends on the spectrum price in relation to other network costs. Comparing recent auctions in different countries we can identify large differences. The spectrum cost per site for the Swedish case is in the same range as the radio equipment whereas in India the spectrum cost per site is as large as the costs for base station sites, see Table 4.4.

Coverage and Capacity of Base Station Sites

The assumptions regarding coverage are shown in Table 4.5. The user demand is satisfied by adding sufficient capacity to each site. When the demand cannot be met with the available amount of spectrum new sites need to be deployed, i.e. the more bandwidth the fewer number of sites. In the analysis we will show how the overall network cost depends on: (i) the amount of available spectrum (for a fixed demand) and (ii) the user demand (for a fixed amount of spectrum). For both the licensed spectrum and the TV white spectrum we assume that we use a LTE type of radio access technology with an average spectral efficiency of 1 bps per Hz. For the capacity estimates we assume three-sector sites and a re-use factor of 1.

User demand

The dimensioning is based on the estimated user demand per area unit (Mbps/km^2). We assume that the data is "consumed" during 8 (equally) busy hours 30 days per month, see Table 4.6.

Costs for radio equipment and base station sites

We can compare mobile broadband systems using TV white space with deployment in the 800 MHz band. Although the uncertainty is high when estimating costs for CR equipment, some insights can be gained if we consider the overall cost structure for the network deployment. In Fig. 4.20 we consider two main components of the cost structure for a radio access network; the radio equipment and "the sites and transmission". In Sweden the cost for deployment of a macro base station site is typically in the range 50–200 k€, we assume a cost of 100 k€ for deployment of a new site. According to Telenor the cost for upgrading existing sites with a fibre connection is estimated to 20 k€ per site [75].

The cost-capacity ratio of commercial radio equipment has improved more than 20 times the last few years. This is illustrated in Fig. 4.20 where HSPA and LTE are compared. For CR we still do not have any cost numbers, in the analysis we assume twice the cost for the same spectral efficiency as LTE, i.e. 20 k€. Factors that may drive costs for CR are: large system bandwidth, additional systems for sensing, interference management, need to add data bases, and no large scale production. Even if the cost for CR equipment would be the same as for standard LTE base stations, the key issue is if new sites need to be deployed or not. In this case the problem is mostly a matter of market entry. In addition to deploying a totally new infrastructure, a new actor needs to invest in and build up marketing, customer base, service and billing platforms.

4.6.3.2 Performance Analysis: Impact of Cost Structure

We have assumed scenarios where a Greenfield and an Incumbent operator deploy networks in order to provide mobile broadband services. Two options are available for the operators; first, it is to run their networks by using licensed spectrum (this means to acquire new spectrum licenses) and second, to use TVWS and only upgrade the network sites with CR equipment. Assuming a fixed demand and varying the amount of bandwidth that each operator gets, we show the impact of this additional spectrum bandwidth on deployment costs.

The more spectrum the less sites are needed. Hence the costs decrease with increasing bandwidth, this is clearly visible for low bandwidths. The impact of spectrum price can be seen for higher levels of bandwidth, see Fig. 4.25. For the low spectrum price levels (European case) a small increase can be observed but for the high price levels (India case) the networks costs increase dramatically. With the used assumption there is minimum for a specific amount of licensed spectrum.

Besides the costs for sites, radio equipment and spectrum the result depends on the demand levels and the assumed coverage areas. Hence, we present a sensitivity analysis where we vary the user demand and the base station coverage. In Fig. 4.26 we illustrate the impact of lower demand and here the same cost minimum can be observed. In Fig. 4.27 we show the cost assuming a smaller coverage area for "high" spectrum prices. In this case a large number of sites are needed and

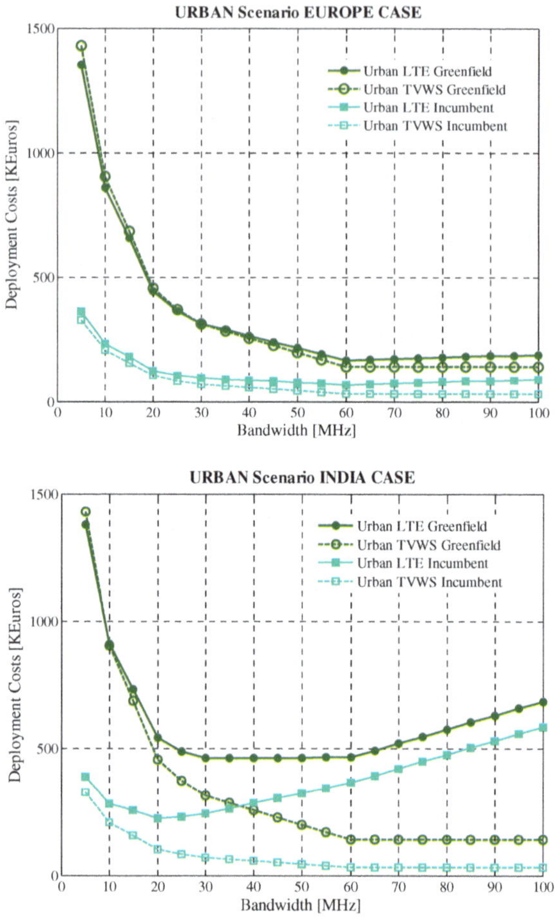

Fig. 4.25 The costs are shown as a function of system bandwidth assuming *low* and *high* spectrum prices (Europe and India respectively) and an urban environment with demand of 200 Mbps/km^2 and a base station coverage area of 1 km^2

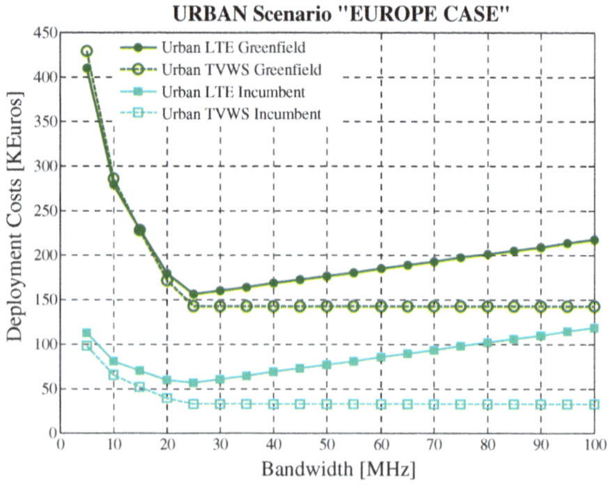

Fig. 4.26 Examples of deployment costs illustrating "Fixed demand and varying amount of spectrum", The costs are shown as function of system bandwidth assuming *low* spectrum prices (Europe) and an urban environment with demand of 50 Mbps/km^2 and *large* base station coverage area (1.0 km^2)

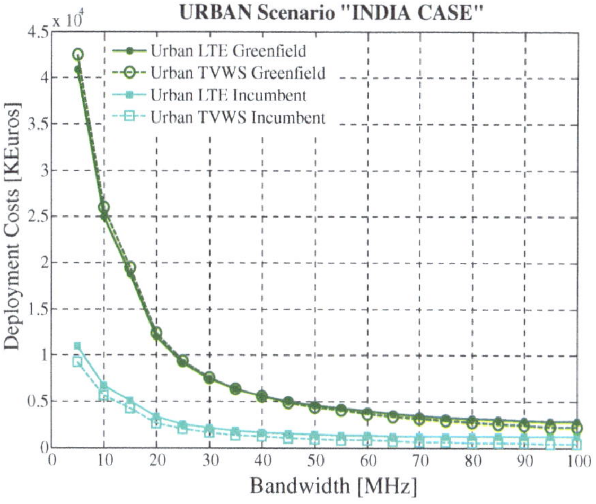

Fig. 4.27 Examples of deployment costs illustrating "Fixed demand and varying amount of spectrum". The costs are shown as function of system bandwidth assuming *high* spectrum prices (India) and an urban environment with demand of 50 Mbps/km^2 and a *small* base station coverage area (0.2 km^2)

hence the site cost is dominating. For lower spectrum prices, the graphs with lower demand levels and smaller coverage areas are similar to Fig. 4.27.

Now we will vary the demand for a fixed bandwidth of 20 MHz. The costs will increase with demand but the interesting thing is to identify the differences between the deployment cases. Figure 4.28 illustrates how a Greenfield operator building up its network from scratch has higher costs than the incumbent operator. The difference is largest for the low demand levels where the incumbent can make use of existing sites. For the assumed levels of site costs, radio costs and spectrum price the Greenfield operator always has higher network costs, even when CR and TV white spaces are used. For the case where the spectrum prices are "high", the situation is different, see Fig. 4.29. Use of TV white spaces (no spectrum cost) results in lower costs for both the incumbent and the Greenfield operator but the incumbent has lower costs.

4.6.4 Uncertainties and Risks

This subsection elaborates on risk and uncertainties in the deployment of new technologies, such as CR. The perspective is techno-economic implying that all parts of the system have to be available in order for the system to function. This is illustrated by the introduction of new standards and the significance of investments in terminals for how it influences the development of the mobile technology system.

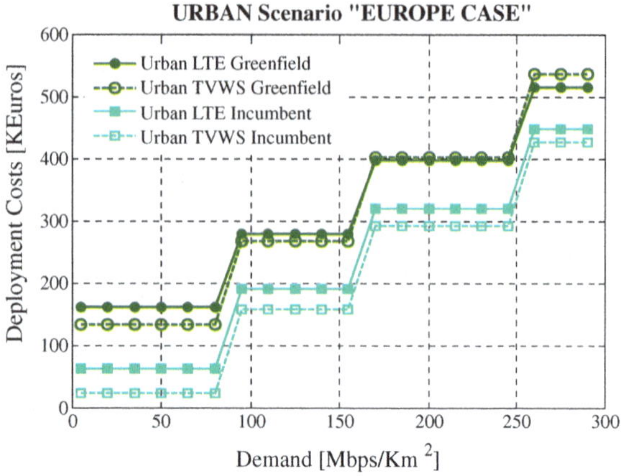

Fig. 4.28 Examples of deployment costs illustrating "Fixed amount of spectrum and varying demand". Network costs as a function of a varying demand in an urban environment assuming *low* level of spectrum cost, 20 MHz of spectrum and coverage area of 1 km^2 per site

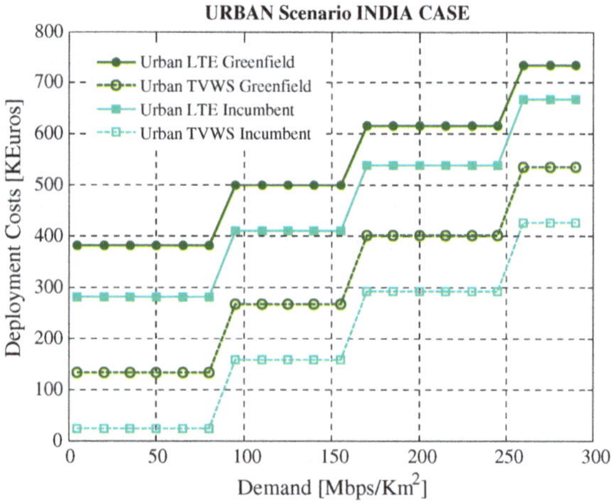

Fig. 4.29 Examples of deployment costs illustrating "Fixed amount of spectrum and varying demand". Network costs as a function of a varying demand in an urban environment assuming the *high* Indian level of spectrum cost, 20 MHz of spectrum and coverage area of 1 km^2 per site

4.6.4.1 It Takes Time to Establish New Mobile Standards on the Market

As mobile communication is a network technology it consists of a number of subsystems. Focus is predominately on network equipment, provided by equipment manufacturers, which have transferred specifications of radio technology standards into the equipment that are manufactured. The advancement of the technology has facilitated multi-band radio enabling operators to easily migrate to new system technologies. But the commercial migration to new technologies requires that end customers have access to appropriate terminals. The historical development of mobile communication has demonstrated that it takes time for new technologies to be established on the market, see Fig. 4.30.

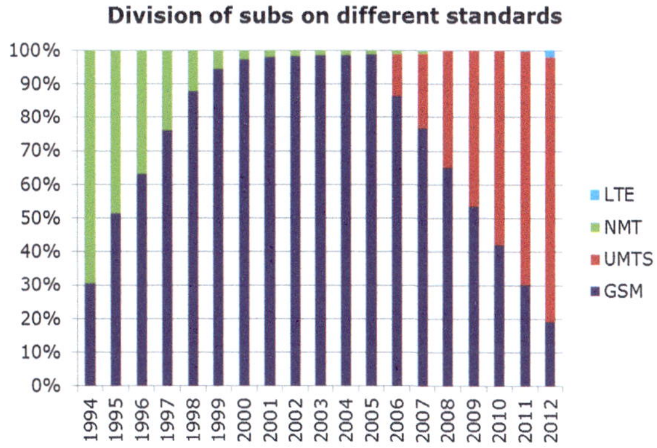

Fig. 4.30 Distribution of the total mobile subscriber base in Sweden. *Source* Svensk telemarknad, PTS

An illustration of this is that NMT, the analogue Nordic Mobile Telephone system, which was launched in the 1980's had its peak in 1995 (measured as sold terminals), 3 years after the official launch of GSM. But as handsets for GSM were not available in commercial volumes when GSM networks were completed it took another couple of years before GSM took off.

UMTS (3G) was initially planned to be launched in Europe in 2001, but the lack of terminals delayed the market introduction and the sales of GSM terminals peaked in 2005. The fact is that it was not until 2009–2010 that 3G made up more than half of the handset market in Sweden. Moreover, TeliaSonera was among the first operators in the world to launch LTE when it opened its network in December 2009, but the inflow of 4G subscribers was minimal due to limited availability of dongles and terminals was an issue for the future .

The initial growth of 4G in Sweden has been slow, although TeliaSonera got competition on 4G in 2012 when a new network opened, and the total share of 4G subscribers where 0.2 % 2011 and 1.9 % 2012.

Although the major focus is on investments (capital expenditures) made by operators the requirement on end-customers is that they have to purchase new terminals in order for a new technology to be established on the market. Based on reported CAPEX made by operator during 2012 and the value of the terminal market, which is derived from the number of sold handsets in Sweden multiplied with the average selling price we can relate these numbers makes it possible to obtain the figure for the total investments. The comparison demonstrates that the consumers' investments in terminals surpass CAPEX provided by operator with four times, see Fig. 4.31. Altogether, the data illustrates that the implementation of new standards is commonly a stretched out process impacted by the introduction of terminals, and determined by the end consumers' willingness to pay for new equipment.

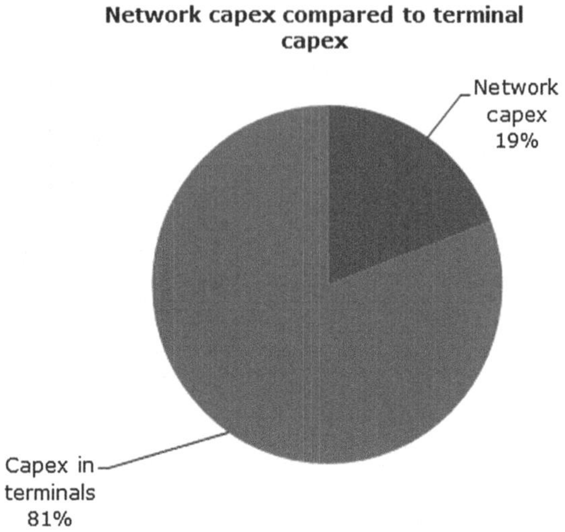

Fig. 4.31 Comparison between network CAPEX and consumers investments of terminals 2012. *Source* operator reports, authors calculations and MTB, the mobile telephone industry

4.6.4.2 Four Factors that Have Implications on CR

Although mobile communication is a technology driven industry investment decisions taken by operators is nowadays governed by financial targets and scrutinized by investment committees and top management in order to safeguard appropriate return on investments. This means that investments and thereby deployment of CR face a number of challenges of which we have identified four factors which we analyse in the following.

Factor 1—It takes time to establish new standards on the market

The introduction of mobile standards, such as GSM, WCDMA, LTE, takes time and the migration from older to newer standards is a stretched out process as the life cycle for older technologies often is prolonged and reach its peak after the new technology has been introduced, as elaborated in the previous section. Given that the mobile industry has matured and operators nowadays are managed with financial targets as a key priority it would be challenging to persuade management to invest and launch CR.

Factor 2—Multiple standards increase complexity

Operators in Europe are currently operating networks with at least three parallel standards—GSM, WCDMA and LTE—which are not optimal for an efficient operator. Although CR could contribute with additional capacity it would add more complexity rather than to streamline the current operation. It also requires a long-term commitment as history show that it takes long time to establish a new standard on the market. This implies that management has to see the merits in CR and be determined that it could contribute with something that the other standards are not able to, which is a challenging task.

Factor 3—Financial commitments calling for additional CAPEX

It requires a financial commitment for operators to deploy a new technology demanding extensive capital expenditures over a number of years. Investments into a new network are irreversible and thereby sunk which implies that management has to be convinced that investments in CR will pay off and deliver a return of investments in line with the financial targets. With declining revenues operators are scrutinizing their investments decision very carefully and make prioritizations meaning that investments in CR come on top of other investments.

The price deflation on network equipment, which has been driven by competition, technology advancement and economy of scale, facilitates for operators to acquire network equipment for the established standards for around EUR 10K per site. The cost for civil works and passive infrastructure makes up the larger part of CAPEX budgets. Given the uncertainty for the volumes of equipment for CR it would rather cost more compared to standardized equipment. This implies that CAPEX budgets has to be extended and accepted in investment committees, which could be challenging as management would rather see higher cash flow than to explore new technologies and increase CAPEX budgets.

Factor 4—Consumers' are the biggest investors

The operator business of today is characterised by standardised products, marketing of services and competition on attracting new customers. The basic principle is to have a large and growing customer base in order to generate a cash flow. The previous section has demonstrated that the end customers' investment in terminals represents the majority of the total investments for mobile systems.

This implies that operators have to persuade the end-customers to not only sign up as subscribers but also to pay for terminals. Although operators could provide various financing options for the acquisition of terminals the end customers has to be convinced by the merits of the offers. The global smartphone trend has demonstrated that economy of scale is essential as its offers customers' good value and enables them to reach internet and unlimited amount of applications while being on the move.

Altogether, this underscores that terminals play a decisive role in the mobile communications system and the availability of terminals that could handle CR is a prerequisite for establishing a business case for the new technology. But given that it will take time before economy of scale could be reached the case for CR could be difficult as end-customers have to be persuaded to pay substantially more for CR terminals compared to standardised smartphones, which are now falling in price.

4.6.4.3 Concluding with an Example

We illustrate the reasoning with a case where we have two operators, of which one is using licensed spectrum and the other is using CR. The case concerns a network for a country with 10 million inhabitants, a mobile penetration rate of 90 % and market share of 30 % for the operator, where the operator using exclusively allocated spectrum has paid the equivalent of EUR 0.50 per MHz/pop for 2 $\times$ 50 MHz, while

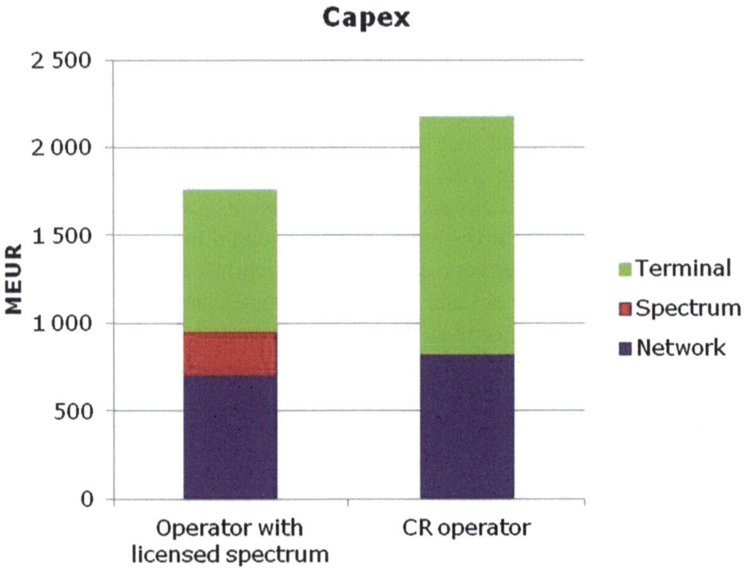

Fig. 4.32 CAPEX for standardized mobile network compared with an operator with CR

the operator with CR has no cost for spectrum. We calculate with 10K sites where half is green field sites and the other half is using existing sites.

Capex for the green field sites is EUR 100K, while the CAPEX for radio equipment is EUR 10K for the standardised radio and EUR 20K for CR. Moreover, we estimate the cost for smartphones to be in the mid of the range of EUR 200–400 while the range for CR capable smartphones is estimated to be EUR 400–800 for smartphones for CR. The aggregated CAPEX is EUR 1760 m for the operator with licensed spectrum and EUR 2175 m for the CR operator (Fig. 4.32).

Altogether, this reinforces the conclusion that the cost for spectrum make up the least part of the CAPEX budget while the cost for terminals is majority of the total investment. This underscores that the success of CR really has to contribute to the consumers benefits and they have to be persuaded that the standardised smartphone is not sufficient and that they rather should choose terminals that have a CR capability. This requires that the price points make sense for the end-consumers and that are persuaded to invest more than what they otherwise would have done. This will be very challenging as the end customers as well as operators are prioritising low cost and low risks.

4.6.5 Summary Assessment of Spectrum Access Options

When we summarize the business feasibility characteristics for CR solutions the result is not that encouraging. The cost of CR currently would be larger than similar commercial LTE or WiFi systems due to larger complexity. In addition the

Table 4.7 Summary of different spectrum access options, selection building on [89]

Aspect	Access using licensed bands	Unlicensed open access	Secondary spectrum access	Licensed shared access
Availability for use	Full	Good	Varying	Full
Radio complexity	LTE type	WiFi type	>LTE	=LTE
Radio cost	LTE type	WiFi Type	>LTE	=LTE
Availability of base station equipment	Standardized, available	Standardized, available	Unclear, low availability	Standardized, available
Availability of user devices/equipment	Standardized, available	Standardized, available	Not available	Standardized, availability
Risk for operators	Low	Medium	High	Quite low

low level of usage and availability of CR equipment contribute to higher risk for operators using this technology. Other solutions for use of non-licensed spectrum band, WiFi and Licensed Shared Access (LSA), make use of existing technology and hence mean lower risk for operators, see Table 4.7.

We can also see that that the value and the usefulness of CR solutions are different for different types of actors. Existing mobile operators (with licensed spectrum too) that use secondary access and CR have an advantage over new actors. First, exiting operators can re-use the existing base station sites whereas a new actor needs to deploy a new infrastructure. Second, a mobile operator using secondary spectrum access as a complementing resource, new actors using CR usually only the secondary access as the main spectrum resources and hence are more vulnerable.

Although a mobile operator may see potential savings in overall network costs a number of potential drawbacks can be identified. The mobile operator may hesitate to include a new type of technology in the networks. The option to use an existing standard (e.g. LTE) in another licensed band may see as a more straight-forward in order reduce the number of standards. In addition, new user terminals and devices with CR need to be developed marketed and adopted by consumers. Since CR equipment is more complex and produced in smaller quantities than existing radio technologies the products will be more expensive which would be a major obstacle.

References

1. Kibilda, J., Nolan, K., Tallon, J., DaSilva, L.A.: Whitespace networks relying on dynamic control channels. In: Proceedings of the 2013 8th International Conference on Cognitive Radio Oriented Wireless Networks (CROWNCOM) (2013)
2. Thanki, R.: The economic significance of licence-exempt spectrum to the future of the internet. Technical report (2012)
3. Cisco Visual Networking Index. Global mobile datatraffic forecast update, 2011–2016. http://www.cisco.com/en/US/solutions/collateral/ns341/ns525/ns537/ns705/ns827/white_paper_c11-520862.pdf (2011)

4. Mancuso, A., Probasco, S., Patil, B.: (draft) Protocol to access white space (PAWS) database: use cases and requirements, Internet Engineering Task Force (IETF) Std., Rev. 09. http://tools.ietf.org/html/draft-ietf-paws-problem-stmt-usecases-rqmts-09 (2012)
5. ECC REPORT 159. Technical and operational requirements for the possible operation of cognitive radio systems in the "white spaces" of the frequency band 470–790 MHz. Electronic Communications Committee (ECC), Technical report. http://www.erodocdb.dk/Docs/doc98/official/pdf/ECCREP159.PDF (2011)
6. Federal Communications Commission (FCC). Unlicensed operation in the TV broadcast bands, ET docket No. 04-186 and 02-380; FCC 10-174, Federal Register vol. 75, No. 233. Federal Communications Commission (FCC), Technical report (2010)
7. Haykin, S.: Cognitive radio: brain-empowered wireless communications. IEEE J. Sel. Areas Commun. **23**(2), 201–220 (2006). http://dx.doi.org/10.1109/JSAC.2004.839380
8. Bahl, P. Chandra, R. Moscibroda,T., Murty, R. Welsh, M.: White space networking with Wi-Fi like connectivity. ACM SIGCOMM Comput. Commun. Rev. **39**(4), 27–38 (2009). http://doi.acm.org/10.1145/1594977.1592573
9. Gur, G., Bayhan, S., Alagoz, F.: Cognitive femtocell networks: an overlay architecture for localized dynamic spectrum access [dynamic spectrum management]. IEEE Wirel. Commun. Mag. **17**(4), 62–70 (2010)
10. Sen, S., Zhang, T., Buddhikot, M.M., Banerjee, S., Samardzija, D., Walker, S.: Dual technology femto cell architecture for robust communication using whitespaces. In: IEEE International Symposium on Dynamic Spectrum Access Networks (DySPAN) (2012)
11. Murty, R., Chandra, R., Moscibroda, T., Bahl, P.: SenseLess: a database-driven white spaces network. IEEE Trans. Mob. Comput. **11**(2), 189–203 (2012)
12. Tallon, J., Forde, T., Doyle, L.: Dynamic spectrum access networks: independent coalition formation. IEEE Veh. Technol. Mag. **7**(2), 69–76 (2012)
13. Wyglinski, A.M., Nekovee, M., Hou, T.: Cognitive Radio Communications and Networks: Principles and Practice. Elsiever Inc. (2010)
14. Sukarevičienė, G., Fomin V.V.: Analysis and modelling of dynamic spectrum access information infractructure (DSA II). In: The Proceedings of the XVIII International Master and Ph.D. Students Conference on Information Society and University Studies (IVUS 2013). Kaunas, Lithuania (2013)
15. Sukarevičienė, G., Fomin, V.V.: Modeling dynamic spectrum access infrastructure, development as a techno-econimic system. In: The Proceedings of 18th EURAS Annual Standardization Conference, Brussels, Belgium (2013)
16. EBU Technology fact Sheet. White space devices. Operating Eurovision and Euroradio (2013)
17. Kokkinen, H.: Fairspectrum provides TV white space database for Europe's first geolocation radio license. Press release. Helsinki, Finland (2012)
18. Sukarevičienė, G.: Developing business model for GDB for the operation of cognitive radio in the TV white space bands. Master thesis. Kaunas, Lithuania (2012)
19. Lyytinen, K., Fomin, V.V.: Achieving high momentum in the evolution of wireless infrastructures: the battle over the 1G solutions. Telecommun. Policy **26**, 149–170 (2002)
20. Smura, T.: Techno-econimic modelling of wireless network and industry architectures. Doctoral dissertation for the degree of Doctor of Science in Technology. Espoo, Finland (2012)
21. ECC Report 185. Further definition of technical and operational requirements for the operation of white space devices in the band 470–790 MHz
22. Gianmarco, B., et al.: The evolution of cognitive radio technology in Europe: regulatory and standardization aspects. Telecommun. Policy **37**, 96–107 (2013)
23. Casey, T.R., Ali-Vehmas, T.: Value system evolution towards a cognitive radio era: implications of underlying market dynamics. In: Proceedings of the 2012 IEEE International Symposium on Dynamic Spectrum Access Networks (2012)
24. Mitola, J., Maguire, G.Q.: Cognitive radio: making software radios more personal. IEEE Pers. Commun. **6**(4), 13–18 (1999)

25. Chapin, J.M., Lehr, W.H.: Cognitive radios for dynamic spectrum access—the path to market success for dynamic spectrum access technology. IEEE Commun. Mag. **45**(5), 96–103 (2007)
26. Manninen, A.T.: Elaboration of NMT and GSM standards: from idea to market. Ph.D. thesis, University of Jyväskylä (2002)
27. Steinbock, D.: Globalization of wireless value system: from geographic to strategic advantages. Telecommun. Policy **27**(3–4), 207–235 (2003)
28. Hillebrand, F.: GSM and UMTS: the creation of global mobile communication. Wiley, New York (2002)
29. Lemstra, W., Hayes, V., Groenewegen, J.: The Innovation Journey of Wi-Fi: The Road to Global Success. Cambridge University Press, Cambridge (2010)
30. Negus, K.J., Petrick, A.: History of wireless local area networks (WLANs) in the unlicensed bands. J. Policy Regul. Strat. Telecommun. Inf. Media **11**(5), 36–56 (2009)
31. Smura, T., Sorri, A.: Future scenarios for local area access: industry structure and access fragmentation. In: Proceedings of the Eight International Conference on Mobile Business, pp. 57–63. Dalian, China (2009)
32. Leiner, B.M., Cerf, V.G., Clark, D.D., Kahn, R.E., Kleinrock, L., Lynch, D.C., Postel, J., Roberts, L.G., Wolf, S.: A brief history of the internet. ACM SIGCOMM Comput. Commun. Rev. **39**(5), 22–31 (2009)
33. Ballon, P., Delaere, S.: Flexible spectrum and future business models for the mobile industry. Telematics Inform. **26**(3), 249–258 (2009)
34. Casey, T.: Analysis of radio spectrum market evolution possibilities. Commun. Strat. **75**, 109–130 (2009)
35. Barrie, M., Delaere, S., Ballon P.: Classification of business scenarios for spectrum sensing. In: Proceedings of the 21st International Symposium on Personal Indoor and Mobile Radio Communications, pp. 2626–2631. Instanbul, Turkey (2010)
36. Medeisis, A., Delaere, S.: High-level scenarios for the future of cognitive radio business. In: Proceedings of the 22nd International Symposium on Personal Indoor and Mobile Radio Communications, pp. 2330–2334, Toronto, Canada (2011)
37. Anderson, P., Tushman, M.L.: Technological discontinuities and dominant designs: a cyclical model of technological change. Adm. Sci. Q. **35**(4), 604–633 (1990)
38. Arthur, W.B.: Positive feedbacks in the economy. Sci. Am. **262**, 92–99 (1990)
39. Fine, C.H.: Clockspeed-based strategies for supply chain design. Prod. Oper. Manage. **9**(3), 213–221 (2000)
40. Senge, P.: The Fifth Discipline: the Art and Practice of the Learning Organization. Doubleday, New York (1990)
41. Porter, M.E.: Competitive Advantage: Creating and Sustaining Superior Performance. Free Press, New York (1985)
42. Stabell, C.B., Fjeldstad, Ø.D.: Configuring value for competitive advantage: on chains, shops, and networks. Strateg. Manag. J. **19**(5), 413–437 (1998)
43. Jacobides, M.G., Knudsen, T., Augier, M.: Benefiting from innovation: value creation, value appropriation and the role of industry architectures. Res. Policy **35**(8), 1200–1221 (2006)
44. Iansiti, M., Levien, R.: Strategy as ecology. Harvard Bus. Rev. **82**(3), 68–81 (2004)
45. Eisenmann, T., Parker, G., Van Alstyne, M.W.: Strategies for two-sided markets. Harvard Bus. Rev. **84**(10), 92–101 (2006)
46. Gawer, A., Cusumano, M.A.: How companies become platform leaders. MIT Sloan Manage. Rev. **49**(2), 28–35 (2008)
47. Strogatz, S.H.: Exploring complex networks. Nature **410**(6825), 268–276 (2001)
48. Anderson, C.: The Long Tail: Why the Future of Business is Selling Less of More. Hyperion Books, New York (2008)
49. Arthur, W.B.: Complexity and the economy. Science **284**(5411), 107–109 (1999)
50. Barabási, A.L., Bonabeau, E.: Scale-free networks. Sci. Am. **288**(5), 50–59 (2003)
51. Faloutsos, M., Faloutsos, P., Faloutsos, C.: On power-law relationships of the Internet topology. ACM SIGCOMM Comput. Commun. Rev. **29**(4), 251–262 (1999)

52. Sterman, J.D.: Business Dynamics: Systems Thinking and Modeling for a Complex World. McGraw-Hill, New York (2000)
53. Rahmandad, H., Sterman, J.: Heterogeneity and network structure in the dynamics of diffusion: comparing agent-based and differential equation models. Manage. Sci. **54**(5), 998–1014 (2010)
54. Wolstenholme, E.: Using generic system archetypes to support thinking and modelling. System Dyn. Rev. **20**(4), 341–356 (2004)
55. Riikonen, A., Kitahara, T., Hämmäinen, H.: Comparison of mobile internet usage and industry structure: Japan vs. Finland. In: 18th Biennial Conference of the International Telecommunications Society. Tokyo, Japan (2010)
56. V. Sridhar., Casey, T., Hämmäinen, H.: Flexible spectrum management for mobile broadband services: how does it vary across advanced and emerging markets? Telecommun. Policy (2012, in press)
57. European Commission. Radio spectrum, a vital resource in a wireless world. http://ec.europa.eu/information_society/policy/ecomm/radio_spectrum/index_en.htm (2011)
58. RSPG10-306. Radio Spectrum Policy Group Report on Cognitive Technologies. EU, Brussels (2010)
59. Lehr, W. Chapin, J.: Role of cognitive radio in the evolution toward converged future networks. In: Second Joint IBBT-MIT Workshop on Cognitive Radio Standardization and Markets, Brussels, 11 Dec 2009
60. Freyens, B.P., Loney, M., Poole. M.: Wireless regulations and dynamic spectrum access in Australia. In: Paper presented at Dyspan (2010)
61. Gonçalves, V., Pollin, S.: The value of sensing for TV white spaces. In: 2011 IEEE Symposium on New Frontiers in Dynamic Spectrum Access Networks (DySPAN), pp. 231–241 (2011)
62. FCC. FCC adopts rules for unlicensed use of television white spaces. Washington D.C., 14 Nov 2008
63. Pawelczak, P., Venkatesha Prasad, R., Xia, L., Niemegeers, I.G.M.M.: Cognitive radio emergency networks-requirements and design. In: 2005 First IEEE International Symposium on New Frontiers in Dynamic Spectrum Access Networks, 2005. DySPAN 2005, pp. 601–606. IEEE (2005)
64. Bourse, D., Agusti, R., Ballon, P., Cordier, P., Delaere, S., Deschamps, B., Grandblaise, D., et al.: The E2R II flexible spectrum management (FSM) framework and cognitive pilot channel (CPC) concept—technical and business analysis and recommendations.In: E2R II White Paper (2007)
65. Anker, P.: Does cognitive radio need policy innovation? Competition Regul. Netw. Ind. **11**(1), 2–26 (2010)
66. Plum Consulting. Valuation of public mobile spectrum at 825–845 MHz and 870–890 MHz. In: A report to the department of broadband communications and the digital economy, 15 Sept 2011
67. Australian Communications and Media Authority (ACMA). Opportunity cost pricing of spectrum. Public consultation on administrative pricing for spectrum based on opportunity cost. http://www.acma.gov.au/webwr/_assets/main/lib310867/ifc12-09_final_opportunity_cost_pricing_of_spectrum.pdf (2009)
68. Doyle, C.: The pricing of radio spectrum, using incentive mechanisms to achieve efficiency, centre for management under regulation, Warwick Business School (2007)
69. Yeo, J.: Valuations in an FCC spectrum auction. University of Minnesota (2009)
70. Ard-Paru, N.: Spectrum assignment policy: towards an evaluation of spectrum, commons in Thailand. Department of Technology. Management and Economics, Division of Technology and Society, Chalmers University of Technology, Gothenburg, Sweden (2010)
71. International Telecommunication Union (ITU). Exploring the value and economic valuation of spectrum. Broadband Series (2012)
72. Buchanan, J.M.: Opportunity cost. In: Durlauf, S.N. Blume, L.E. (eds.) The New Palgrave Dictionary of Economics Online, 2nd edn. Palgrave Macmillan. 03 June 2009. doi:10.1057/

9780230226203.1222. http://www.dictionaryofeconomics.com/article?id=pde2008_O000029 (2008)
73. Zander, J.: On the cost structure of future wireless networks. In: Proceedings of IEEE VTC'97. Phoenix, AZ, 5–7 May 1997
74. Markendahl, J., Mölleryd, B.G., Mäkitalo, Ö., Werding, J.: Business innovation strategies to reduce the revenue gap for wireless broadband services. Commun. Strat. **75**, 35–56 (2009)
75. Markendahl, J.: Mobile network operators and cooperation—a tele-economic study of infrastructure sharing and mobile payments services. Ph.D. dissertation, Royal Institute of Technology, Stockholm. http://www.impgroup.org/dissertations.php (2011)
76. Johansson, K.: Cost effective deployment strategies for heterogeneous wireless networks. Ph.D. dissertation, Royal Institute of Technology, Stockholm (2007)
77. Marks P., Viehoff, I., Saadat, U., Webb, W.: Study into the use of spectrum pricing. Appendices 1 and 2: Case Studies, Prepared by NERA & Smith for the Radiocommunications Agency. http://www.ofcom.org.uk/static/archive/ra/topics/spectrumprice/documents/smith/smith1.htm (1996)
78. Sweet, R.I., Viehoff, D., Linardatosc, N., Kalouptsidis. C.: Marginal value-based pricing of additional spectrum assigned to cellular telephony operators. Inf. Econ. Policy **14**(3), 371–384 (2002)
79. Marks, P., Pearson, K., Williamson, B., Hansell, P., Burns, J.: Estimating the commercial trading value of spectrum. A Ofcom report by Plum Consulting. www.plumconsulting.co.uk/pdfs/Plum_July09_Commercial_trading_value_of_spectrum.pdf (2009)
80. Mölleryd, B.G., Markendahl, J., Mäkitalo, Ö.: Spectrum valuation derived from network deployment and strategic positioning with different levels of spectrum in 800 MHz. In: Proceedings of 8th Biennial and Silver Anniversary ITS Conference. Tokyo (2010)
81. Mölleryd, B.G., Markendahl, J.: Valuation of spectrum for mobile broadband services—engineering value versus willingness to pay. In: Presented at 22nd European Regional ITS Conference Budapest. http://www.econstor.eu/bitstream/10419/52146/1/672544792.pdf, 18–21 Sept 2011
82. Markendahl, J., Mölleryd B.G.: Mobile broadband expansion calls for more spectrum or base stations—analysis of the value of spectrum and the role of spectrum aggregation. Int. J. Manage. Netw. Econ. **2**(2), 115–134 (2011)
83. Mölleryd, B.G., Markendahl, J.: Valuation of spectrum for mobile broadband services—the case of India and Sweden. Telecommun. Policy (2014)
84. Kronander, J., Nekovee, M., Sung, K.W., Zander, J., Kim, S.-L., Achtzehn, A.: QUASAR scenarios for white space assessments and exploitation. In: URSI EMC Conference. Worclaw (2010)
85. Zander, J., et al.: On the scalability of cognitive radio: assessing the commercial viability of secondary spectrum access. IEEE Wirel. Commun. Mag. **20**(2), 28–35 (2013)
86. Quasar deliverable D5.1. Model integration and spectrum assessment methodology. http://www.quasarspectrum.eu/deliverables.html (2011)
87. Markendahl, J., Gonzalez-Sanchez, P., Mölleryd, B.G.: Impact of deployment costs and spectrum prices on the business viability of mobile broadband using TV white space. In: 7th International Conference on Cognitive Radio Oriented Wireless Networks (Crowncom) (2012)
88. Markendahl, J., Mäkitalo, Ö.: Analysis of business opportunities of secondary use of spectrum, the case of TV white space for mobile broadband access. In: 22nd European ITS Conference (2011)
89. Zander, J., Mähönen, P.: Riding the data tsunami in the cloud—myths and challenges in future wireless access. IEEE Commun. Mag. **51**(3), 145–151 (2013)
90. Sutton, P., Nolan, K., Doyle, L.: Cyclostationary signatures for rendezvous in OFDM-based dynamic spectrum access networks. In: 2nd IEEE International Symposium on New Frontiers in Dynamic Spectrum Access Networks (DySPAN 2007), pp. 220–231 (2007)
91. Sutton, P.D., Nolan, K.E., Doyle L.E.: Cyclostationary signatures in practical cognitive radio applications. IEEE J. Sel. Areas Commun. **26**(1), 13–24 (2008)

92. Sutton, P., Lotze, J., Lahlou, H., Fahmy, S., Nolan, K., Ozgul, B., Rondeau, T., Noguera, J., Doyle, L.: Iris: an architecture for cognitive radio networking testbeds. IEEE Commun. Mag. **48**(9), 114–122 (2010)
93. Gummadi, R., Wetherall, D., Greenstein, B., Seshan, S.: Understanding and mitigating the impact of RF interference on 802.11 networks. ACM SIGCOMM Comput. Commun. Rev. **37**(4), 385–396 (2007). http://doi.acm.org/10.1145/1282427.1282424
94. Reed, J.H., Bostian. C.W.: Understanding the issues in software defined radios. Virginia Tech (unpublished)

Chapter 5
Impact Assessment of CR Policy and Regulation

Leo Fulvio Minervini and Peter Anker

Abstract This chapter looks at emerging issues related to carrying out Impact assessment (IA) for identified combinations of techno-economic circumstances and conditions of CR deployment. The aim of such analysis would be to aid the policy discussion and development, by recognising the most attractive and beneficial combinations of regulatory provisions to form the basis for the ultimate CR regulatory framework. Section 5.1 provides an overview of IA and offers perspective on existing IA guidelines in the case of CR policy. Section 5.2 discusses the alignment of regulation and technology, applying an actor-centric approach. It highlights that successful introduction of CR will require alignment between the characteristics of CR and the regulatory regime under which CR will operate. Section 5.3 discusses role of spectrum regulation and argues that more relaxed spectrum regulations would trigger generation of well suited and flexible services, as they could reduce market entry barriers and allow more service providers to access the spectrum resources. Then, Sect. 5.4 describes a study on IA of Dynamic Spectrum Access (DSA). The introduction of DSA has been challenged by several technical, economic and regulatory factors. The authors develop a framework that combines system dynamics modelling (top-down approach) and Bayesian network data analysis (bottom-up approach) for analysing current mobile markets and their future evolutions possibilities. This is followed by Sect. 5.5 that looks at the matter of type conformity assessment for future CR/SDR apparatus, which would be an important consideration for placing equipment on the market. Then Sect. 5.6 analyses reasons of rather sluggish pace of CR innovation, with the aim of suggesting a range of suitable policies to boost further and more fertile developments of CR technology. The chapter is concluded by Sect. 5.7 that offers spectrum

L. F. Minervini (✉)
University of Macerata, Macerata, Italy
e-mail: leofulvio.minervini@unimc.it

P. Anker
Ministry of Economic Affairs and Delft University of Technology, Delft, The Netherlands
e-mail: P.D.C.Anker@minez.nl

A. Medeisis and O. Holland (eds.), *Cognitive Radio Policy and Regulation*, Signals and Communication Technology, DOI: 10.1007/978-3-319-04022-6_5,

policy analysis from both positive and normative perspectives. It proposes an "agreement framework", which could be used as reference template against which future policy analysis could be carried out in similar cases, with regard to emerging CR applications and CR technology in general.

5.1 Impact Assessment of CR/SDR Policy: Overview and Guidelines

Peter Anker[1] and Leo Fulvio Minervini[2]

[1]Ministry of Economic Affairs and Delft University of Technology, Delft, The Netherlands
[2]University of Macerata, Macerata, Italy

5.1.1 Impact Assessments: An Overview

Impact assessment (IA) is the formal analysis of the potential effects of new policies before their adoption. It is used as an aid to policy-making, since IA aims at forecasting the socio-economic impacts of policy proposals in terms of costs, benefits and risks. Therefore, in many countries IA provides a ground work for evidence-based policy-making. It also helps to communicate to stakeholders the evidence upon which legislative or regulatory changes are proposed. Indeed, there has been a wide diffusion of IA to increase the ability of governments to produce high quality regulation (or, as frequently referred to, better regulation). For instance, since 2003, the European Commission has employed an integrated IA system for estimating ex ante the impacts of its policy and regulatory proposals in economic, social and environmental terms [1]. Moreover, research has explored various areas where IA has become relevant and shed light on different rationales for the existence of IA, e.g., improvement of regulatory quality, improvement of market competition as well as of regulatory competition or cooperation, creation of conditions for policy learning, extension of the range of policy options, inclusion of stakeholders' opinions [2].

Notwithstanding increasing use of IA for policy making, IA is a relatively novel tool. Discussions around implementation and use of IA have stimulated an ongoing debate, that has brought to the fore a number of issues, including: different definitions of IA; various views on the relevant role of IA in the policy cycle; gaps between IA rationale(s) and practices; heterogeneous approaches in North America and Europe. Specifically, Torriti ([2], p. 243) noted that "North American authors support a prevalently economic rationale for decision-making, whereas their European colleagues seem more inclined to the view that the problems related with EU policies and regulations cannot be solved solely by adopting cost effective models" (see also [3, 4]).

The rest of this Section is organized as follows. The next subsection will provide a brief description of the European Commission's guidelines for IAs. Then, Sect. 5.1.3 will turn to the CEPT Electronic Communications Committee's guidelines for IAs regarding spectrum matters. Section 5.1.4 will reflect the outcome of discussion in COST-TERRA with regard to IA of CR/SDR regulation.

5.1.2 The European Impact Assessment Guidelines

In 2009, the European Commission adopted a new set of IA guidelines [5], which is offered for Commission staff preparing IAs. However, those guidelines present a number of answers to many relevant questions for any IA exercise. Therefore, this subsection will provide a brief summary of a few high-level questions addressed there. The interested reader is encouraged to refer to the full EC document.

The EC guidelines highlight that IA is "a process that prepares evidence for political decision-makers on the advantages and disadvantages of possible policy options by assessing their potential impacts" ([5], p. 4). In particular, IA is defined as a set of logical steps to be followed in the preparation of policy proposals. Six key analytical steps are identified and further considered:

(1) Identify the problem;
(2) Define the objectives;
(3) Develop main policy options;
(4) Analyze the impacts of the options;
(5) Compare the options;
(6) Outline policy monitoring and evaluation.

Those fundamental steps should consider a few issues which characterize any IA: the nature and scale of the problem at hand; its likely evolution; the stakeholders affected by it and their views; the objectives to be set to address the problem; the main policy options for reaching those objectives; the likely economic, social and environmental impacts of the identified options; the relative effectiveness, efficiency and coherence of different options in solving the problems; last not least, the organization of future monitoring and evaluation.

5.1.3 Impact Assessment for Spectrum Policy

The Electronic Communications Committee (ECC) released *Guidelines for the implementation of impact assessment in relation to spectrum matters* in 2008 [6]. The ECC guidelines took into account existing EC guidelines, and developed them specifically for spectrum policy making. The ECC proposed that IAs will normally involve seven stages:

(1) identify/describe the issue/problem(s);
(2) describe the policy/measure and identify the objectives;
(3) identify and describe the regulatory options;
(4) determine the impacts on all stakeholders, including relevant spectrum incumbents;
(5) determine the impact on competition—if relevant (cf. previous stage);
(6) assess the impacts and choose the best option;
(7) outline policy monitoring and evaluation.

The ECC guidelines examine each stage in detail, thus providing an operational guide for IAs.

The activities to be carried out at each stage are shaped according to a few principles, which include the following ones: (i) IAs provide a framework for weighing up the costs and benefits of the options; (ii) they aim to consider a wide range of options, including not regulating or status quo in regulatory measures; (iii) IAs should take into account the whole value chain and knock-on effects across the relevant spectrum users as well as other sectors, in order to minimize any unintended consequences of decisions; (iv) IAs should be guided by the principle of proportionality and aim to have a low level of uncertainty ([6], pp. 3–4).

The guidelines also address some misapprehensions about IAs. Here, it seems interesting to note, first, that IA is not concerned solely with commercial or monetary considerations, to the exclusion of social or public policy goods; and, second, that IA does not comprise quantitative cost-benefit analysis, to the exclusion of other analytical tools ([6], pp. 4–5).

5.1.4 Impact Assessment for CR Policy

IA was the focus of one of COST-TERRA Working Groups, namely WG4. WG4 worked on carrying out IA for identified combinations of techno-economic sets of CR/SDR deployment rules, with the aim of identifying the most attractive combinations to form the basis for the ultimate CR/SDR regulatory framework with any variations therein. Other COST-TERRA WGs and special interest groups (SIGs) also provided inputs to discussions and constructive feedback on WG4 activities. During COST-TERRA meetings, IA procedures for CR policy were discussed and there was a general agreement on the relevance of EU and ECC guidelines for CR policy. Therefore, COST-TERRA did not elaborate a set of IA guidelines to deal with CR policy.

Nevertheless, the discussion emphasized a few ingredients that IA for CR policy should use in producing effective IAs. In particular, the following ones were suggested:

(i) Definition of problem/issue(s) should consider whether markets or regulatory failures exist, and, if such failures exist, they should be brought to the fore and discussed in IAs;

(ii) A baseline scenario should be included for comparing policy options;
(iii) Qualitative and, as far as possible, quantitative analyses of alternative policy options (incl. sensitivity analyses and risk assessments) should be considered;
(iv) An appropriate time horizon should be selected for IAs;
(v) Objectives considered in IAs should be SMART, that is, Specific, Measurable, Achievable, Realistic, and Time-dependent.

5.2 Aligning the Regulatory Environment with the Technology, an Actor-Centric Approach

Peter Anker

Ministry of Economic Affairs and Delft University of Technology, Delft, The Netherlands

5.2.1 Introduction

Already nearly 15 years ago the concept of CR was proposed by Mitola and Maguire as a promising technology to deliver personalized services to the user through the most efficient radio resource available [7]. Since then the concept of CR has been further explored and the importance of CR for efficient use of the radio spectrum has gained momentum [8, 9]. Significant efforts are put in the development of various aspects of CR. Trials with the commercial use of CR are on-going but are mainly limited to the TV broadcasting bands. There is still no commercial use of CR.

One of the main reasons for the lack of practical and commercial use of CR technology is uncertainty about the regulatory model. Although there are possibilities to use CR under the current radio spectrum management regime, the current regulatory model is not conducive for dynamic access of spectrum made possible by cognitive technology. Regulatory provisions are needed to align the regulatory model with the new capabilities of CR technology of flexible and more efficient utilisation of the radio spectrum [10].

The dilemma that governments are facing since the liberalization is that prevailing policy suggests a technology neutral assignment of radio spectrum, while enabling the deployment of a specific technology, i.e., CR technology, is of public interest to achieve more efficient utilisation of the radio spectrum. It appears that in this light, regulation to allow the deployment of a specific type of CR technology in parts of the radio spectrum that would otherwise be underutilised or not used at all is justified [11].

As CR encompasses a very versatile set of technologies, the subsequent challenge governments are facing is the choice among some of the more fundamental features of CR technology, such as the technology used to make a CR aware of its radio environment and the band in which the CR is allowed to operate. Their choices will need to be well informed as their choices play a pivotal role in the business models of the entrepreneurs. The way governments allocate the use of radio spectrum to particular radio communication services on the (inter)national level and assign the rights to use the radio spectrum on the national level is determining the viability of the business case for particular radio communication products and services. In this respect there is the issue of 'the chicken and the egg': certain types of radio spectrum rights assignment facilitate certain types of usage, while certain types of perceived usage will require a particular type of assignment. In other words, entrepreneurs are reluctant to invest in new products and/or services based on CR technology because of the degree of regulatory uncertainty and regulators cannot provide this certainty because it is uncertain if their choices will support a viable business case.

This section proposes a way forward to deal with this dilemma by explaining the relationship between the fundamental choices regulators will have to make and possible business cases for the introduction of new products and/or services based on CR technology. The section starts with an introduction on the regulatory environment and the relationship between the regulatory regime and possible business cases. It will address the changes that will have to be made to allow CR technology and more dynamic forms of spectrum access and the relationship between the regulatory regime and possible business cases. It is followed by a description of the basic technological solutions that are possible for CR and the relationship between the CR capabilities and possible business cases. This exploration is used to assess the impact of a chosen regulatory environment and associated CR technology on possible business cases.

5.2.2 Regulating CR

CR is often associated to unlicensed secondary white space access to spectrum. However, there are more regulatory regimes possible under which the CR application can have access to spectrum [10, 12]. White space access means that CR applications are granted access to white spaces of spectrum as long as the conventional (primary) user is not using it. The white space users operate on a secondary level of usage of the spectrum. Therefore this type of sharing is also referred to as vertical sharing. This secondary usage does not necessary have to be unlicensed. It may be restricted to a closed user group. Restricting access to a closed user group can be arranged through licensing.

Secondly, CR technology can be used to pool spectrum between a number of users or user groups. Spectrum pooling is the situation in which a common "*pool of spectrum*" is shared among multiple users [13]. Access to the pool may be restricted to a (licensed) closed group of users or the pool might be open to all under certain use restrictions.

All users that share a pool of spectrum have the same rights to access the spectrum. Therefore this kind of sharing is also referred to as horizontal sharing.

Table 5.1 Four different regulatory scenarios for implementing DSA

	Horizontal sharing (spectrum pooling)	Vertical sharing (white space access)
Property rights regime (closed user groups)	Spectrum owners dynamically share spectrum	Owners of the spectrum grant specific CR's access to their white spaces
Open access regime (commons)	All users dynamically share spectrum on an equal footing	CR's dynamically access white spaces from incumbent users

This leads to four different scenarios for the implementation of DSA. The different scenarios are summarized in Table 5.1.

Apart from an overall need for more flexibility, the changes that are needed within the regulatory regime will be different within each regulatory regime under which dynamic spectrum access is realised. In the following two subsections these changes as needed to implement dynamic spectrum access are further explored for the different scenarios.

5.2.2.1 Dynamic Spectrum Access in an Open Access Regime

In an open access regime, any user can obtain access to spectrum under certain specified conditions. These conditions will have to be clearly defined to limit the interference level. In the vertical sharing regime, a commons is created by giving devices access to the unused parts of the spectrum of licensed users. This type of sharing is also referred to as Opportunistic Spectrum Access. In this case, the rules for spectrum access will have to guarantee that the interference to the primary user(s) of the band is kept below an acceptable level. The spectrum regulator will need to define an acceptable level to detect and protect incumbent users.

The definition of an appropriate level is not an easy task. If the level is too restrictive the potential gains of Opportunistic Spectrum Access are marginal, while a level that is too permissive may affect the Quality of Service of the primary user. The regulator will have to cooperate with industry to set a realistic level, which is based on the state of the art of technology. The level will have to be re-assessed if the primary user changes its technology. In the case of a true commons in which a frequency band is dynamically shared among all users, there is less need for involvement by the spectrum regulator. The main task of the regulator is in that case to designate a band for such purposes.

The regulator can also support OSA by providing information on the use of the band that will be dynamically shared between primary users and OSA devices.

5.2.2.2 Dynamic Spectrum Access in a Property Rights Regime

A property rights regime is based on the introduction of spectrum usage rights. These property rights go a step further than the licenses of today. They are used to create a market for spectrum in which these rights can be sold, leased and rented. The spectrum regulator will have to define these rights, with as few restrictions as possible. A number of countries have already introduced the possibility of secondary trading. However, in most cases there is an approval mechanism involving the authorities before trading may take place. This kind of barrier induces a delay before a trade can take place and thus makes real-time trading impossible. Hence, this barrier will have to be removed to exploit the full potential of dynamic spectrum access. Trading based on a much shorter time basis may make the market for spectrum more fluid. A central entity (a spectrum broker) could be used to facilitate this spot market.

A spectrum market can only function if information about the actual ownership of the spectrum property rights is readily available to facilitate trading. The regulator is ideally positioned to perform the task to keep a record of the ownership of these rights. Inclusion of monitoring information about actual usage of spectrum can further facilitate trading by giving more insights in the possibilities for trading and secondary usage.

5.2.2.3 Enforcement and Dispute Resolution

To successfully introduce dynamic spectrum access, there must be some assurance for the incumbent users of the spectrum that their usage will not be subject to (harmful) interference. This means that there is a need for a dispute resolution mechanism. To ease the settlement of disputes, it may be necessary to introduce a unique identifier for all CRs that is send alongside with the message with all radio transmissions. This will require that regulators are actively involved in the development and/or standardisation of CR technology [14].

A related point is that regulators will have to be very active in enforcement, especially in the start-up phase of the use of CR technology. This will provide the necessary confidence to existing users of the band that all efforts are taken to prevent CRs from inducing harmful interference and at the same time it will provide useful information to the industry to further develop their product [10].

5.2.2.4 Conclusions

The role of the regulator and the necessary conditions in the various regulatory regimes are outlined in Table 5.2.

5.2.3 The Impact of Regulations on the Business Case

The way in which the regulatory regime allows access to spectrum will greatly influence the business opportunities. This subsection gives an overview of the impact of the regulatory regime on the business opportunities.

Table 5.2 Necessary conditions for DSA in various regulatory regimes

Regulatory regime	Necessary conditions
Open access	Strict protection rules needed to keep the interference to the primary users at an acceptable level
	Rules to promote fair sharing of spectrum resources among OSA devices
	Availability of information on primary use to detect and protect incumbent users
Property rights	Well defined exclusive licenses granted to primary users or brokers
	As few usage restrictions as possible
	No barriers to instant trading
	Electronic information about ownership and actual usage should be available

5.2.3.1 White Space Access

In the white space access regimes the CR devices will always have to respect the needs of the primary user. White space access is only possible as long as there is no need for the spectrum by the primary user and no interference is created to the primary user.

This sets limitations to the business case for unlicensed white space access with an unrestricted number of devices. There will never be a guarantee that a CR device can have access to a white space and there is always the possibility that a CR device has to cease its operation because a primary user wants access to the spectrum. This makes this regulatory regime less suitable for time critical CR applications.

Restriction of access to white spaces to a specific user group provides the possibility for active coordination between the incumbent user and the secondary (cognitive) user about the likelihood of interference, and on guarantees about access to spectrum. Restricted access may also increase the level of trust for the incumbent user and may make them more willing to share their white spaces with a known and trusted CR user.

5.2.3.2 Spectrum Pooling

In case spectrum is pooled between a number of users or user groups, CR technology is used to dynamically share the spectrum resources. Pooling of spectrum in a closed user group between spectrum owners is only a viable option if the various owners are not in direct competition with each other. This is for instance the case if spectrum is used for company internal purposes, such as fixed links or private mobile radio. It is also possible to pool spectrum between various owners which have a completely different service, e.g., between a terrestrial service and a satellite service. Coordination between various owners will be easier if there already is a relationship whereby spectrum is shared at present. This will increase the level of trust and will make it easier to come to an agreement.

CR technology can also be used to pool spectrum between unlicensed applications. Knowledge of the radio environment is in this case used to realise a fair distribution of access to spectrum between the devices.

Table 5.3 Impact of the regulatory regimes for spectrum access on the business case for CR applications

	Horizontal sharing (spectrum pooling)	Vertical sharing (white space access)
Open access (commons)	Fair distribution of spectrum access	No guarantees for spectrum access, i.e., less suitable for time critical applications
Closed user group (licensed)	Increased level of trust	
	More certainty about access to spectrum	
	CR user groups not in direct competition with each other	Possibility for active coordination. More guarantees for spectrum access

5.2.3.3 Conclusions

The regulatory regime has a huge impact on the Business Case for CR. Each regulatory regime will facilitate a different kind of CR applications and/or service offerings. A mixture of these regimes will be necessary to unlock the full potential of CR technology in increased spectrum efficiency. The impact on the business case of the regulatory regime under which the CR application will operate is summarised in Table 5.3.

Especially the use of CR technology in a closed user group can help to bring this technology further for two reasons. First, restricting access to a controlled group may increase the level of trust between the users who share the spectrum. Second, restricted (licensed) access can provide certainty about access to spectrum over a longer period of time needed to recover the investments to be made in CR technology.

5.2.4 The Impact of CR Capabilities on the Business Case

The fundamental difference between a CR and a conventional radio is that a CR uses information of the radio environment to select and deploy the most appropriate communications profile, such as frequency band, access technique and modulation method. There are various techniques possible to obtain information about the radio environment.

The regulator will have to make fundamental choices about the radio environment in which the CR will operate and on the way in which the CR collects information of the radio environment. Each of them will have different implications for potential CR applications and the magnitude of the required investments.

5.2.4.1 The Radio Environment

CR technology is proposed to improve utilisation by using spectrum that is allocated but actually not used at a given time and location. The question is whether there is enough capacity in these unused spaces that can be made available to support the underlying business case for CR technology and if the business case is solid enough to recoup the necessary investments in this new technology.

The ease of making unused spectrum available for cognitive use depends on the characteristics of the incumbent user. It is easier to find a white space if conventional user(s) and usage is relatively static than when conventional users are mobile and/or their usage fluctuates.

Moreover, the fact that large parts of the spectrum are not utilised does not imply that an attractive business case for the remaining unused parts exists. The fact that in rural areas GSM spectrum is underutilised does not necessarily mean that there is a viable business case for these unused GSM frequency channels, at least not for mobile communications. The business case for the exploitation of these white spaces will have to be distinctively different from the business case of the conventional user.

5.2.4.2 Sensing

In its basic form a CR senses the radio environment to acquire information on the local usage. The CR device relies thereby on its own judgment of the local use of the spectrum to transmit over sections of the spectrum that are considered free. No matter how good the sensing technology is, a system that only relies on its own judgment to obtain information about spectrum usage might come in a situation where it inadvertently is not able to detect usage of a radio channel. This means that with a CR based on sensing alone, there is always the possibility of interference to the conventional users of the band. To limit this risk, restrictions on the output power of the CR devices will have to be set. As a consequence, the CR can only be used for applications which use low power in relation to the incumbent usage.

Sensing can be used without the need for coordination with the "outside world". Hence, sensing can be used for stand-alone applications, whereby there is no need for investments in the roll-out of associated infrastructure.

The probability of finding a white space that can be utilised depends on the activities of the incumbent user(s), the range of frequencies which is sensed and the number of active white space devices. Sensing will have to take place over a sufficiently large frequency range to support the capacity needed by the CR application. Sensing becomes more challenging, and more expensive, when a wider range of frequencies and/or a wider range of conventional user applications are to be taken into account. At the current state of technology and field experience on sensing, a case-by-case approach will be required which takes into consideration the existing spectrum usage. Hence, for new CR regulations to be meaningfully applied, i.e., before making available a band for white space devices, an assessment should be made of the amount of white spaces that can be made available against the capacity needed for the introduction of the application that uses these white spaces.

Sensing can be made more reliable by cooperation between the sensing devices [15]. Cooperation can improve the probability of detection and reduces the detection time and thus increase the overall agility of the system. Drawbacks are the need for a common signalling channel between the devices and the additional overhead needed to exchange sensing information over this channel.

Especially the need for a signalling channel makes this coordinated approach complex. The cognitive devices become part of a network. This makes this coordinated approach especially feasible in applications where the CR device is already part of a (local) network, e.g., in-house networks. Coordination is a less attractive option for stand-alone CR applications.

5.2.4.3 Database

A second option is to get information about the local use of the spectrum from a database. Such a database should contain the relevant information on the frequencies that can be used at a certain location as well as the applicable restrictions. The database will have to be kept up-to-date, which makes this option especially suitable in cases where spectrum usage of the conventional user(s) does not change frequently, e.g., in a broadcasting band or a band for fixed satellite communications.

The restrictions for the CR application imposed by the use of a database are twofold. First of all, the CR device needs to be aware of its geographical location. This information can be programmed in the device during the installation of the CR device for fixed applications. Mobile CR devices will need a means to acquire that information, for instance by incorporating radio navigation in the terminal. However, the use of radio navigation will be difficult for indoor applications.

Secondly, the CR device will need to have access to this database on a regular basis. Access to the database is easier to arrange if the CR device is already part of a network than for stand-alone CR applications. The rate at which the CR devices have to obtain updated information on the local radio environment depends on the rate at which the information on the incumbent user may change and on the degree of mobility of the CR device.

5.2.4.4 Cognitive Pilot Channel

Coordination between CR devices can be realised through a so-called Cognitive Pilot Channel (CPC). A CPC is a dedicated carrier providing information about the availability of spectrum and possibly usage restrictions to the CR devices in a certain area. The CPC can be used to (1) give general—local—information on the availability of white spaces in relation to the service to be protected, or (2) to coordinate the use of the spectrum resources by the CR devices competing for spectrum access or (3) a combination of both [16].

The first option requires that the CPC broadcasts information on channels that are available and possibly the associated use restrictions, unless these restrictions are already known beforehand by the CR device. The second option is more complex because there is also a need for the network to know which channels are actually used by the CR devices and therefore there is a need for a feedback channel.

Implementation of a CPC will require a radio-infrastructure to support the CPC. The CPC can be provided by a dedicated, autonomous network, but this will require substantial investments. The necessary investments can be lowered if the CPC uses a logical channel within an existing network, e.g., within a mobile network. Standardisation activities in this field are on-going (see Sect. 1.3).

Because a CPC can provide real-time information, a CPC is highly suitable in cases where spectrum usage of the user(s) with which the band has to be shared is more dynamic. In this case, the network will need to have up-to-date information of the spectrum usage of all user(s) at all times.

5.2.4.5 Conclusions

The means a CR uses to acquire information on the radio environment has a significant impact on the business case for CR applications. An outline of the main conclusions of the impact of the CR technology on the CR applications, and thereby on restrictions for a viable business case, are given in Table 5.4.

An apparent difference between sensing on the one hand and a database or Cognitive Pilot Channel on the other hand, is that the latter two will require investments in infrastructure. This means that sensing can be used for stand-alone applications, whilst the other options are better suited for the delivery of services with an associated infrastructure roll-out, i.e., sensing can be used in a business case based on the sales of equipment whereas the database and CPC are better suited for a service provider driven business case based on the sales of a service. In that case there will be a direct relationship between the service provider and the customer. This relationship is necessary to recoup the investments in infrastructure.

Of course, it is always possible to use a combination of techniques. Especially a combination of database access and sensing seems promising. The database can be used to protect existing services with which the band is shared and sensing can be used to assess whether the opportunity is really available or already in use by another CR device.

Another possibility is the use of a local CPC (or so-called beacon) to reduce some of the drawbacks of sensing, especially the complexity and associated costs of sensing devices. A relatively complex master device can be used to process the sensing results of a range of locally connected devices. The master device decides based on this information on what channel the connected devices may operate and sends this information to these devices over a local beacon. This solution can only be used if these devices form a local network. The relatively expensive master device acts as an intelligent central node for the relatively cheap connected devices.

Table 5.4 The impact of the CR technology on the business case for CR applications

	Implication to potential CR applications	Remarks
Sensing	Low power in relation to the primary user Sensing over a relatively small band sets limits to the data transfer capacity available Wide band sensing increases the capacity available, but is more complex and expensive Can be used for stand-alone applications	There remains a potential for interference to the conventional user
Database	Can be used for applications which need a higher power CR device needs to be aware of its location Application needs a connection to the database on a regular basis	Only useful in bands with relatively static conventional users Costs of database service will have to be recovered
Cognitive pilot channel	Can be used for applications which need a higher power CR device is part of a network	Can also be used for more dynamic conventional use Large scale deployment more expensive than a database

5.2.5 *Aligning the Regulation and Technology, an Actor-Centric Approach*

Although there are possibilities to use CR under the current radio spectrum management regime, regulatory provisions are needed to align the regulatory model with the new capabilities of CR technology of flexible and more efficient utilization of the radio spectrum [10].[1]

In this subsection it is proposed to use an actor-centric approach to deal with this issue of alignment. After all, CR is a technology to share spectrum among various users. The various users of the spectrum, the industry that has to develop the equipment and the government that has to provide the necessary regulations will have to coordinate to come to a successful exploitation of CR. The actors involved in this coordination will all have their own objectives and incentives.

This subsection will offer an explanation of methodology to analyze the alignment between a new technology and the regulatory environment within which it will be introduced. Evidence for the relevance of this approach may be found in the historic discourse provided in Sect. 1.1 of the first chapter, which analysed the coordination of radio spectrum use in the past and the development of radio spectrum regulations resulting from those coordination efforts. This approach will be then applied in Chap. 7 to carry out the analysis of the so far best known intended use of CR technology: white space access in the TV bands and, based on that analysis, proposing some forward looking recommendations.

[1] This subsection also reflects on [12, 13].

5.2.5.1 Two Levels of Alignment

Various contributions have been made on the need to adapt the regulatory framework to the new capabilities of CR [10, 17]. While alignment between new technologies, such as CR, and the associated regulations is an important prerequisite, it is not enough to assure a successful introduction of this new technology. There are numerous examples on the introduction of new technologies were the necessary alignment between the technology and the regulations was in place but the market for the provisioning of products and services based on this new technology did not mature.

Our analysis of the underlying causes is that firms will only decide to invest in new products and/or services if they can expect a future return. These investment decisions are driven by three major considerations: (1) the prospective demand and willingness to pay for new products and/or services; (2) the magnitude of the investments required; and (3) the degree of risk or uncertainty involved.

The profile of the business case, in terms of depth of investment and the recovery period required, will influence the ability to obtain the necessary (external) funding. As such the business case is especially challenging for service provisioning that requires a huge, upfront investment, e.g., an infrastructure roll-out to provide mobile telephony. In these cases the right to exploit the radio spectrum or any other infrastructure over a significant period of time and on an exclusive basis will contribute to the willingness of firms to invest as it reduces the uncertainty, which may make the business case more viable [11].

Although the regulator can't do much about technological and market uncertainties as such, the regulator plays a crucial role. The regulator should create a regulatory environment in which these uncertainties are lowered to an acceptable level for commercial applications to emerge. This environment should, among other things give clear directions on the expectations of CR technology [18].

However, in setting up institutional arrangements, governments will steer technology and possible business cases in a certain direction. It was shown [19] that the specificities of the entry and authority rules will favor certain types of usage over other types of use.[2] This is also true the other way around; certain types of perceived usage will require particular entry and authority rules.

Hence, decisions made by governments on the market design and associated regulations will have an influence on the viability of possible business cases. For example, decisions made in spectrum policy on the amount of spectrum allocated, whether the spectrum is made available on a license exempt basis or not, the number of licenses issued, the roll-out and other obligations attached to the licenses and the award mechanism for the licenses (e.g., an auction or a beauty contest) will all influence the required investments and the possibilities to exploit a certain business case. This is quite well demonstrated by mobile communications

---

[2] Ostrom made this observation in the investigation of common pool resources. The problems associated to infrastructures are quite similar [16]. The latter source argues that infrastructures

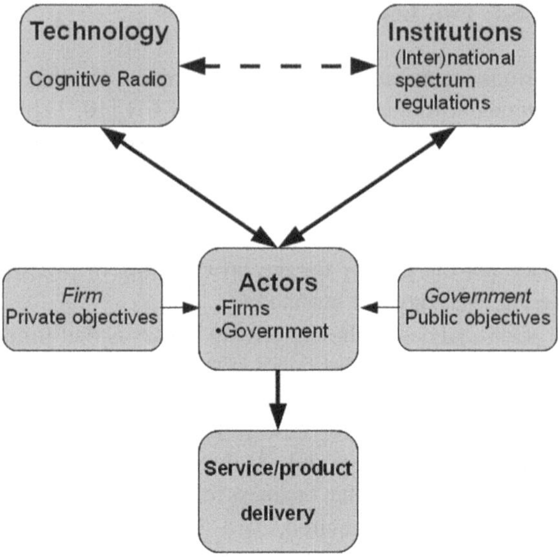

Fig. 5.1 Two levels of alignment

(GSM) which could flourish under a strict licensing regime and Wi-Fi that could develop under a license exempt regime.

Governments will need to be very well informed to make the right decision in order to let the intended business case flourish. Lessons learned from the past seem to suggest that a too "pushy" approach from governments may be counterproductive and retard or stall technological development [20]. Governments will need to take decisions that are not only in line with their own goal(s), but also make it possible for entrepreneurs to realize their goals. After all, it is through the actions of the firms, individually and collectively, that the governmental goals will be realized. This is illustrated in Fig. 5.1.

The government and the entrepreneurial firm have different objectives. In a somewhat simplistic view of the world, since the liberalization governments have, above all, an objective of economic efficient use of spectrum.

This is accompanied by societal objectives, such as universal service delivery, and in some cases also by industry policy. Governments rely on a market design and associated regulations to serve this mixture of economic and societal objectives. In the case of mobile communications, radio spectrum policy is used to create a market for mobile telephony. Specific auction rules may be used to allow new entrants and to influence the number of players on the market. Specific obligations are attached to the licenses to serve societal objectives, e.g., a coverage obligation.

(Footnote 2 continued)
(including energy, communication, transport, and postal services) can be perceived as common pool resources providing essential services to society.

Firms, on the other hand, have a completely different objective. They want to invest in (new) technology to develop products and services with the aim to maximize profit. The government and the firm are highly interdependent in the realization of their objectives. The institutional arrangements that are set up will have to provide certainty to entrepreneurial firms to invest in new technology and the exploitation thereof. If, as a result of profit maximization considerations, firms decide not to use the system as intended, the government fails in realizing its governance objectives.

Use of the new technology in such a way that both the government and the entrepreneurs can realize their goals is what we call a "sweet spot". A "sweet spot" is only possible if the use of certain technology and the associated institutional arrangements are aligned in such a way that both the intended business opportunity and the public objectives can be realized. The finding of "sweet spot" for CR technologies shall be further discussed in Chap. 7 of this book.

5.3 Inter-Operator Spectrum Sharing: From Techno-Economic Enablers to Real Market Show Stoppers

Michal Szydelko

EIT Plus Co.Wrocław, Poland

During study on the inter-operator spectrum sharing scenarios in HSPA networks [21], key modifications to the market and business models were identified, which were seen required by the spectrum sharing scenario implementation in future networks [22]. Due to the fact, that the spectrum usage in the current markets is subject to various constraints coming from regulatory, standardisation bodies, as well as hardware limitations point of view, various market players and their positions within described spectrum sharing scenarios were considered in order provide full picture of this complex environment. Spectrum resources were considered as goods being required for the mobile network subscriber value delivery. Therefore, spectrum valuation aspects as well as the end user requirements and expectations were incorporated in the discussion.

Aim of the study was to evaluate, whether the spectrum sharing extended by appropriate novel resource allocation techniques, might be able to generate new services for mobile subscribers, at the same time enabling new revenue streams for mobile network operators. Considering various aspects of the end user value perception, possibility of cost and complexity of the services delivery has been also addressed. Described inter-operator spectrum sharing was found to be an opportunity being able to open new markets and generate new mobile services. Based on system level simulation results [21] and using game theory to model cooperation among spectrum users, it was observed that there is a sharing gain achieved in terms of the total sum rate of the cell goodputs of both Mobile

Network Operators (MNO) participating in the game. Nevertheless, in the referred solution, which was based on the spectrum valuations being modelled by the cell specific buffer states, it was further observed that a load imbalance between sharing participants is the key aspect of the analysis. When the network load imbalance was high enough, the less loaded network was not able to gain from the participation in the sharing. This meant that any risk-free network operator might not be attracted by such cooperation mechanism. For that reason, it was found that appropriate motivation for potential players has to be formulated in order to stimulate their willingness to participate in such sharing mechanism. In order to find solution for the identified concern, we can think of the mechanism, where the sharing MNO is modifying his utility function in order to improve the sharing outcome (e.g., from proportional fair to throughput maximisation).

Based on the telecom market analysis, it was not difficult to notice, that fruitful deployment of the cellular network depends on the availability of the spectrum resources, which are scarce resource and can be allocated only by the national regulatory agency. Due to limited availability of this medium, in most cases the allocation is based on the long-term auctioning process, which aims to maximise the revenue from the spectrum and allocate it to the network operator, who values particular spectrum band the most. For all the reasoning mentioned earlier, the auctioned spectrum bands in many markets have generated bids, which were much higher than expected by national regulators. Simple conclusion is that in order to be able to provide mobile services, one has to consider high investment to obtain the spectrum resources.

It comes at no surprise that after successful acquisition of spectrum band, every mobile network operator tries to cover the market as wide as possible (in terms of the amount of the subscribers) in order to compensate expenses from the spectrum resources acquisition, by possibly largest revenue flow. What it means, is that MNOs are not focusing on too granular user definitions and are targeting their offer at high population of users. In such case, it is not very likely that the offer will cover very specific end-user requirements.

From the economical point of view, the most optimal offer creation process shall be constructed from as little building blocks as possible. At the same time, this process should allow to cover the market as wide as possible. In other words, for the Operational Expenses (OPEX) reduction, company would be interested in maintaining as little product lines as possible and at the same time, for the revenue maximisation purposes, the company's goal would be to generate as many product variants as possible, in order to satisfy possibly widest range of customers.

In relation to the cellular networks and mobile services, this can be translated into an offer, which is constructed from limited number of basic services (limitation of costs of services provisioning and maintenance), being able to attract certain population of subscribers. Due to the granularity of the service offer, it is likely to happen, that it will not be possible to offer sufficiently large number of various subscription plans and respective mobile services to certain, well defined group of end users having specific demands, or not willing to pay for the subscription which is not suited for their needs and expectations.

Trying to analyse this problem from the standpoint of the new market entrant, it is felt that this situation might give the opportunity for new mobile services creation, but cannot attract the current MNOs due to relatively low (in reference to their expectations) revenues forecast. What is the most important observation, is that in contrary to the MNOs, the presented case might be highly attractive for new market players (e.g., virtual operators, mobile service providers, etc.), who's cost structure is much less complicated and which is not being affected by high investments in the spectrum and infrastructure. The enabler for this to happen is the modification of the spectrum access regulations for bands that are not yet allocated, as well as for the re-use of spectrum resources, which is already in the possession of the MNOs (e.g., short-term auctioning, leasing, etc.).

Looking back at the referred simulation results [21], the observation on the sharing gains as a function of the load imbalance has to be highlighted, i.e., the higher packet load imbalance was considered between operators, and the higher sum rate throughput gains were observed. That brought us to the conclusion, that the most optimal spectrum utilisation (irrespective of the radio access technology consideration) shall be met in case of services generating highly uncorrelated (in terms of the generated data traffic) packet load over particular areas in case of shared spectrum or full network sharing.

Thinking of the relaxed spectrum regulations, which would allow dynamic (i.e., variable in time) and flexible (i.e., variable amount) spectrum allocation grants, well suited and flexible services would emerge on the market being provided by new market entrants, as there would be much lower entry barrier, i.e., no need for long-term spectrum band acquisition. For the current market players owning already acquired spectrum bands, it open new revenue generation possibilities, at the same time, allowing them to decide on the competitiveness of the emerging offer being delivered via their radio access network. The proposed model differs from the Mobile Virtual Network Operator (MVNO) business case, which allows new market players not owning spectrum bands, but does not allow spectrum allocation flexibility and its dynamics—only medium to long-term contracts are considered.

From the current market stage, novel spectrum sharing techniques are expected to arise on the market, gaining from the opportunistic spectrum availability. Frequency, time and space specific network capabilities boosts coming from spectrum sharing will materialise only in case of the matching UE needs, as well as available scheduling grants. From the business point of view, authors are of the opinion that the current market of mobile services is too generous. With the users expecting flat rate subscription, network operator agrees to offer cell's peak rate (of course with no guarantees). Spectrum sharing will make the cell's peak rate even higher, but it is not guaranteed that the revenue will increase for the operator(s).

The missing element is the definition of the Service Level Agreement (SLA) being signed between market players who are providing radio access and the mobile services. SLA specifies the radio access bearer as well as the services, including definition of the bearer throughput, its availability and quality of service, which should be specified independent from the network provider and RAT being used.

Moreover, SLA might cover the traffic volume being guaranteed for the service provider, possibly being time and geographical location specific. Depending on the volume of the SLA and its parameters, the consideration of the service provider requirements in certain radio access network might be high, requiring appropriate network planning and dimensioning actions, or capacity extensions.

In conclusion, it is suggested that more relaxed spectrum regulations, which would allow more service providers to access the spectrum resources, would trigger generation of well suited and flexible services to emerge on the market, as the consequence of much lower entry barrier.

5.4 Introduction of DSA: The Role of Industry Openness and Spectrum Policy

Arturo Basaure[1] and Varadharajan Sridhar[2]

[1]Aalto University, Espoo, Finland
[2]Sasken Communication Technologies, Bangalore, India

5.4.1 Introduction

Dynamic Spectrum Access (DSA) aims to improve the spectrum efficiency by accessing dynamically the spectrum resources. Despite large efforts in R&D, these technologies have not been successfully introduced into the mobile market. Several technical, economic and regulatory challenges have been identified for this slow deployment. In practice, a dynamic management of the spectrum involves most of the telecommunication layers and players; and not only end user terminals as Mitola originally suggested. This involves a wide restructuring of the mobile industry. In addition, under an unclear evolution, several standards for DSA are currently under development, such as those related to IEEE and ETSI organizations.

This section aims to analyze how industry openness and spectrum policy affect the introduction of such technologies taking while taking into consideration their main challenges. Industry openness refers to entry and exit barriers of the industry, and spectrum policy refers to the number of spectrum holders and the type of licensing. We consider the deployment of DSA technologies in two opposing scenarios: end-user centric devices and mobile operator centric devices.

5.4.2 Methodology

This section combines top-down and bottom-up approaches. The bottom-up approach is used for analyzing the current data on market structure through Bayesian network analysis. The top-down approach analyzes the future impact of DSA technologies on different markets through System Dynamic modeling.

System Dynamics analyzes different organizational systems as a whole with the objective to understand their dynamic behavior, and the relations between different factors. We support the main assumptions of this model with country data analysis, using a Bayesian network. While System Dynamic modeling describes the relation of different variables within time, our Bayesian network describes the conditional probabilities between variables in one point of time. Thus, in this section we use a Bayesian analysis as an input for modeling the dynamic behavior of the whole system characterized by feedback loops between variables within time.

5.4.3 *Bayesian Analysis of Mobile Market Structure*

This analysis utilizes a diverse collection of variables to define the level of openness of the mobile industry. Variables that can potentially describe the level of investment and return of the industry that act as entry barriers are: mobile average revenue per user (ARPU), cellular investment per capita, investment as percentage of revenue (investment/ARPU). Variables that describe entry barriers related to customers are: churn rate (monthly %), mobile price (average price of one minute in USD), mobile penetration (%), prepaid ratio (% from total subscription). Other variables that explain entry barriers related to operators are: termination rate (in USD) and network operator—service operator separation (yes or no).

Regarding spectrum policy, we utilize variables that are related with the spectrum licensing and policies. The most relevant variables describing spectrum policy are: technology neutrality policy (yes or no, according to law), technology neutrality of the market (yes or no, according to the market), spectrum concentration index (HHI[3]), market share concentration index (HHI), spectrum reselling rights (yes or no, the possibility of trading spectrum) and the number of mobile network operators (MNOs). Since spectrum is a heterogeneous resource, we use another variable to analyze the impact of the concentration in lower frequency bands (below 2.0 GHz); which are usually considered as more valuable for mobile operators.

A Bayesian network[4] describes the conditional dependencies between variables and therefore it shows which variables are more adequate to describe the industry openness and the spectrum policy. Pearson correlation detailed graphically by the color and numbers in the arcs (Fig. 5.2b). Blue indicates a positive correlation and red indicates a negative correlation. Arrows depict the dependence between variables. Figure 5.2b describes the causal relation between variables related to the spectrum policy from 24

[3] HHI stands for Herfindahl—Hirschman Index.

[4] The utilized Bayesian network was implemented with the help of BayesiaLab 5.1 software, which includes machine learning algorithm functionality.

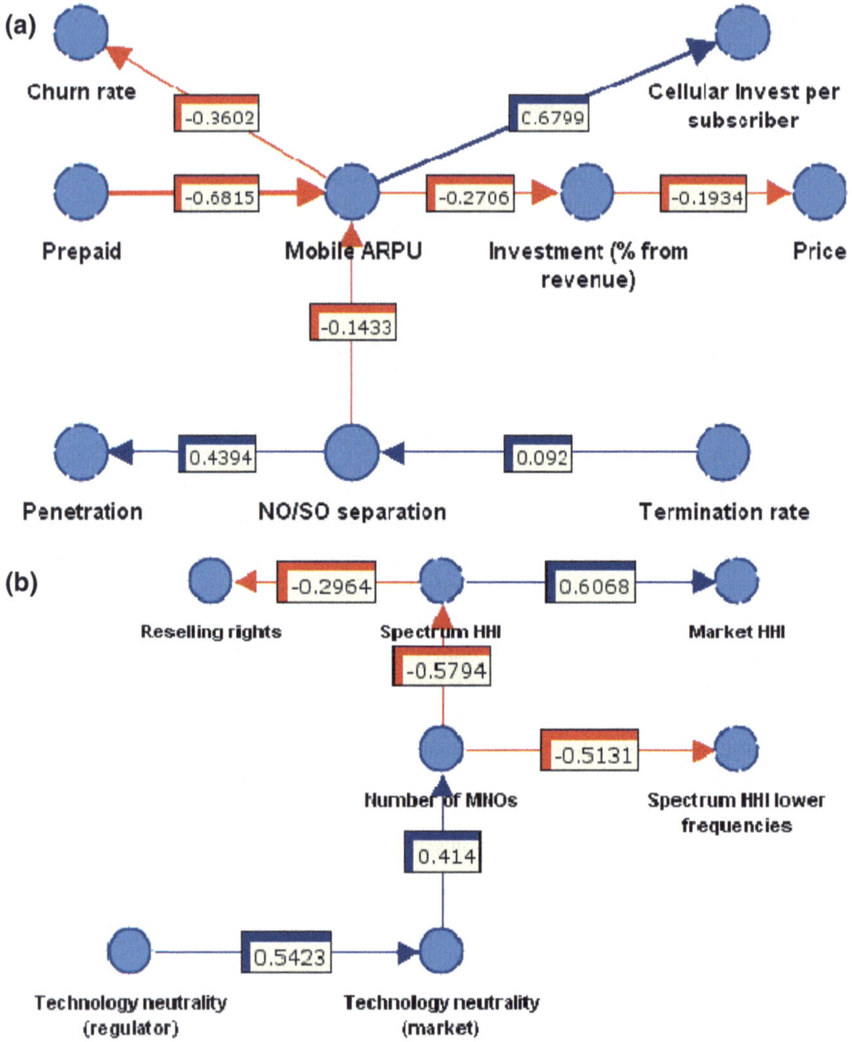

Fig. 5.2 Industry openness (**a**) and spectrum policy (**b**) variables and their causalities explained by a Bayesian network. Numbers in arcs indicate Pearson correlation

selected countries[5] [23–25] and Fig. 5.2a explains the same relation for the industry openness from 37 selected countries.[6] These models use a confidence level of 95 %.

[5] Australia, Austria, Belgium, Brazil, Canada, China, Chile, Czech Republic, Denmark, Finland, France, Germany, India, Italy, Japan, Netherlands, New Zealand, Norway, Spain, Sweden, Switzerland, Turkey, United Kingdom and United States.

[6] Australia, Austria, Belgium, Brazil, Canada, Chile, China, Czech Republic, Denmark, Estonia, Finland, France, Germany, Greece, Hungary, Iceland, India, Ireland, Israel, Italy, Japan, Korea

From Fig. 5.2b, we can conclude that spectrum policy is best described by the market share HHI rather than the initial spectrum allocation or other regulatory decisions (spectrum decentralization index $\approx$ market HHI). From Fig. 5.2b we can conclude that the level of industry openness can be best described by the price level and the cellular investment per capita, which best explain the level of entry and exit barriers (industry openness $\approx$ 1/(price index* cellular investment per capita).

5.4.4 *System Approach for Understanding the Introduction of DSA Technologies*

In the following subsection we use a system dynamics to analyze the introduction of DSA technologies, using two approaches: competition with high network externalities and competition with predator–prey substitution.

The network externality competition model [26] describes a struggle between two homogeneous networks to dominate the market without space for competitors. In this case, the network effect is higher when the success of competing technologies depends on compatibility issues, such as the spectrum availability for sharing, compatible services and service providers, critical mass of terminals in the market, etc. The predator–prey substitution model describes competition based on the substitution effect using the Lotka-Volterra predator–prey equations [27]. This model describes a substitution model, in which a new technology competes against an older technology of a saturated market.

We base our assumptions on the previous work. Regarding spectrum policy, it is suggested that a centralized allocation of spectrum does not incentivize transactions in a secondary market since it does not provide room for improvements in the original allocation, while a market driven decentralized allocation incentivizes further improvements through a secondary spectrum market [28]. Regarding industry openness, it is observed that the unbundling favored by regulators to incentivize competition can have a negative impact on the investments due to an increase in the intensity of price competition [29]. Thus industry openness can be described by price to quality ratio.

Figure 5.3a depicts a system dynamic diagram for a network effect competition model. This model assumes that when an industry is open, low prices disincentivise the investment for centralized technologies favoring the usage for new technologies, such as end-user centric devices [30]. In addition, end user centric devices need a decentralized spectrum policy and a high level of standard cooperation. Thus, the inability to agree on standards or the existence of closed systems has a negative impact for reaching critical mass [31].

(Footnote 6 continued)
(South), Luxembourg, Mexico, Netherlands, New Zealand, Norway, Poland, Portugal, Slovak Republic, Slovenia, Spain, Sweden, Switzerland, Turkey, United Kingdom and United States.

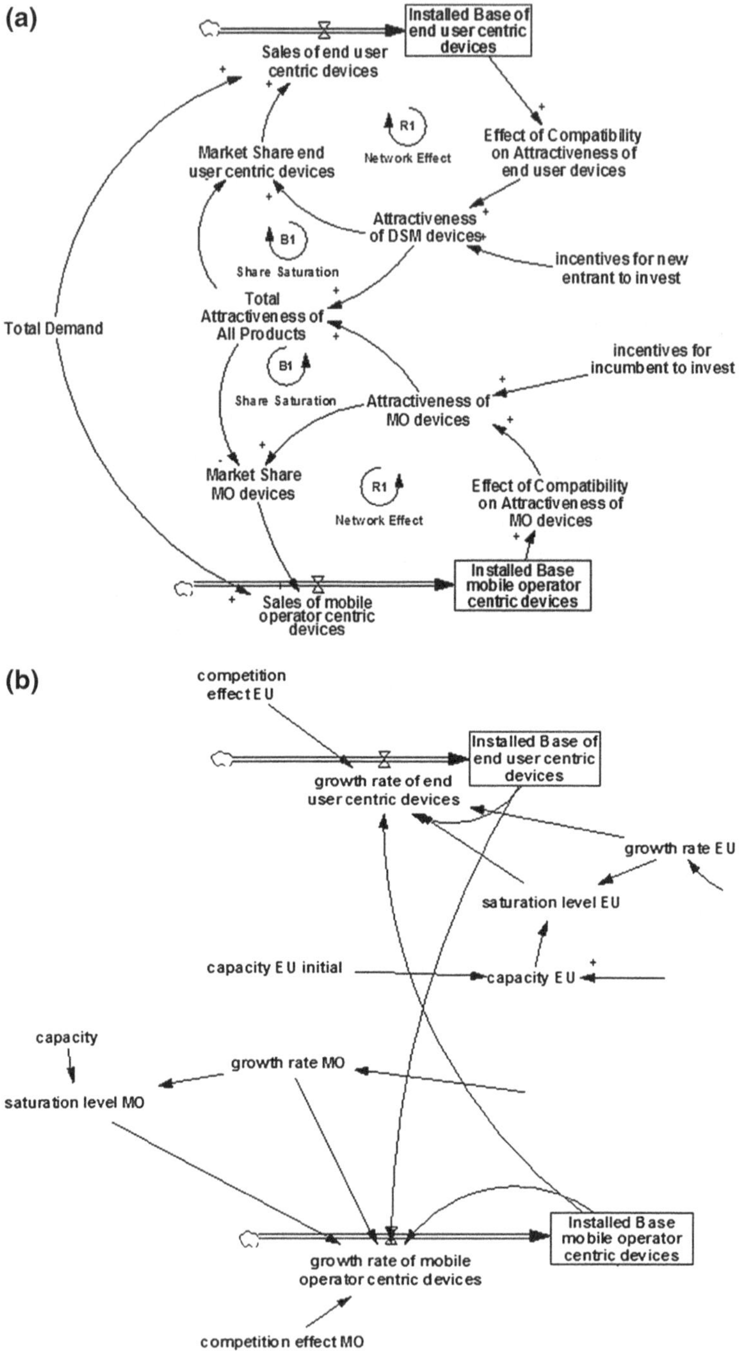

Fig. 5.3 System approach to describe the introduction of DSA technologies considering the network effect (**a**) and the predator–prey competition model (**b**)

The predator–prey competition model (Fig. 5.3b) considers the end user centric devices as predator and mobile operator centric devices as prey. It is suggested that the rate of technology adoption is directly proportional to the expected profitability and a decreasing function of the size of the investment [32].

From this perspective, a high level of standard cooperation increases the expected profitability and decrease the level of investments. Therefore, we assume that standard cooperation positively affects the adoption of new end user devices (saturation level) as well as the incentives for investments.

5.4.5 Results of Simulations

The results of simulation of Fig. 5.4a show how the introduction of end-user centric devices can be successful under a minimum required HHI and openness of industry. If these factors do not reach the required level, the industry locks into a centralized management of spectrum, based on current mobile operator centric devices, see Fig. 5.4b–d.

Figure 5.5 shows the results of the simulation using the predator–prey competition model.

The results show that in this competition model, end user devices have a slower diffusion than in the competition model with strong network effect, but have higher chance of success, because of its predator behavior. Figure 5.5a indicates that end user device diffusion is slightly faster when spectrum is decentralized and industry is open. Figure 5.5d additionally suggests that when spectrum is decentralized, the introduction of end user devices opens the industry. If the spectrum is centralized, the industry continues with the domination of mobile-operator centric device, even though a predator–prey competition model allows certain level of coexistence with end-user centric devices.

5.4.5.1 Conclusions

This section explored the introduction of DSA technologies by analyzing different mobile markets in a combined approach consisting of Bayesian network data analysis and System Dynamic modeling.

The two different competition models analyzed in this section show significantly different results. Under the presence of high network externalities, end-user centric devices will dominate only under an open industry and decentralized spectrum policy. Under a predator–prey model, a decentralized spectrum policy should be enough to drive an end user device scenario. Using the concepts of modular and integral design [31], a modular standard would show a predator–prey behavior. This means that a modular design should augment the substitution process while decreasing the network externalities.

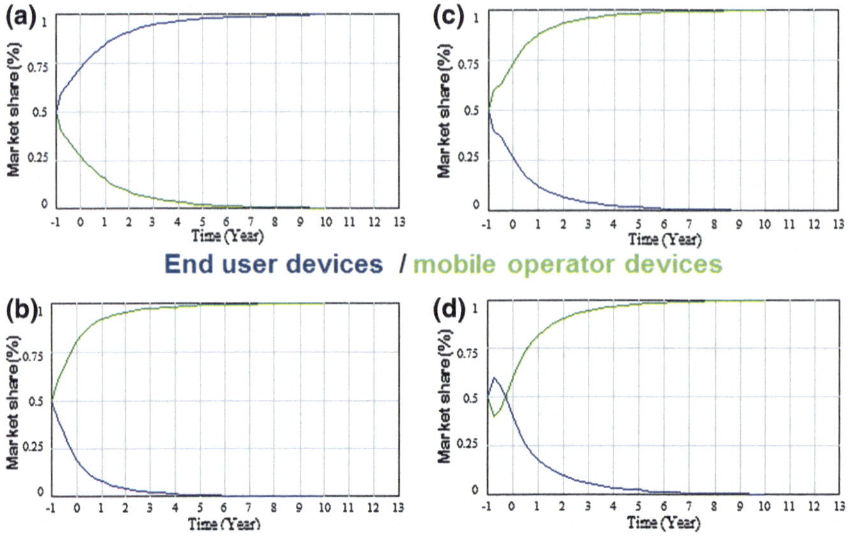

	A	B	C	D
HHI	0.2 (decentralized)	0.6 (centralized)	0.2 (decentralized)	0.6 (centralized)
Industry openness	0.6 (open)	0.4 (closed)	0.4 (closed)	0.6 (open)

Fig. 5.4 Results for system simulation of competition with network effect

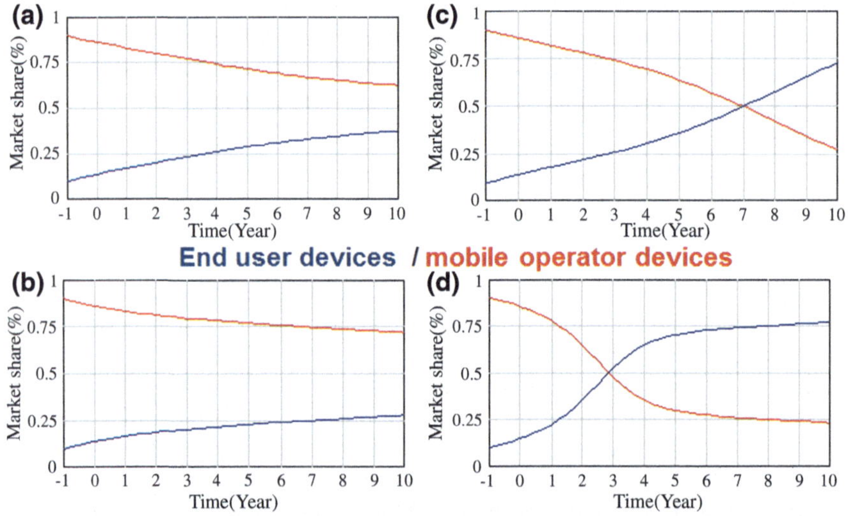

	A	B	C	D
HHI	0.6 (centralized)	0.6 (centralized)	0.2 (decentralized)	0.2 (decentralized)
Industry openness	0.9 (open)	0.3 (closed)	0.9 (open)	0.3 (closed)

Fig. 5.5 Results for system simulation of competition with substitution effect

Table 5.5 Strategies for industry players in different DSA scenarios

Regulation favors	Incumbent operator	New entrant or challenger operator	Spectrum holder (other than mobile operator)
End-user centric devices	Take active role in DSA offering new services	Early adoption of DSA Focus on innovation	Sell spectrum in the market. Consider becoming a new player
Mobile operator centric devices	Utilize DSA to decrease costs and increase efficiency	Cooperation with incumbent operators. Offer compatible services	Share spectrum with operators

Finally, this section gives valuable insights to regulators to understand the current type of policy in practice in their countries and the future consequences of their decisions. It also gives a global overview to different stakeholders on how to deal with DSA technologies as ICT evolves, see Table 5.5.

Regulators should analyze if their current regulation in terms of industry openness and spectrum decentralization is suitable to enable future innovation in a DSA scenario. At the same time, regulators should study the most appropriate mechanism to allow spectrum sharing and trading in their regulations. Incumbent operators should check their current level of cooperation and prepare strategies for spectrum sharing. In this way, they can take the most appropriate decision when investing in these technologies. Other spectrum holders should think on new business opportunities to actively drive spectrum sharing rather than taking passive role. New entrants and challenger operators should try to build a competitive advantage through early adoption of DSA technologies.

5.5 European Market Access and Compliance Regulation for CR/SDR[7]

Dirk-Oliver von der Emden
Federal Office of Communications OFCOM, Biel, Switzerland

5.5.1 *Introduction*

So far the debate on the regulatory dimension of the deployment of CR/SDR systems focused mainly on issues related to spectrum access regulation. But prior to any transmission these systems would have to be lawfully placed on the market.

[7] The views expressed here are solely those of the author in his private capacity, and do not necessarily reflect the views of OFCOM.

Hence the pertinence to analyse the hypothetical impact of present-day regulation on market access and compliance of CR/SDR.[8] One of the purposes of this contribution is to raise the awareness to the fact that the requirements—both essential and administrative—flowing from market access and compliance regulation should be taken into consideration early in any undertaking aiming ultimately at the deployment of CR/SDR systems.

The present contribution will centre on the regulatory framework for market access and compliance of radiocommunication equipment implemented in EU/EEA/EFTA countries[9] (hereafter "Europe"). The R&TTE Directive[10] is the regulatory centrepiece of this framework for placing on the market, free movement,[11] and putting into service of radio equipment and telecommunications terminal equipment in Europe.[12]

The R&TTE Directive puts into effect the core objective of free movement of goods in the Single Market.[13] As a consequence, if a radiocommunication apparatus is compliant with the provisions of the R&TTE Directive it can be lawfully placed on the market in EU/EEA/EFTA countries even if this equipment cannot be operated in any of those countries.[14] In other words, placing on the market must be permitted despite the existence of interdictions and restrictions on putting into operation in some or all of the EU/EEA/EFTA countries. One notable curiousness of this regime is thus that an end-user in an EU Member State cannot infer the possibility to use a radiocommunication equipment simply from the fact that it is lawfully sold in that country.[15] Accordingly, the R&TTE Directive foresees that additional mandatory information has to be provided on or with the equipment by

---

[8] The European Telecommunication Conformity Assessment and Market Surveillance Committee (TCAM) has for long recognised that CR/SDR will have an impact on market access and conformance regulation. It launched initiatives analysing this possible impact and contemplates adapting said regulation to the new realities which CR/SDR could bring about.

[9] EU Member States, EEA EFTA countries (Island, Norway and Lichtenstein), and Switzerland (the transposition of R&TTE in this country is expressly foreseen in a mutual reconnaissance agreement with the EU).

[10] Directive 1999/5/EC of the European Parliament and of the Council of 9 March 1999 on radio equipment and telecommunications terminal equipment and the mutual recognition of their conformity ("R&TTE Directive").

[11] See the rulings of the European Court of Justice (ECJ) in Joined Cases C-388/00 and C-429/00 Radiosistemi [2002] ECR I-5845, Case C-14/02 ATRAL [2003] ECR I-4431, and Case C-132/08 Lidl Magyarország Kereskedelmi [2009] ECR I-3841.

[12] Accordingly, the approach followed in the United States—whose paradigms differ markedly from those of the European approach—will not be addressed here. For an overview, see Annex 2 to [17].

[13] Considerings (12) and (32) of the R&TTE Directive.

[14] Article 8(1) of the R&TTE Directive.

[15] See Sect. 9. "Possibility to place products on the market in the Community, which cannot be used in the Community" of "Interpretation of the Directive 1999/5/EC" under http://ec.europa.eu/enterprise/sectors/rtte/documents/interpretation_en.htm [Accessed 31 October 2013].

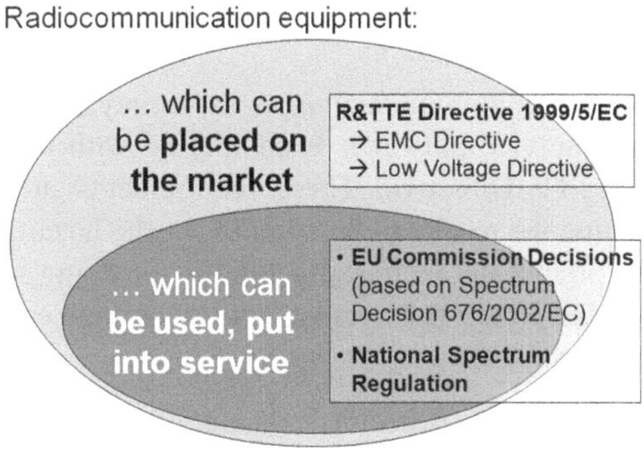

Fig. 5.6 In Europe, a radiocommunication apparatus operating lawfully needs to be lawfully placed on the market.

the manufacturer or importer when differing restrictions on the use of the apparatus apply in some countries or geographic areas[16] (Fig. 5.6).

5.5.2 *Selected Aspects of European Market Access and Compliance Regulation*

Dwelling on the details of the R&TTE Directive would be of no avail as the Commission has made a proposal for a revision of this Directive [33].[17] The discussion in this section will concentrate on some fundamentals which are likely to remain unchanged[18] and are of relevance for developers of CR/SDR. The intent of this overview is to emphasize that various approaches are open to CR/SDR developers (manufacturers, programmers, or integrators) when going to market in Europe. Several approaches lay at hand of equipment manufacturers in order to induce for their products a presumption of compliance with the technical requirements laid down by the Directive. In theory, there never is only one unique technical solution which has to be used in order to meet the essential requirements set out in the Directive.[19]

[16] Article 6(3) of the R&TTE Directive.

[17] Formally the proposed directive should repeal the R&TTE directive (see draft Article 50) but in effect it will be a revision of the latter as far as radiocommunication equipments are concerned.

[18] For further particulars readers are referred to European Commission, Guide to the R&TTE Directive 1999/5/EC, available under http://ec.europa.eu/enterprise/sectors/rtte/files/guide2009-04-20_en.pdf [Accessed 31 October 2013].

[19] Considering (27) of the R&TTE Directive: "whereas compliance with such harmonised standards gives rise to a presumption of conformity to the essential requirements; whereas other means of demonstrating conformity to the essential requirements are permitted".

5.5.2.1 Essential Requirements

A radiocommunication equipment may be placed on the European market and put into service only if it is complying with material requirements laid down in the R&TTE Directive. These requirements are mandatory. Essential requirements define the results to be attained, or the hazards to be dealt with, but do not specify or predict the technical solutions for doing so [34].

The R&TTE Directive lays down three essential requirements that are of public interest:

- Protection of health and safety of the user and any other person, based on the protection requirements of the Low Voltage Directive;
- Protection requirements with respect to electromagnetic compatibility contained in the Electromagnetic Compatibility Directive; and
- Effective use of the radio spectrum/orbital resource so as to avoid harmful interference.[20]

The R&TTE Directive also empowers the European Commission to stipulate that some products fulfil additional—or "elective"—essential requirements in addition to the three above-mentioned mandatory ones. By decision the Commission can mandate that certain functions have to provided.[21] Up-to-now the only additional requirements that have been mandated aim at ascertaining access to emergency services by particular types of equipment pursuant to Article 3(3)(e) of the R&TTE Directive.[22]

5.5.2.2 Administrative Requirements

In addition to the requirements pertaining to their qualities, radiocommunication equipment must comply with some formal requirements. To list a few:

- Application of the adequate conformity assessment procedure[23]
- Marks and inscriptions[24]
  - Conformity marking ('CE' Mark)
  - Identification of the notified body, if applicable
  - Class identifier, if applicable
  - Batch and/or serial number
  - Name of the manufacturer or the person responsible for placing apparatus on the market

[20] Respectively Articles 3(1)(a), 3(1)(b), and 3(2) of the R&TTE Directive.

[21] Article 3(3) of the R&TTE Directive.

[22] See relevant Commission decisions under http://ec.europa.eu/enterprise/sectors/rtte/documents/legislation/decisions/ [Accessed 31 October 2013].

[23] Article 10(1) of the R&TTE Directive.

[24] Article 12 of the R&TTE Directive.

- Notifying authorities of the placing on the market of certain types of radio-communication equipment[25]
- User information[26]
  - Intended use
  - Declaration of conformity
  - Identification of the countries where use of the equipment is permitted, where appropriate
  - Possible restriction to the use of the equipment.

It is worth highlighting that some administrative requirements could be subject to notable changes (which includes abolition) when the Radio Equipment Directive will enter into force. Nevertheless, two of them deserve a succinct development:

(1) **'CE' Mark**
The CE marking symbolises the conformity of the product with the applicable requirements imposed in Europe. It is affixed under the responsibility of the manufacturer, his authorized representative or the person responsible for placing the apparatus on the market in Europe. The CE marking affixed to products is a declaration by the person responsible that (i) the product conforms to all applicable Community provisions, and that (ii) the appropriate conformity assessment procedures have been completed.[27] (Fig. 5.7) European market surveillance authorities must presume the conformity of CE marked products with the applicable requirements and are not allowed to restrict their placing on the market unless they can demonstrate noncompliance on the basis of evidence.

(2) **Conformity assessment procedures**
The manufacturer is responsible for assessing the conformity of his product with the essential requirements of the R&TTE Directive or for having it assessed by a third party (generally, an accredited laboratory). He has to prepare technical documentation providing evidence that the apparatus complies with the essential requirements. This includes evidence that the apparatus complies with the relevant harmonised standards or, if harmonised standards are not used or used only in part, a detailed technical justification. Once the product successfully passes the conformity assessment procedure, the manufacturer does not need obtaining further approvals from any authority. The R&TTE Directive identifies several conformity assessment procedures for radio equipment including a transmitter (see Table 5.6). One procedure (1) in principle does not require the involvement of an accredited laboratory, whereas in the two other ones the laboratory's assessment can include either (2) its opinion on compliance with the essential requirements based on the technical documentation drawn by the manufacturer, or (3) its assessment of the

[25] Article 6(4) of the R&TTE Directive.

[26] Article 6(3) of the R&TTE Directive.

[27] Article 12 of the R&TTE Directive.

Fig. 5.7 The CE marking must be affixed visibly, legibly and indelibly to the product and have a height of at least 5 mm

Table 5.6 Overview of the procedures for conformity assessment of radio equipment including a transmitter

Conformity assessment procedure	Condition for application	Role of the accredited laboratory (if applicable)
Internal production control and specific apparatus tests	Radio equipment including a transmitter complying in full with harmonised standards and harmonized standard comprising complete test suites	*Involvement not mandatory*
Technical construction file	Radio equipment including a transmitter not complying or only partially complying with harmonised standards	*Involvement mandatory* Opinion on the conformity of the equipment based on the review of the technical construction file established by the manufacturer
Full quality assurance	Any equipment covered by the R&TTE directive	*Involvement mandatory* Certification of the manufacturer's quality system

manufacturing process. Attention is drawn on the fact that different conformity assessment procedures may have to be used for each essential requirement.

The complexity of the procedures increases from 1 to 3. In case of full compliance with a harmonised standard, the assessment procedure is least burdensome for the manufacturer. When however a harmonised standard does not exist, has not been fully followed, or the test suites in a harmonised standard are incomplete, a manufacturer needs to demonstrate more extensively how the requirements of the Directive were met. In other words, the presumption of conformity with the corresponding essential requirements bestowed on products manufactured in compliance with harmonised standards[28] translates in a simpler conformity assessment procedure.

[28] Article 5(1) of the R&TTE Directive.

5.5.2.3 Harmonised Standards

The technical specifications of products meeting the essential requirements set out in the directives are laid down in harmonised standards. Harmonized standards are a particular form of European Standard (EN) and can only be produced by the three recognized European Standards Organizations (European Committee for Standardization (CEN), European Committee for Electrotechnical Standardization (CENELEC) and European Telecommunications Standards Institute (ETSI). The European Commission mandates harmonised standards from these standardisation organisations.[29]

Essential requirements define the results to be attained, or the hazards to be dealt with, but do not specify or predict the technical solutions for doing so [34]. This flexibility allows manufacturers to choose other ways to meet the requirements. Technical solutions laid down in harmonized standards can be used to meet the essential requirements of the R&TTE Directive, but they are not mandatory. Application of harmonised or other standards remains voluntary, and the manufacturer may always apply other technical specifications to meet the essential requirements [34]. However, as stated previously, if a type of radiocommunication apparatus is covered by a harmonised standard, abiding to the specifications and test suites devised in it provides the simplest route to market. A third party assessment is considered necessary where products are not manufactured in compliance with harmonised standards, in absence of such standards, or if the essential test suites in these standards are not complete or missing.

5.5.3 Issues Pertaining to CR/SDR Arising from Actual Market Access and Compliance Regulation

CR is believed to include necessarily SDR functionalities [35].[30] SDR is defined as radio equipment (including a transmitter) in which the RF operating parameters comprising *inter alia* frequency range, modulation type, and/or output power can be set or altered by software, or the technique by which this is achieved [36]. Furthermore, it is imaginable that signal processing could be handled over general purpose processors, rather than done using special purpose chips, such as application-specific integrated circuits (ASIC) [37]. Accordingly, it would be in the normal course of events that CR/SDR devices in use would be reprogrammed, i.e., functions would notably be changed, reconfigured without modifying the hardware, after first placement on the European market.[31]

---

[29] Article 2(h) of the R&TTE Directive.

[30] In other words, SDR technology is a precursor and an enabler of CR technology.

[31] ETSI uses the term Reconfigurable Radio Systems (RRS): Such systems exploit the capabilities of reconfigurable radio and networks and self-adaptation to a dynamically changing

It is assumed that the R&TTE Directive will apply to CR/SDR devices as they usually will be transmitting radiocommunication equipment. Accordingly, all the different possible stages of configurability of such a device would *in theory* have to fulfil the requirements of the R&TTE Directive. But the paradigms of the R&TTE Directive do not appear to be well-suited to field-programmable CR/SDR equipment. This is particularly true when one entity manufactures the hardware, other ones develop software steering the operational parameters for spectrum use, and the software reconfiguration is subject to no safeguards. Dealing with this type of technology (pertaining foremost to the issue of who should be responsible for overall R&TTE compliance) will be less straightforward than in the case of 'regular' devices [38].

The matter of the need of an evolution of the regulatory scope of the R&TTE Directive in order to accommodate CR/SDR has been under study by European market surveillance authorities for many years. A first set of questions flows from the uncertainty whether the R&TTE Directive provides certain categories of CR/SDR with loopholes that impede the achievement of the intended goals of the directive. If existing, could these gaps simply be closed by means of new interpretations of the R&TTE Directive's current text or is there a need for modified or new provisions? The second battery of questions results from the suspicion that the R&TTE Directive in its present-day reading could not be applied in a workable manner to some categories of CR/SDR.

The solutions springing from these reflections found their concretisation in the actual proposal for the Radiocommunication Equipment Directive [33] that should supersede the present-day R&TTE Directive in the course of 2014 (Fig. 5.8).

5.5.3.1 Essential Requirements

Market surveillance authorities apprehend that the potential advantages of flexibility (adjustment of parameters generated by a combination of hard- and software modules following a reconfiguration of the software) could increase the risk either that equipment which is or could become non-compliant would be placed on the market or that compliant equipment on the market would be rendered non-compliant afterwards by reconfigured software.[32] Pieces of equipment that lawfully displayed a CE mark when they were first placed on the market could sometime become non-conformant due to *ex post* modifications to the operating parameters of the equipment—and display of the CE mark would actually become illegal.

(Footnote 31 continued)
environment (ETSI TR 102 802 *Reconfigurable Radio Systems (RRS)*; *Cognitive Radio System Concept*).

[32] First cases were reported by market surveillance authorities where wrong or old firmware installed by the supplier on request of customers or directly by the end user appeared to disable the Dynamic Frequency Selection (DFS) mechanism of Wireless Access Systems (WAS) operating in the 5 GHz range. This requirement flows from the necessity to prevent undue interference to meteorological radars by WAS [66].

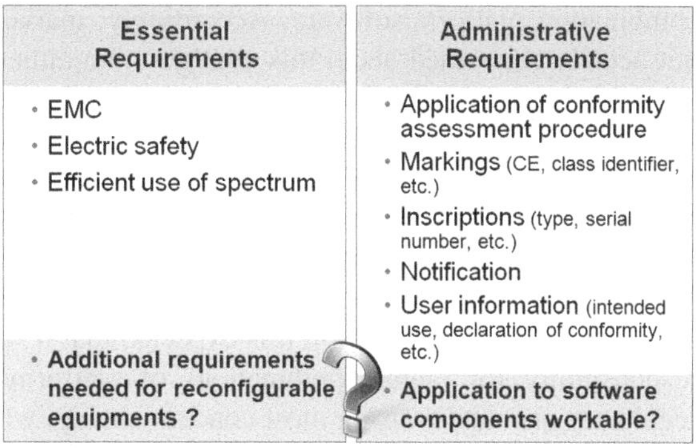

Fig. 5.8 Selected questions on the adequacy of the R&TTE Directive with upcoming CR/SDR systems

Specifically, operating parameters could contravene with regulatory conditions of use of spectrum following a reconfiguration. Conformity with the essential requirement of efficient use of spectrum is resource-consuming to enforce. Indeed it is often difficult for authorities to find non-compliant equipment put into use and generating harmful interference. The potential damage provoked by such interference is easy to visualise in the scenario of jammed air traffic management, public safety, or security services' frequencies.

The analysis over the last years has shown that no supplementary mandatory essential requirement is needed. However, additional (or "elective") essential requirements might have to come into operation for some configurations of CR/SDR.

CR/SDR operating under the control of a network

Under this configuration radio transmitters can only transmit under the control of a network and thus do not need any technical adjustment by the user (who may not even be given the opportunity to undertake them). This configuration does not really raise concerns of market surveillance authorities as the conformity of all hardware and software (initial or updates in the course of the life cycle) is likely to be monitored and controlled by the operator of the network. Control of both the reconfigurable platform and software is centralised (though it is unlikely that it will be in the hands of the manufacturer who is principally responsible for the equipment's compliance according to the R&TTE Directive). Unlawful equipment would be prevented from operating by the operator and could thus not create much harm.

CR/SDR where reconfiguration can be undertaken autonomously

In order to facilitate the legal handling of this type of CR/SDR, regulators envisage classifying them in two categories [39]:

- *Vertically integrated CR/SDR*

All hardware and software during the whole life cycle are controlled by the manufacturer who can ensure that software is only loaded through well-controlled mechanisms. This means that one entity could be held responsible of the combination platform/software. Accordingly, market surveillance authorities are not acutely concerned about this configuration either.

- *Uncontrolled CR/SDR*

In this case the products of independent hardware and software providers are combined. Independent companies develop and sell hardware and CR/SDR software separately—it may be the intention of the hardware manufacturer or due to his lack of caution to implement safeguards preventing the free installation of software. In this configuration it is expected that it will be very complex to assign responsibility for faulty combinations of platform/software. Consequently, the regulatory questions raising most concern emerge where software is developed by an entity other than the manufacturer of the hardware.[33]

In order to mitigate the concerns of market surveillance authorities with respect to uncontrolled CR/SDR, different ideas were brought forward. Discussions produced a scheme to include a new provision for an additional—or "elective"—essential requirements in the proposal for the Radio Equipment Directive. This provision empowers the Commission to require "tamper-proofness" from certain CR/SDR:

- Security and integrity requirement (hindrance of inappropriate downloads): Ensuring that only compliant combinations of software and hardware come together and that the equipment only accepts authorized software [39, 40].[34]

Under this scheme the manufacturer can be obliged to make sure that only particular types of software (those registered—similarly to an AppStore) would be obtainable [38].

An unmet demand concerned the ready availability of information which would notably assist market surveillance authorities in order to determine the respective responsibilities of persons brought in association with an non-compliant CR/SDR equipment:

- Traceability requirement: Histories of (i) software changes/versions and (ii) reconfigurations (logs) [39].

Harmonised Standards

It is reminded that the simplest way for manufacturers to prove compliance is to apply a harmonised standard, if available.[35] Harmonised EN Standards define one

[33] As seen previously, difficulties may also occur when several versions of firmware exist, some of which causing the equipment to contravene with an essential requirement.

[34] See [33], drafts of Article 3(3)(g) and considering (17): "The user, the radio equipment or a third party should only be able to load software into the radio equipment where this does not compromise the subsequent compliance of the radio equipment with the applicable essential requirements.".

[35] Although manufacturers always have the choice of involving a notified body, they may self-certify against the relevant harmonised standards that references complete test suites and make an EU Declaration of Conformity: harmonised standards give a presumption of conformity with the R&TTE Directive for the equipment to be placed on the market in the EU.

of the possible technical specifications as the one where, if complied with, a device certainly meets an essential requirement of the R&TTE Directive.

But standardisation of efficient spectral utilisation[36] by CR/SDR will be addressed by European standardisation bodies[37] gradually. For example, the European Commission has only recently mandated ETSI to produce a harmonised standard on reconfigurable radio systems [41]. The mandate's most tangible part addresses white space devices operating in the UHF TV band and getting access to spectrum through a geo-location database.[38]

"Harmonising essential requirements and making them mandatory by directives is appropriate only where (…) a wide range of products [is] sufficiently homogenous, or a horizontal hazard identifiable, to allow common essential requirements. The product area or hazard concerned must also be suitable for standardisation" [34]. Furthermore, once the specifications are drafted, it must also be ensured that most of the specifications of test procedure aiming at ensuring the compliance of CR/SDR devices should be included in the harmonised standard ("Essential Radio Test Suite") [42].

When moving from the research and development phase to commercial deployment, the lack of harmonised standards which complicates the placement of innovative products on the market and the unavailability of suitable spectrum allocations and associated conditions of use creates legal uncertainty. This can deter potential investors in technology [43]. Thus, the R&TTE Directive appears to be less suited to allow the placing on the market of products based on fundamentally new radio technologies not yet covered by harmonised standards. Indeed, in the absence of harmonised standards, the manufacturer has to consult a notified body for placing a product on the market. For both of them there is no certainty when attempting to establish under these circumstances the conformity of radio equipment with the essential requirements of the R&TTE Directive [44]. Another obstacle in the standardisation process for innovative technologies is that ETSI's work is more accessible to larger market players. Small and medium enterprises (SMEs) and societal stakeholders are underrepresented in the European standardisation process [45].

Nevertheless, the future may bring interesting developments in favour of CR/SDR as the Commission has plans to allow references in public procurement of ICT to ICT

---

[36] For example harmonised standards dealing with spectrum access of one type of proto-CR/SDR should include specifications for the exchange of information between a devices and a database, ensuring that the devices will be connected with the relevant database, on the geo-location systems, on the need for the devices to obtain the authorisation to emit from the database.

[37] Though it is admitted that Harmonized Standards (HS) for Cognitive Radio are being developed by ETSI. For an overview of the ETSI Technical Committees and their responsibilities relevant to CR/SDR.

[38] Actually, the master–slave model with geolocation database as described in the mandate is at best a rudimentary type of CR/SDR system with a network-centric "intelligence": it rather is a network with a basic automated frequency assignment method.

standards developed by other standards development organisations than European Standardisation Organisations, provided that these standards comply with quality criteria [45].

5.5.3.2 Administrative Requirements

The prime issue is the presumption of conformity associated with the CE mark and the declarations of conformity. The CE mark is placed on a radiocommunication apparatus by the manufacturer after its conformity has been assessed. In the case of CR/SDR, software patch may be loaded into the radio equipment that compromises the radio equipment's subsequent compliance with the applicable essential requirements. Under these circumstances, is it still fair that the manufacturer (who has affixed the CE mark and underwritten the declaration of conformity at the origin) is held responsible for any non-conformity, as prescribed by one of guiding principles of the R&TTE Directive?

The discussions about the applicability of the R&TTE Directive to dematerialised "components" of radiocommunication equipment like software (especially if it would be programmed by another entity then the integrator manufacturing the equipment) are also intricate. This is especially true for the application of administrative requirements like marking and user information,[39] whose application to over-the-air reconfiguring software opens up many regulatory questions. Another captivating question is whether every combination of hardware and software will need to undergo conformity assessment. Or would authorities be tempted to distinguish between routine updates and more substantial software updates?

The current Directive, however, was not written with software in mind. It may therefore have to be clarified in future how objectives like traceability, marking and user information are to be achieved for software should it be subject to these requirements [46]. As far as CR/SDR is concerned, it is likely that with respect to a number of administrative requirements, market surveillance authorities will have to demonstrate some flexibility. In particular, it is doubtful whether compliance with administrative regulations of software "components" of reconfigurable systems at all times and under all circumstances would be straightforward.

It is worth drawing the readership's attention on the fact that it is likely that the Radio Equipment Directive will compel CR/SDR manufacturers or software developers to make available information on the compliance of intended combinations of radio equipment and software in order to facilitate competition and provision of software by independent parties.[40]

---

[39] Respectively Articles 12 and 6(3) of the R&TTE Directive.

[40] See [33], drafts of Article 4 and considering (19).

5.5.3.3 Responsibility Ascription

Where more than one manufacturer produces components, each of them can assume responsibility for its own component according to the R&TTE Directive.[41] The company that integrates these components into an equipment will warrant that the new product is also compliant with the requirements of the R&TTE Directive. Often, it may not be practical to perform a assessment of the requirements on the module alone and a complete assessment only takes place after integration.

Today most SDR implementations remain under the control of a single manufacturer. Many base stations and handsets include already proto-SDR technology and some operating parameters are implemented in software. Yet, the hardware and the related software are typically highly optimized. Sometimes the upgrade of the software accommodating different standards may actually not be foreseen. A third party would hardly be in a position to tamper with this software.

What is at stake is to establish responsibilities if the hardware-software combinations do not adhere to the regulation (due to whatever cause[42]). The R&TTE Directive, which assumes that a single legal entity designs the equipment and ensures its compliance once and for all,[43] is not well adapted to address the flexibility where equipment can be reconfigured during operations by users and/or an entity other than the initial manufacturer [43]. It was not drafted with software in mind (Fig. 5.9).

Market surveillance authorities have an interest in rules clearly ascribing responsibilities in the case of non-compliance. The question for them is how to materialise this desire with regard to wireless systems dynamically reconfiguring and upgrading purely by software means. At present the approach where the combination of hard- and software that are produced by different legal entities implies that each legal entity is responsible for its own product is rather rejected.

---

[41] ETSI has produced Guides to the application of harmonized standards to multi-radio and combined radio and non-radio equipment: ETSI TR 102 070-1 *Electromagnetic Compatibility* and ETSI TR 102 070-2 *Effective use of the radio frequency spectrum.*

[42] The lurking complexity can be visualised by means of a very few examples: (a) A user deliberately downloads reconfiguration software in defiance of the use intended by the software programmers and/or the hardware manufacturers; (b) Several third party software applications run in parallel on a hardware platform, creating thus a multitude of combinations which could cause non-compliant behaviour of the radiocommunication equipment; (c) In case of concomitance of several software versions, which software version (the latest?) is used for compliance assessment; (4) Availability of older firmware abiding by specifications of outdated versions of a harmonised standard but no longer compliant with the essential requirements.

[43] According to the R&TTE Directive, manufacturers have the sole and full responsibility (sometimes taken over by the importer) of ensuring through testing that their products are compliant to the applicable directives. The liability of the manufacturer (or the importer) hinges on the CE mark and on the declaration of conformity: the responsibility for an equipment is assigned to the entity affixing the CE mark—who's also the "declarer".

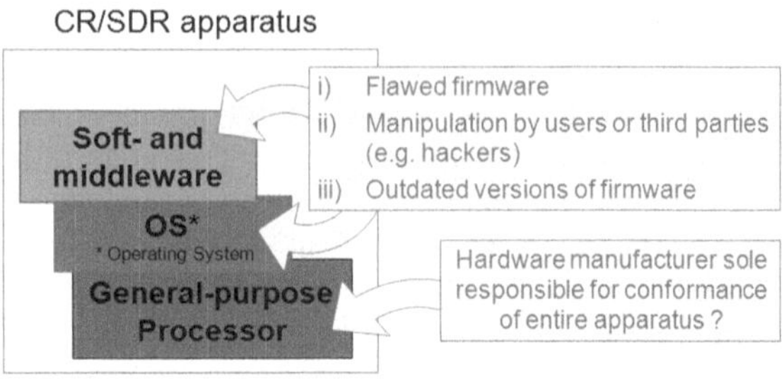

Fig. 5.9 Issues arising from reconfigurability having an impact on responsibility ascription

A rule where the person who puts a product into service must assume the responsibilities instead of the manufacturer (compliance with the requirements and accomplishment of the conformity assessment) would only be workable with professionals. For example, nowadays the system integrator for fixed link systems assembled on-site is responsible for ensuring compliance of the system with the Directive when the system is brought into service. The same applies for products manufactured for own use. In the case of mass-market pieces of equipment the users usually do not know—and can mostly not know—the technical specifications and internal operation/design of their device. They cannot determine whether their actions lead to R&TTE incompliant situations. Moreover, such users cannot be expected to have access to or utilise test equipment to assess conformity of their device.

Presently, apparatus which at the time of supply has provision for later user-added components that fall under the R&TTE Directive but are otherwise not covered by the Directive (e.g., computers without an integral modem and/or wireless capability) should not be marked according to the Directive. One can very well imagine programmable CR/SDR devices where the hardware is not specific to any particular radio technology: amongst other things a software would be needed to create a radiocommunication equipment [47]. Under the current interpretation of the R&TTE Directive this piece of hardware would not have to abide by the compliance provisions of the directive.

Another potential evolution is associated with Open-source software (OSS)[44] developers working in the wireless space and not affiliated with device manufacturers. These developers are already at work now [48]. It is feared by market surveillance authorities that user-modifiable code (which is a subset of OSS) will make it difficult to identify the "author(s)" of a non-conform product (software) and to establish when modifications of software leading to an irregular situation were made.

[44] Open-source software (OSS) is computer software that (i) is available in source code form and where (ii) the provision of the source code occurs under a "public" software license. This means there is a freedom to run the program, for any purpose and a freedom to study how the program works, modify it, and release the improvements to the public.

Fig. 5.10 Key points for certain categories of CR/SDR under the current regime of the R&TTE Directive

- Additional essential requirements?
  - E.g. network, equipment integrity
    - Protection against unauthorised programming (e.g. hacking)
    - Recording of configuration history ("Reconfiguration Controller")
    - Autonomous downloading of updated software (i.e. "patches")
- Additional, modified administrative requirements?
  - Less stringent provisions for SDR/CR devices "notably modified" (updated) by manufacturer or network operator?
- Review of responsibility ascription?
  (Relevant only for cases of reconfiguration using third party software)
  - Extent of duties of equipment, hardware manufacturers established relatively clearly
  - What about radio software developers (IT industry)?
    - Acceptance of responsibilities for their products similar to manufacturers?
    - Enforceability by market surveillance authorities?

5.5.4 Outlook

It emerges from the above discussion that the R&TTE Directive in its present-day form is challenged primarily by CR/SDR with very specific features ("uncontrolled autonomously reconfigurable") (Fig. 5.10).

CR/SDR with such features could become a possible far-reaching problem at earliest in the medium-term. The challenges foreseen in this contribution for market access and compliance regulation in Europe are currently mostly theoretical in nature. In particular, the border cases identified would prove critical only if a market for user-reconfigurable devices starts to form.

Some emerging problems are tackled in the proposal for the Radio Equipment Directive. Yet it would be inefficient and ineffective to anticipate already now in detail any imaginable issue (e.g., rogue software loaded on mass market consumer equipment) that might arise when CR/SDR becomes pervasive.

Consequently, the most reasonable approach under the present circumstances is to maintain the present-day responsibility-ascription scheme also for CR/SDR products. The future structure of the market deploying CR/SDR may require adjustments but there is no urgency to modify this scheme in anticipation. Furthermore, for the time being it is realistic to command that CR/SDR meet the essential requirements under all circumstances.

Finally, the future will also show whether European standardisation organisations, and in particular ETSI, manage to draw up harmonised standards which aim at ensuring that CR/SDR meet essential requirements set out in mandates of the European Commission. Will SMEs and academia participate actively in and provide their innovative technology solutions to these standardisation efforts?

Acknowledgments My gratitude goes to Mr Lucio Cocciantelli, Head of Market Surveillance at OFCOM, for his valuable insights. However, any errors remain the sole responsibility of the author.

5.6 Helping Innovation of CR

Arturas Medeisis[1] and Leo Fulvio Minervini[2]

[1]Vilnius Gediminas Technical University, Vilnius, Lithuania
[2]University of Macerata, Macerata, Italy

5.6.1 Introduction

This section analyses possible reasons of rather sluggish pace of CR innovation with the aim of suggesting a range of suitable policies to boost further and more fertile developments of CR technology.

Thanks to its advanced features of environmental awareness and, ultimately, propensity for autonomous decision making, CR represents a significant evolutionary step from traditional radiocommunications systems. The autonomous, cognitive re-configuration of CR opens up opportunities for new business models in the wireless communications marketplace built on the novel utility profiles of CR, as was discussed in second chapter. Yet this also means that fledgling CR innovation must overcome significant technological and other challenges on its road to practical implementation. If not addressed properly and quickly, these challenges may fester and become "reverse salient" barriers [49] in the composition and functioning of an eco-system of CR innovation and thus restraining the impetus of CR development.

Therefore this section sets out to explore the technology-push and demand-pull processes [47] as applied to CR [50], and then tries to identify and discuss the barriers that may be stalling CR innovation and how they might be reduced or overcome.

5.6.2 CR Innovation: Technology-Push and Demand-Pull

In the context of modern wireless markets, it can be argued that CR represents an important new option that contributes to a variety of competing technological solutions. Therefore successful implementation of CR technologies should become a matter of heightened attention by regulators, who might need to address the situation in order to provide opportunities for CR to succeed. This attitude of "creating windows of opportunity for new radio services and applications"[45] is not something new, in fact the very same stance was quite often taken by regulators over past decades, whenever allocating spectrum to broad swath of untried

[45] This visionary expression is credited to Mr. Reiner Liebler of German regulatory agency for posts and telecommunications (RegTP, later BNetzA), at the time of his leadership of the CEPT's Working Group Frequency Management, which he chaired between 1998 and 2003.

technologies and systems, many of which eventually flopped in the marketplace, such as e.g., Terrestrial Flight Telephony Systems[46] or quixotic Meteor Scatter Applications[47], to name but a few. In fact one could argue, that European regulators have witnessed so many technology innovation proposals that went awry, that they became increasingly reticent whenever asked to take bold decisions for promoting new types of wireless systems. However the CR technology is different in that it embodies not a specific system or technology, but rather a family of technologies, a new paradigm of wireless networking and innovation. Therefore of all proposals, this should really deserve a closer consideration as it might be laying foundation for new wave of unconstrained wireless innovation for the years to come.

The evolutionary perspective leads us to consider two complementary yet distinct strategic forces shaping the dynamics of innovation and impacting the transfer of technology from the research labs to the market. The first of these forces can be described as the "technology-push", which explains technology transfer as motivated by means. In this process, the sheer technological superiority of the innovation compared with traditional technologies dictates its broad acceptance by an industry. A second contributing force is characterized as a "demand-pull" or "market-pull", the intensity of a market proposition and a commercial promise of a new technology [47]. It may be hypothesized that the halting dynamics of CR innovation may be indicating some barriers that inhibit the workings of one or both of these forces.

The main impetus of the classical technology-push is built on the premise of the technological soundness and superiority of new innovative solutions compared with existing state-of-the-art technologies. Normally this requires a clearly formulated technological concept and an initial working prototype that can pass the elaborate testing of the market and convince stakeholders of the emergence of a new, dominant technological design [51, 52]. Such scenario, however, is made much more complicated in the case of CR due to the principal multi-dimensionality of this concept as a family of technologies and their inherent complexities, which are likely to require some phased implementation (see discussion on this in Sect. 2.8). So for the

[46] TFTS was allocated frequency bands 1670–1675 MHz/1800–1805 MHz by CEPT in 1997 (cf. ERC/DEC(97)08) and envisaged to provide voice communication to passengers on planes flying over the European continent. A great effort was put into establishing the system: from allocating necessary frequency bands to carrying out a meticulous planning of terrestrial base stations and their frequency assignments to provide suitable pan-European coverage for air traffic. However after brief period of limited deployment the system was deemed a fiasco and rolled down, the allocation of frequency bands was cancelled in 2003 (cf. ECC/DEC(03)03).

[47] The Meteor Scatter Application system was a land mobile system working in the range 30–50 MHz promoted by industry in 1990s with the aim of providing low bit-rate pan-European coverage for truck fleet management and similar applications, by using the phenomenon of (very weak) reflection of radio waves from ionised gas trails of microscopic meteorites that constantly bombard the Earth's atmosphere. The ultimate regulatory recommendation allowing the use of such systems was taken in CEPT in year 2000 (cf. ERC/REC(00)04), however practical implementation of such systems never took off.

development of CR technology, it can be suggested, and duly observed, that the technology-push may be happening along two paths.

The first path is an incremental process of wireless innovation by equipment vendors whereas CR-related technologies and use cases of various stripes are making their inroads into wireless marketplace. Characteristically, these innovations might not even be consciously associated to the "making of CR", as often happens in the CR application areas other than the DSA. Nevertheless, all these incremental innovations would eventually add up to creating the critical knowledge base that would propel the CR technology to the centre-stage of wireless innovation and provide it with the sense of maturity and status of de facto dominant industrial concept. At that moment the "paradigmatic" switch would occur toward the embracing new technology by means of industry consensus [51].

If looking at the situation today, one may observe that the process of incremental innovation does take place, as evidenced by attention to CR technology from existing wireless players. However, the traditional operators may be tempted to act with great caution to avoid disturbing the status quo. Therefore, it is likely that these operators would proceed in carefully measured steps to ensure that any realized technological gains are harnessed as part of the toolbox of existing wireless service offerings or through a carefully screened set of CR use cases that may be of interest to the incumbents [53].

The second path is through a standardization process in which the incumbent wireless stakeholders as well as CR proponents without current stakes in the wireless industry but wishing to enter the field, are pushing CR technology to the position of a recognized industry standard by means of standardization processes that involve formal Standards Development Organizations (SDO), such as IEEE or ETSI. It is important to note that the formal standardization process might be an effective avenue of technology-push toward gaining market recognition of the disruptive aspects of CR. However the interests of the different lobby groups, and then individual companies deeper down inside the respective camps, would often clash making the standardisation process lengthy, perilous and, sometimes, inconclusive.

This situation may be observed clearly in the case of CR. Standardization efforts were initiated in several SDOs, such as the IEEE Standards Committee DySPAN (former DySPAN-SC), IEEE 802, ETSI Technical Committee RRS, ITU-R and others. Of these, the IEEE SC DySPAN takes the most holistic approach. However, even there (or especially there because of the attempted wholeness of consideration), the standardization process is excruciatingly slow because it needs to reconcile technological advancement with business and policy considerations [54].

The market-pull of an innovative technology may be described as a gravitational force generated by market players that appreciate the commercialization prospects of the new technology. The question of a credible business case is of paramount importance when attempting to understand the gravitas of the market-pull.

So far the main focus of business forecasting in the field of CR was firmly concentrated on the application areas of DSA. A few early examples of CR technology road-mapping exercises [55, 56] highlight the potential for business propositions of CR in such scenarios. DSA should enable nearly instant access to radio spectrum usage gaps and transcending complex and cumbersome traditional administrative spectrum allocation procedures. This offers an attractive conceptual proposition for the many companies wishing to enter the fray of high-profile and profitable wireless businesses.

As regards the other application areas of CR family and noting their inevitable incremental implementation, they would seldom offer any distinctive business case in its own right, or significant changes to the primary business case of the wireless player. Therefore such incremental innovations would be relegated to the niche of process optimisation rather than generating their own commercial value. For example, the Self-Organised Networking solutions, which might be clearly attributed to be a sub-class of CR technologies, are making steady inroads into modern cellular mobile networks. Yet they do not change anything in the prime business model of the cellular operators and therefore do not command special attention other than being seen as yet another technological gimmick helping to deploy networks faster and manage them with lesser effort.

Alas, even the great promise of DSA has failed so far in generating the necessary commitment and investments sufficient to overcome the challenges of CR innovation. Other than the baffling complexity of implementation that deters the prospective interest, another reason for lacking attention might be indeed doubt in monetisation of whatever achievable DSA benefits [57].

5.6.3 Future Development Options

The previous analysis described a situation in which the innovative development of CR faces an uphill technological battle toward market recognition. This situation may lead to a standstill, described as a chicken-and-egg dilemma in which vendors wait for large operators to announce support for CR technology as an indication of sufficient volume potential, whereas the operators are reluctant to support new technology unless it is standardized and embraced by the manufacturers as a pre-requisite of acceptable pricing for mass-market devices [58].

This situation is suggestive of (market and government) failures to provide the necessary testing ground for the trial-and-error dynamics required for efficient evolutionary processes [59]. Thus, the next issue to consider could be the type of regulatory intervention that might be considered appropriate to facilitate the innovative process of CR.

The barriers in the CR innovation path prevent any substantial opportunity for CR to quickly push the market into a Schumpeterian cycle of destabilization and a subsequent innovation leap. Accordingly, governmental policy to clear that path seems crucial. Policy failure (or, in some cases, policy absenteeism) is as important as market failure [59].

Table 5.7 Regulatory policies to assist the innovative dynamics of CR development

Innovation force	Assistive regulatory policy examples	Implementation status
Technology-push	Promoting development of CR standards	Work of ETSI TC RRS and IEEE SC DySPAN
	License and technical conditions for spectrum access conducive to implementation/experimentation of CR technologies	e.g., new proposals of Pluralistic Licensing concept for licensed bands, or ISM-Advanced concept for unlicensed bands, see Chap. 7
	Allocating dedicated exclusive spectrum band as testing and development incubator for CR-enabled applications	New proposal
Market-pull	Technology-neutral liberal spectrum licensing (i.e., with right of change of radio service/application)	European WAPECS initiative
	Governmental support to chosen business applications of CR technologies	Development of ASA/LSA concepts and regulatory framework

So what kind of policies might help spurring technologically neutral CR innovation (because it is not yet clear, and probably will never become definitive, how the CR framework of technologies will look like), while avoiding earlier pitfalls of making over-confident bets on new wireless technological propositions?

With reference to previous discussion, it may be argued that CR innovation would be boosted if both the technology-push and the market-pull forces might be allowed to unleash their full potential and dynamics. Interestingly, it was previously already observed that the incremental innovation is more likely to respond to demand-pulls than technology-pushes, and non-incremental innovation is more responsive to technology-pushes [60]. This implies that, by providing unrestricted working of both forces, the CR eco-system would be able to develop in any imaginable way. In other words, the regulatory policies should be designed so as to establish those "windows of opportunity" through creation of some kind of conditions where CR innovation could flourish in a kind of controlled learning environments. Table 5.7 outlines a set of possible regulatory options that could help improving the innovative dynamics of CR.

All regulatory options mentioned in Table 5.7 contribute to creation of liberalised spectrum access conditions, which foster wireless innovation in general, and CR as the most prominent case of such wireless innovation.

It may be also noted that it is important to reduce the risk that CR testing may cause direct disturbances of existing markets. Therefore exiting incumbent users, traditional wireless systems, should be protected by either defining clear frequency boundaries, such as with proposed dedicated CR band, or establishing spectrum access rules that are conducive to CR deployment while ensuring the reasonable degree of protection to incumbents, such as with proposed Pluralistic Licensing scheme [61], for further discussion on this see Sect. 7.1.

Fig. 5.11 Working of CR innovation forces in liberal CR-friendly market

The ultimate aim of such policies should be to create certain "safe havens" where CR technology could evolve and mature in a kind of learning platform that provides the necessary freedom for experimentation. This would effectively remove the identified barriers and would allow more practical experience to be gathered in using this technology, as illustrated in Fig. 5.11 [50].

It is notable that many of the identified regulatory options are already being implemented or in the process of consideration. An example of more radical intervention, which was not yet seriously considered by regulators, would be allocating a designated band for CR-enabled wireless applications. On several earlier occasions providing dedicated frequency bands for innovative technologies has proven to be a wise choice that established the technological trajectory and provided the necessary regulatory certainty for innovating companies to concentrate their focus and investments [62]. Successful examples of aiding innovative ideas by allocating a designated frequency band include the allocation by the FCC of a spectrum for cellular telephony in 1970 that led to the first commercial deployment of a cellular system in 1983 by Bell Labs and the designation of the 2.4 GHz ISM band for spread spectrum technologies in 1985 that paved the way for widespread WiFi systems.

5.6.3.1 Conclusions

The process of CR innovation is slowly progressing within a complex environment shaped by the combined workings of technology-push and market-pull forces. However, these forces are being stifled by reverse salient barriers that inhibit the development and dissemination of CR technologies and applications.

The provided analysis supports the notion that an effective means of overcoming these extant barriers and revitalizing the innovation process for CR technologies could be for governments to provide some kinds of "windows of opportunity" for CR technologies, i.e., certain spectrum access conditions conducive to deployment of CR or outright safe heavens, such as dedicated licensed or unlicensed frequency bands where CR technologies could mature through repetitive learning cycles of trials and errors.

Providing the CR innovation community with an open and unrestrained testing grounds would represent a plausible solution for effectively removing the identified innovation barriers: "Governments can [...] encourage innovation in two ways: they can implement measures that reduce the private cost of producing innovation, technology-push, and they can implement measures that increase the private payoff to successful innovation, demand-pull" [47].

This is not to suggest a mere "engineering" approach to CR innovation, where CR will be able to flourish and be adopted widely as soon as some spectrum will be made available to experiment with CR. Technology-push has been crucial, but demand-pull issues are also relevant. Barriers on both sides need to be reduced to capture the effective essence of CR. Therefore, regulatory intervention seems appropriate to provide conditions for finding out whether stalling CR innovation is due to barriers on the technology-push side, or rather on the demand-pull side. In the end, it might be found that CR is not as beneficial (nor disruptive) as a decade of research has tried to suggest—but at least there will have come a time when innovators should not bother with CR anymore, as trials might prove they have been erring.

5.7 CR Policy Analysis: "Agreement Framework" and its Implementation

Leo Fulvio Minervini[1] and Arturas Medeisis[2]

[1]University of Macerata, Macerata, Italy
[2]Vilnius Gediminas Technical University, Vilnius, Lithuania

5.7.1 Policy Issues

This section will propose a method for structured representation and analysis of CR policies, by considering regulation of White Space Devices (WSD) as a representative example discussed in the first chapter. The aim of the proposed methodology would be to create a general framework for aiding an assessment of existing and future CR policies. Such framework would be especially useful as basis for organizing productive debate around any contemplated regulatory changes.

We adopt the methodology of positive and normative policy analyses elaborated in [63] with regard to environmental policy. The viewpoint of positive analysis allows focusing on how things stand with regard to CR, as well as on how CR systems are described and modeled. The normative analysis offers focus on how things should be, in order to implement meaningful CR policy, according to a few (explicit) value judgments on critical aspects. Drawing a clear-cut line between positive and normative analyses is normally not possible. Robert and Zeckhauser recall that "any normative or prescriptive analysis necessarily includes positive analysis, plus values" [63]. They also propose a few key elements of positive and normative analyses, which can be seen as the fundamental ones for policy discussion: "Any positive analysis will tend to include elements of scope, model and estimation, though often these elements are either implicit of undifferentiated. Likewise, normative analysis will additionally include elements of standing, criteria and weights, whether or not these distinctions are recognized" [63].

5.7.2 *Positive CR Policy Analysis*

As a general observation, CR policy falls in the area of wireless communications business. Hence, discussion is immediately placed within the context of using a valuable natural resource (i.e., radio spectrum). Today CR technology seems reaching its maturity, but it has not yet crossed the commercial deployment milestone. However, solid evidence of maturity can be derived from the recent increase in the number of pilots (such as pilots using white space in the and the UK) and appearance of new cases and applications, most notably the LSA[48] concept.

Based on this premise, we outline some criterions that, on one hand, describe technology maturity, and, on the other hand, establish an applicable time horizon for our analysis. Generally speaking, technology can be considered mature when real-life field pilots (test deployments) are taking place, and business proposition looks promising, i.e., cost-benefit analyses support the case either for-profit or non-profit deployment.

As regards time horizon, it is proposed to consider a short-to-medium time frame, some 3–5 years. This would correspond rather well with the usual cycle of regulatory policy developments (i.e., time usually elapsing from novel regulatory proposals to effective legislation being in place).

With those criterions in mind, we elaborate on the "taxonomy of disagreement" and the three elements of positive analysis [63]—i.e., scope, model and estimates—which we apply to the case study of WSD policies.

Building on the taxonomy of disagreement, potential disagreement areas will be outlined, as pertaining to a particular consideration aspect. However, it should be noted that this does not presume that disagreement shall be always present. To the contrary, such taxonomy may just provide a helpful check-list of issues, where the agreement could be thought by different stakeholders when developing new policy, thus building ground for well-informed and compromise-driven decision making process.

5.7.2.1 Scope of Application

In the case of policy analysis of WSD applications, the scope of analysis can be narrowed with regard to sector/area, therefore considering the wireless (mobile) data market segment, predominantly in city areas with high population and business user densities—i.e., where the demand for mobile services is highest and puts great pressure on mobile network resources.

Primary stakeholders in such case can be grouped as follows:

- mobile operators, who are interested in offloading the excess data capacity overloads from their Wide Area Networks (3G/4G);

[48] That is, Licensed Shared Access, see Sect. 2.6.

- mobile network users, who are interested in getting higher data throughputs (and, probably, reduced or zero fees) at their dominant "hang-out" spots (such as home or office);
- equipment vendors.

So when analyzing the scope, one could consider where the *potential disagreement* areas might emerge. For instance, disagreement might be on the most appropriate level of decision and policy making (i.e., global, regional or national level?). Also, there might be different views on the list of stakeholders involved. The degree of disaggregation might bring to a clash those who favor application-tailored policies against those who favor CR generic policy. Last but not least, strategic considerations might be relevant: for instance, what is the role of CR as possible catalyst of wireless innovation and enabler of new paradigms of spectrum management?

5.7.2.2 Model of Policy Relationships

We identify the following institutional actors in modeling the CR policy interaction area:

- the National Regulatory Authority (NRA), i.e., the governmental agency/ department dealing with radio spectrum management;
- geolocation database (GDB) operator(s), or in more general sense—spectrum broker(s);
- telecommunication service providers;
- end users;
- equipment vendors.

The general routes for policy impact may be depicted as shown in Fig. 5.12. Note that large arrows indicate strong policy impact connections, while line-arrows indicate weak impact connections.

For instance, the NRA has strong policy impact on operation of spectrum broker(s) and service providers through issuing of spectrum access rules (incl. GDB operation rules as relevant). It has somewhat lesser impact on equipment vendors/manufacturers (e.g., through endorsing specific standards or type approval norms) and end users (e.g., in cases of market interventionary measures such as service price control, or licensing process conditions).

With regard to CR policy model, *potential disagreement* areas might involve, for instance, the following considerations:

- Should the GDBs be only facilitators of opportunistic access?
- Are the links among stakeholders appropriate?
- Is uncertainty taken into account? If yes, how? For instance, although policy provisions for CR are provided, it is possible that the technology does not progress into the commercial deployment phase.

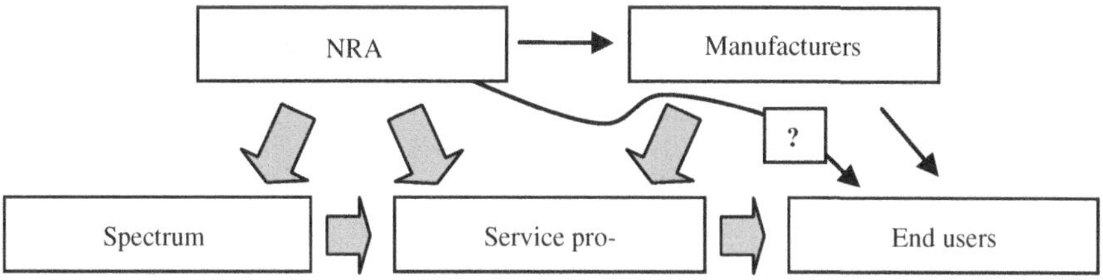

Fig. 5.12 The policy relations model for CR deployment (in TVWS)

5.7.2.3 Estimates

This part of policy analysis is concerned with estimates of model parameters—e.g., (comparative) indication of strength/extent of those identified policy impact links (cf. Fig. 5.12). Although CR has been discussed for years, there is still little work on estimation of model parameters. Quantitative research, applied to CR, is still limited, especially if one considers that CR is still almost confined to laboratories. It may be of great relevance, for CR implementation, to illustrate the business impact of considered cases in a quantitative manner, based on thorough cost-benefit analyses. Calculations of CAPEX, OPEX, turnovers, GDP impact, etc. could enrich qualitative analysis and provide solid grounds for policy making.

With regard to estimates, *potential disagreement* areas might involve, for instance: (i) valuation of the (economic) impact on the constituents of a proposed model; (ii) estimation and quantification of model parameters; (iii) uncertainty (e.g., relevance of the pace of learning).

5.7.3 Normative CR Policy Analysis

The framework that we propose for normative CR policy analysis focuses on three elements: namely, standing—i.e., who counts?; criteria—i.e., what counts?; and weights—i.e., by how much?

Implementation of CR devices is not expected to radically change the configuration of stakeholder groups and, in general, the approach used for normative policy valuations. However, we consider that CR will—and, perhaps, should—have an impact on the relative strength of stakeholders, as well as on what counts (and by how much). Consumers and traditional business operators will not disappear from the list of major stakeholders. However, the most innovative business operators and flexible (ready-to-adapt) consumer groups should receive more consideration—and weight—from a normative perspective. At the same time, while innovation and consumer welfare growth should remain crucial criteria for assessment, issues of spectrum efficiency, QoS, innovation and investment are likely to become even more complex and valuable than in the past.

Development and successful implementation of CR devices may have a considerable impact on the current wireless ecosystem and its associated spectrum management regime. Traditionally, regulation and policy have been, to a large extent, network-centric and most of commercial wireless systems have developed around networks. However, a more CR-friendly spectrum management regime might enhance developments toward more device-centric (self) regulation and policy. Indeed, CR is likely to deliver its promises to an extent which critically depends on the ability of the wireless environment to dynamically change and adapt to new conditions in the realm of technology, policy and economic welfare.

In this subsection we propose an application of the framework for normative positive analysis to the case of WSDs. We will consider, in turn, the three elements of normative analysis [63] and we will also attempt to figure out, with regard to each element, areas of potential disagreement.

5.7.3.1 Standing

In the analysis of standing, three groups of stakeholders can (and—we believe—should) be seen as the major ones: operators; equipment vendors; and mobile device users.

It may be noted that large traditional operators were initially skeptical towards new CR technologies: CR was considered potentially disruptive to their business, as it would allow breaking their oligopoly on providing of wireless data services. However, the emergence of traffic off-loading concept—which could make good use of white spaces—might be changing their stance, by offering mobile operators some tangible business gains. Another potential advantage for mobile operators may be found in leveraging the joint benefits of WSD in TV bands with the use of the LSA solutions in the higher frequency bands (see the discussion of that concept in Sect. 2.5).

As regards the equipment vendors, in the short-to-medium term most of the large ones will be focusing their attention on roll-out of LTE technologies, and the complete overhaul/modernization of cellular networks that is often associated with introduction of LTE.[49] Hence any entries into the new and technologically challenging WSD niche could mostly be anticipated from small vendors seeking entry into the market.

Mobile users *per se* are not likely to care much about technology development, as they seem to be already spoilt by burgeoning supply of telecom/data services in various forms, especially in cities where the use of WSDs would be a very critical option due to spectrum overload. However, CR policy should consider the impact of technological change on mobile users' welfare, taking into account impacts on operators' as well as on equipment vendors' markets.

Potential disagreement areas with regard to standing might concern the role of licensing as guarantee of spectrum quality, and the use of GDB platforms as universal spectrum broker (including the extension of their scope across various bands, i.e., beyond TV bands).

[49] LTE represents a paradigmatic shift to an "all-IP" solution compared against the circuit-switched paradigm of previous 2G/3G network designs. Therefore it makes sense for operators to combine introduction of LTE with converting the rest of their network core to a new IP-based platform.

5.7.3.2 Criteria

In the context of this section, we are looking at the market domain through the prism of the service providers' business proposition to the end user. The underlying problem is that so far there is still no visible or obvious commercial benefit from exploiting the WSD, as compared with other already existing wireless data communication solutions—be it 3G/4G or Wi-Fi. The off-loading of traffic from the macro-network to small cells is, essentially, a pure internal technological efficacy improvement by operators, which does not directly translate into offering innovative/value added service to end users. Any (marginal) benefit of small cells (e.g., by bringing in more paying customers through more competitive payment plans with "home zones" with low or zero tariffs) could be achieved by using femto-cells of the same technology/frequency band as used in the macro network.

This means that we are excluding from the analysis various market players that the service provision segment could be made of. Such simplification is warranted because the future structure and revenue sharing models of service provision in the field of CR are not yet clear (and may take various complex forms, as discussed, e.g., in [64, 65]). In other words, that is one relevant area of *potential disagreement* as regards criteria. Disagreement about criteria may also concern, for instance, interference potential to incumbent users, in terms of interference to TV reception and interference to co-secondary users (such as users of wireless microphones).

5.7.3.3 Weights

The matter of respective weights may be considered through direct extension of the above debate on criteria. Thus we may move on to direct consideration of *potential disagreement* areas as regards weights, that is, by how much or to what degree the identified criteria should count. One such area might concern the relative importance and pervasiveness of the incumbent users, i.e., the primary users that need to be protected by WSD operation, say for TV bands:

- What is the proportion of households that still rely on over-the-air TV reception, as opposed to cable TV, IPTV, or satellite TV reception? (This question would have an obvious national context, as the situation will be different in each country.);
- How many of wireless microphones are there, and what should be the proportionate level of their protection?

5.7.3.4 Conclusions

The analysis carried out in this section is an attempt to build a structured policy consideration framework in the discussion of the future of CR. It is also an attempt to highlight considerations of "values" and to delineate areas where agreements

Table 5.8 The "agreement framework" for building and analyzing CR policies: WSD example

Positive analysis: Scope

• global, regional, or national level of CR policy most appropriate?
• degree of policy detail: application-tailored policies, or CR-generic policies?
• strategic considerations: i.e., the role of CR as catalyst of innovation and new spectrum management paradigm?

Positive analysis: model

• role of GDB as the only (prime) facilitator of opportunistic access?
• what links inside the model and their respective strength?
accounting for uncertainty: e.g., policy provisions for CR are provided, but technology does not progress to commercial phase

Positive analysis: estimate

• quantification and valuation of model parameters
• quantifying and valuing (economic) impact on the constituents of the model
• uncertainty parameters, including the pace of technological evolution

Values analysis: standing of stakeholders

• role and scope of licensing
• role/value of GDB platforms as universal spectrum broker across various bands (i.e., beyond TV bands)

Values analysis: criteria

• extent of market impact
• realistic interference potential to incumbent users:
–interference to TV reception
–interference to co-secondary users: wireless microphones

Values analysis: weights

• relative importance and pervasiveness of the incumbent users:
–proportion of remaining over-the-air TV users
–extent of the wireless microphone use, proportional level of their protection

should be thought in order to avoid conflicts. With this attitude in mind we propose building the "framework for policy agreement", as depicted in Table 5.8.

Note that here we take a more positive (non-conflicting) stance as compared to the approach of building the "taxonomy of disagreement" by [63]. We believe that positive attitude is appropriate for the case of CR policy discourse: we largely deal with the green-field situation where differences in opinions and values, held by the concerned stakeholders, are not yet deeply ingrained nor prominent.

To conclude, the above described framework of agreement may be used as reference template against which future normative policy analysis could be carried out in similar cases, with regard to emerging CR applications and CR technology in general. This kind of analysis may be also useful as part of the bigger picture concerned with spectrum management (i.e., the debate whether CR should be reflected as a salient entity in the ITU Radio Regulations and similar normative acts).

References

1. Torriti, J., Löfstedt, R.: The first five years of the EU impact assessment system: a risk economics perspective on gaps between rationale and practice. J. Risk Res. **15**(2), 169–186 (2012)
2. Torriti, J.: Impact assessment in the EU: a tool for better regulation, less regulation or less bad regulation? J. Risk Res. **10**(2), 239–276 (2007)
3. Jacob, K., Hertin, J., Hjerp, P., et al.: Improving the practice of impact assessment, project no. 028889, Evaluating Integrated Impact Assessments (EVIA), Mimeo (2008)
4. Renda, A.: Impact Assessment in the EU: The State of the Art and the Art of the State. Centre for European Policy Studies, Brussels (2006)
5. EC-European Commission: Impact Assessment Guidelines, Sec (2009) 92. EC, Brussels (2009)
6. ECC–Electronic Communications Committee: Guidelines for the implementation of impact assessment in relation to spectrum matters. Report 125, Liége, ECC (2008)
7. Mitola, J., Maguire, G.Q.: Cognitive radio: making software radios more personal. IEEE Pers. Commun. Mag. **6**(4), 13–18 (1999)
8. Haykin, S.: Cognitive radio: brain-empowered wireless communications. IEEE J. Sel. Areas Commun. **23**(2), 201–220 (2005)
9. RSPG11-392 Report on Collective Use of Spectrum (CUS) and Other Spectrum Sharing Approaches. EU Radio Spectrum Policy Group, Brussels (2011)
10. Anker, P.: Does cognitive radio need policy innovation? Compet. Regul. Netw. Ind. **11**(1), 2–26 (2010)
11. Lemstra, W., Anker, P., et al.: Cognitive radio: enabling technology in need of coordination. Compet. Regul. Netw. Ind. **12**(3), 210–235 (2011)
12. RSPG10-306 Radio Spectrum Policy Group Report on Cognitive Technologies. EU, Brussels (2010)
13. Lehr, W., Jesuale, N.: Spectrum pooling for next generation public safety radio systems. In: DySPAN 2008, 3rd IEEE Symposium on New Frontiers in Dynamic Spectrum Access Networks 2008, IEEE (2008)
14. Atia, G., Sahai, A., Saligrama, V.: Spectrum enforcement and liability assignment in cognitive radio systems. In: DySPAN 2008, 3rd IEEE Symposium on New Frontiers in Dynamic Spectrum Access Networks 2008, IEEE (2008)
15. Mishra, S.M., Sahai, A., Brodersen, R.W.: Cooperative sensing among cognitive radios. In: IEEE International Conference on Communications, 2006. ICC'06, vol. 4, pp. 1658–1663. IEEE (2006)
16. Bourse, D., Agusti, R., et al.: The E2R II flexible spectrum management (FSM) framework and cognitive pilot channel (CPC) concept–technical and business analysis and recommendations. E2R II White Paper (2007)
17. RSPG10-348 Final RSPG Opinion on Cognitive Technologies. EU Radio Spectrum Policy Group, Brussels (2011)
18. Ankre, P.: Cognitive radio, the market and the regulator. In: IEEE Symposium on New Frontiers in Dynamic Spectrum Access Networks, 2010, IEEE (2010)
19. Ostrom, E.: Governing the Commons: The Evolution of Institutions for Collective Action. Cambridge University Press, New York (1990)
20. Haug, T.: A commentary on standardization practices: lessons from the NMT and GSM mobile telephone standards histories. Telecommun. Policy **26**(3–4), 101–107 (2002)
21. Szydełko, M., Byrka, J., Oszmianski, J.: Dynamic valuation function based definition of the primary spectrum user in colocated cellular networks. In: Proceedings of 7th International ICST Conference on Cognitive Radio Oriented Wireless Networks and Communications (CROWNCOM 2012), pp. 297–302, June 2012

22. Szydełko, M.: Business model analysis for spectrum sharing with the spectrum broker. In: Proceedings of IEEE Symposium on New Frontiers in Dynamic Spectrum Access Networks (DySPAN 2012), pp. 355–365, 16–19 Oct 2012
23. OECD Communications Outlook. OECD Publishing. http://dx.doi.org/10.1787/comms_outlook-2011-en (2011)
24. Yearbook of Statistics: Telecommunication/ICT Indicators 2002–2011. ITU-D, Geneva (2012)
25. Global Wireless Matrix 2011. Bank of America, Merrill Lynch, Industry Overview. 28 Sept 2011
26. Sterman, J.D.: Business Dynamics: System Thinking and Modeling for a Complex World. McGraw-Hill, USA (2000)
27. Pistorius, C., Utterback, J.: A Lotka-Volterra Model for Multi-mode Technological Interaction: Modeling Competition, Symbiosis and Predator Prey Modes. Sloan School of Management, MIT, WP # 3929 (1996)
28. Sridhar, V., Casey, T., Hämmäinen, H.: Flexible spectrum management for mobile broadband services: how does it vary across advanced and emerging markets?. Telecommun. Policy **37**(2–3), 178–191 (2013)
29. Davies, J., Howell, B., Mabin, V.: Telecom regulation, regulatory behaviour and its impact–a system view. Commun. Strateg. **70**, 145–168 (2008)
30. Bauer, J.M.: Regulation, public policy, and investment in communications infrastructure. Telecommun. Policy **34**, 65–79 (2010)
31. Funk, J.L.: Standards, critical mass, and the formation of complex industries: a case study of the mobile internet. J. Eng. Technol. Manag. **28**(4), 232–248 (2011)
32. Mansfield, E.: Technical change and the rate of imitation. Econometrica **29**(4), 741–766 (1961)
33. European Commission: Proposal for a Directive of the European Parliament and of the Council on the harmonisation of the laws of the Member States relating to the making available on the market of radio equipment, COM(2012) 584 final ("Radio Equipment Directive") (2012)
34. European Commission: Guide to the Implementation of Directives Based on the New Approach and the Global Approach (2000)
35. Prasad, R.V., Pawelczak, P., Hoffmeyer, J.A., Berger, H.S.: Cognitive functionality in next generation wireless networks: standardization efforts. IEEE Commun. Mag. **46**(4), 72–78 (2008)
36. Electronic Communications Committee (ECC) Report 80. Enhancing Harmonisation and Introducing Flexibility in the Spectrum Regulatory Framework, Oulu (2006)
37. Merino Artalejo, M. F.: Market Impact of SR: Benefits and Barriers. Massachusetts Institute of Technology (MIT), USA (2002)
38. Technopolis Group: Impact assessment concerning a proposed mandatory registration system in the scope of directive 1999/5/EC–final report (2009)
39. Telecommunication Conformity Assessment and Market Surveillance Committee (TCAM), Document (31)21. Chairman's summary of the TCAM Workshop 'Software Defined Radio in the context of the revision of the R&TTE Directive' of 1 October 2010. https://circabc.europa.eu/w/browse/3ea3bbdb-2cec-4a30-aa15-c97b3400a456. Accessed 31 October 2013
40. TCAM, Document 18 (04) 25. Application of the Directive to Software Defined Radio (SDR) (2004)
41. European Commission, Document M/512 EN. Standardisation Mandate to CEN, CENELEC and ETSI for Reconfigurable Radio Systems. http://ec.europa.eu/enterprise/standards_policy/mandates/database/index.cfm?fuseaction=search.detail&id=515#. Accessed 31 October 2013
42. R&TTE Compliance Association (R&TTE CA) Document TGN 09 Rev1. Guidance on Essential Requirements, Essential Radio Test Suites & HS Requirements & Conformance Test specifications Table (HS-RTT) (2007)
43. European Commission: Second progress report on the operation of the directive 1999/5/EC, COM (2010) 43 final (2010)

44. Radio Spectrum Policy Group (RSPG): RSPG Opinion on Streamlining the Regulatory Environment, RSPG 08-246 final (2008)
45. European Commission: Proposal for a Regulation of the European Parliament and of the Council on European Standardisation and amending Council Directives 89/686/EEC and 93/15/EEC and Directives 94/9/EC, 94/25/EC, 95/16/EC, 97/23/EC, 98/34/EC, 2004/22/EC, 2007/23/EC, 2009/105/EC and 2009/23/EC of the European Parliament and of the Council, COM (2011) 315 final (2011)
46. TCAM, Document 28 (09) 06: Working document for problems definitions and objectives regarding the R&TTE Directive 1999/5/EC (2009)
47. QinetiQ: An Evaluation of Software Defined Radio–Main Document (2006)
48. Software Freedom Law Center: FCC Rules on FOSS and Software-Defined Radio (2007)
49. Hughes, T.P.: The evolution of large technological systems. In: Bijker, W.E., Hughes, T.P., Pinch, T.J. (eds.) The Social Construction of Technological Systems: New Directions in the Sociology and History of Technology. MIT Press, Cambridge (1987)
50. Medeisis, A., Minervini, L.F.: Stalling innovation of cognitive radio: the case for a dedicated frequency band. Telecommun. Policy **37**(2–3), 108–115 (2013)
51. Teece, D.J.: Profiting from technological innovation: Implications for integration, collaboration, licensing and public policy. Res. Policy **15**(6), 285–305 (1986)
52. Suarez, F.F.: Battles for technological dominance: an integrative framework. Res. Policy **33**(2), 271–286 (2004)
53. Fitch, M., Nekovee, M., Kawade, S., Briggs, K., Mackenzie, R.: Wireless service provision in TV white space with cognitive radio technology: a telecom operator's perspective and experience. IEEE Commun. Mag. **49**(3), 64–73 (2011)
54. Granelli, F., Pawelczak, P., Prasad, R.V., Subbalakshmi, K., Chandramouli, R., Hoffmeyer, J.A., Berger, H.S.: Standardization and research in cognitive and dynamic spectrum access networks: IEEE DySPAN-SC efforts and other activities. IEEE Commun. Mag. **48**(1), 71–79 (2010)
55. Casey, T.: Analysis of radio spectrum market evolution possibilities. Commun. Strateg. **75**(3), 109–130 (2009)
56. Chapin, J.M., Lehr, W.H.: Cognitive radios for dynamic spectrum access-the path to market success for dynamic spectrum access technology. IEEE Commun. Mag. **45**(5), 96–103 (2007)
57. Nguyen, T., Zhou, H., Berry, R.A., Honig, M.L., Vohra, R.: The impact of additional unlicensed spectrum on wireless services competition. In: Proceedings of the IEEE Symposium on New Frontiers in Dynamic Spectrum Access Networks (DySPAN 2011), pp. 146–155 (2011)
58. Fomin, V.V., Medeisis, A., Vitkute-Adzgauskiene, D.: In search of sustainable business models for cognitive radio evolution. Technol. Econ. Dev. Econ. **18**(2), 230–247 (2012)
59. Metcalfe, J.S.: Evolutionary economics and technology policy. Econ. J. **104**, 931–944 (1994)
60. Dosi, G.: Sources, procedures, and microeconomic effects of innovation. J. Econ. Lit. **26**(3), 1120–1171 (1988)
61. Holland, O. et al.: Pluralistic licensing. In: 6th IEEE Symposium on New Frontiers in Dynamic Spectrum Access Networks, Washington, USA, October 2012
62. Anker, P.: Cognitive radio-the need for coordination: lessons from the past. In: Pres. 3rd Meeting of COST Action IC0905 (TERRA), Brussels, 20–22 June 2011
63. Robert, C., Zeckhauser, R.: The methodology of normative policy analysis. J. Policy Anal. Manag. **30**(3), 613–643 (2011)
64. Delaere, S., Ballon, P.: Multi-level standardization and business models for CR: the case of the cognitive pilot channel. In: Proceedings of the IEEE Symposium on New Frontiers in Dynamic Spectrum Access Networks (DySPAN 2008) (2008)
65. Fomin, V.V., Medeišis, A., Vitkutė-Adžgauskienė, D.: In search of sustainable business models for cognitive radio evolution. Technol. Econ. Dev. Econ. **18**(2), 230–247 (2012)
66. ECC Report 192 on the Current Status of DFS (Dynamic Frequency Selection) in the 5 GHz Frequency Range (soon to be published)

Chapter 6
Case Studies for Advancing CR Deployment

Dariusz Więcek and Fernando Jose Velez

Abstract This chapter shows selected practical cases studies dealing with advancing CR deployment. Section 6.1 provides details about TV White Space spectrum estimation methodology based on ITU GE06 Plan rules for cases of countries where high levels of TV interference exist. In addition, an example of addressing the practical co-existence of TV white space devices with incumbent applications in the UHF TV band is presented in this section as well. Section 6.2 looks at the practical possibilities of deploying CR systems in the ISM bands, including a techno-economic viability study of CR solutions in factory scenario. Finally, Sect. 6.3 describes a concept and provides an in depth analysis of using CR technologies in medical environments.

6.1 Utilisation of White Space Devices in TV Bands

6.1.1 Methodology of White Space Estimation in TV Bands Based on ITU GE06 Rules

Dariusz Więcek

National Institute of Telecommunications, Wrocław, Poland

This section presents methodology of Television White Space (TV WS) estimation taking into account rules for broadcasting services published in the GE06 ITU

D. Więcek (✉)
National Institute of Telecommunications, Wrocław, Poland
e-mail: D.Wiecek@itl.waw.pl

F. J. Velez
Instituto de Telecomunicações–DEM, Universidade da Beira Interior, Covilha, Portugal

A. Medeisis and O. Holland (eds.), *Cognitive Radio Policy and Regulation*, Signals and Communication Technology, DOI: 10.1007/978-3-319-04022-6_6,

Agreement [1]. This method was used for preparing maps of TV WS availability in Poland, which were presented originally at COST-TERRA Workshop [2] as well as in CEPT work, which resulted in their inclusion in the ECC Report 185 [3].

6.1.1.1 Introduction

During two sessions of the ITU Regional Radiocommunication Conference RRC'04–RRC'06 technical principles for establishing TV Plans in the VHF and UHF TV bands (174–230 MHz and 470–862 MHz) were prepared. At the RRC Plans for the Digital Terrestrial Television (DTT) Assignments and Allotments were also developed. Rules for changing the Plans and future DTT coordination procedures were written in the GE06 Agreement [1].

Due to lot of incompatibilities between entries in the GE06 Plan (i.e. Assignment and Allotments data planned for inclusion into Plan) many such interference problems were solved by bilateral or multilateral Agreement among interested Administrations. This caused the situation where high interference levels among DTT stations in the UHF band exists in real operational conditions—which cannot be neglected in many countries.

Such "interference limited conditions" (where reception of the DTT should be analysed, taking into account existing interference coming from other DTT stations) is studied and considered, for example during DTT stations coordination procedures between Administrations—when one Administration wishes to include in the Plan new DTT Assignment or change its technical details. Such approach is useful also as a basis of the TV WS availability estimation methodology presented in this chapter.

6.1.1.2 The Methodology

Taking into consideration current UHF TV interference situation, protection requirements can be established by the general GE06 Agreement rules [1] used in case of analysis of new DTT station entry (new planned Assignment in Broadcasting Service). In such situation White Space Device (WSD) can be treated as an additional (much like as new DTT station) transmitter which could be analysed as a "new entry" into the Plan based on the GE06 technical rules. Protection criteria for DVB-T are based on GE06 allotments/assignments parameters (relevant to RN/RPC) for 95 % of location, receiving antenna at 10 m. a.g.l. with directional antenna (for fixed reception) or omnidirectional antenna (for portable receptions). In the analysis, interferences calculations are performed using propagation method ITU-R P. 1546 at 1 % of time 50 % of locations [4]. Aggregate interferences from WSDs should not exceed more than existing interferences from other DTT stations. Such calculation can be performed using GE06 Allotment or Assignment Plan entries data, however it is better to use real operational transmitter characteristics due to fact that such transmissions data can differ from the data written into Plan. In the case of this section, real calculation of DVB-T transmitter data of all DTT multiplexes in Poland were used, as well as, known technical data of transmitters of neighbouring countries.

In Poland fixed reception type (RPC1 GE06) is used for coverage and interference calculation in case of operational DTT Plans. Taking this into account the technical conditions rules based on the GE06 for calculation were as follows:

- Maximum permissible interfering level at DTT coverage areas: $\sim$ 56 dBmV/m (depending on exact TV channel number, using GE06 field strength corrections);
- Protection Ratio = 21 dB, as defined for GE06 Reference Planning Configuration (RPC1);
- Location probability correction = 13 dB (used for fixed reception RPC1 in case of 95 % of locations);
- Aggregation of WSD interference = 10 dB;
- Receiving antenna discrimination = −16 dB (Fixed antenna discrimination);
- Permissible interfering field strength $\sim$28 dBmV/m in 1 % of time at 10 m a.g.l.

The technical data were used for calculations of size of the buffer zones around DTT service areas generating exclusion areas due to protection of co-channel TV reception. For adjacent N + 1 and N-1 DTT channels the protection areas were the same as the adjacent channels DTT service areas—which means that no adjacent WSDs operations were allowed in the N + 1, N-1 DTT service areas.

6.1.1.3 WSD Channels Availability Calculations Example

TV White Space channels availability calculation based on the methodology presented in the previous subsection was performed. Figure 6.1 shows the calculation done at around 600,000 points (each point of territory within raster of 1 km × 1 km) within whole territory of Poland at UHF frequency range (470–790 MHz). The calculation was performed using the Digital Elevation Map of Poland and neighbouring countries. Protection of all relevant TV transmitters was taken into account.

The colors on the map represent the number of available channels for WSD operation without causing unacceptable interference (for 20 dBm EIRP portable WSD terminal with transmission at 1.5 m a.g.l.). The picture presents one example of such calculations. Further examples can be found in the ECC Report 185 [3].

6.1.1.4 Final Remarks

The methodology presented in this section was used for preparation of the TV WS maps with channels availability for different WSD type configurations and frequency ranges in Poland. Most important results were included in the ECC Report [3]. The method could be used also for TVWS channel availability estimation in

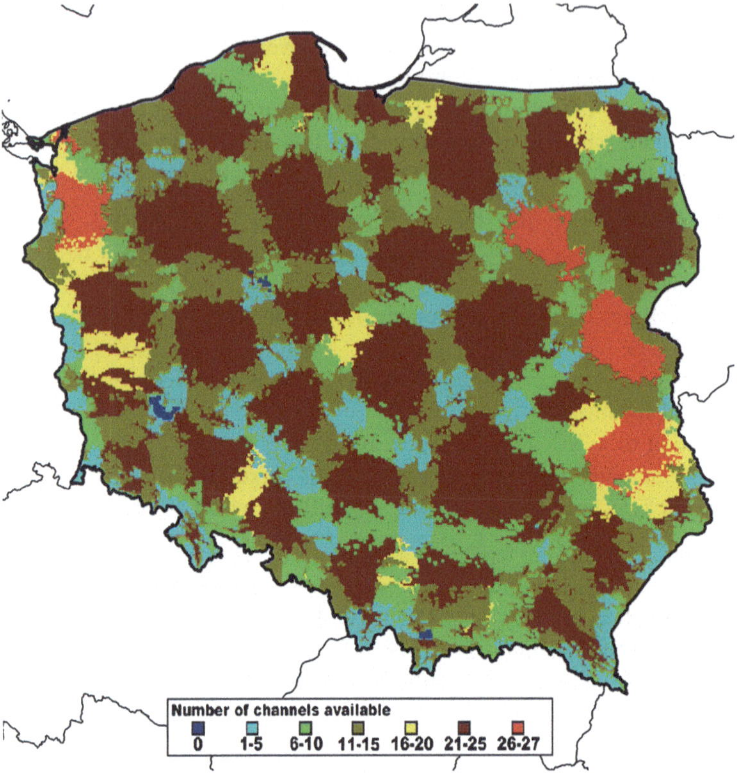

Fig. 6.1 TV white space channels availability calculation at around 600,000 points in Poland

other GE06 countries (ITU Region 1 and part of Region 3). Required DTT protection level may be established individually in the countries depending on local conditions (DTT type of reception, number of multiplexes, modulation etc.). Such approach is interesting especially in countries where high levels of interference exist which cannot be neglected, i.e. mainly in countries with high-power and high-elevated DTT transmitters.

6.1.2 TVWS Coexistence with Incumbents

Jarkko Paavola,[1] Juha Kalliovaara[2] and Jussi Poikonen[3]

[1] Turku University of Applied Sciences Turku, Finland
[2] University of Turku, Turku, Finland
[3] Aalto University, Espoo, Finland

Fig. 6.2 Considered TVWS CR license area

6.1.2.1 Test Environment

A TV White Space (TVWS) test bed has been set up in the WISE project [5], which is part of the Finnish Tekes Trial technology programme [6]. This test bed consists of the following components:

- A commercial level DVB-T test network with three transmitters;
- The Turku University of Applied Sciences radio laboratory equipped with e.g. white space radios, and spectrum analysis and measurement equipment.
- A complementary simulation environment developed by University of Turku and Aalto University.
- A full CR license for the frequency range 470 MHz -790 MHz issued by the Finnish Communications Regulatory Authority (FICORA). This radio license requires a geolocation database to be used to determine available frequencies and maximum transmission power for white space devices operating within the test network. The Turku TVWS testbed utilises a geolocation database maintained by Fairspectrum Ltd.

The CR license of the above outlined test network covers an area of approximately 40 km × 40 km over the Turku region as illustrated in Fig. 6.2.

The test environment has been utilised in interference measurements in order to evaluate maximum transmit power limits for cognitive devices while protecting incumbent operations. Incumbents to be protected in the frequency range 470–790 MHz are TV receivers and PMSE (Programme Making and Special Events) equipment including for example wireless microphone systems.

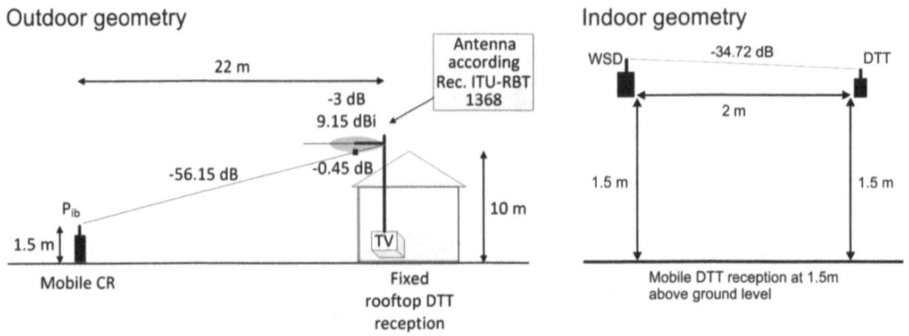

Fig. 6.3 Reference geometries reproduced in the outdoor and indoor measurement campaigns. Adapted from [9]

6.1.2.2 Interference Measurements and Simulations

Measurement campaigns were performed in the TVWS test network to quantify the sensitivity of digital TV reception to interference from nearby WSD devices. Specifically, test TV receivers and WSD transmitters were set up according to outdoor and indoor reference geometries proposed by the ECC [7] to represent worst-case interference scenarios where an interfering WSD is located relatively close to a TV receiver. These geometries are illustrated in Fig. 6.3.

The main objective of the measurements was to determine the protection ratios between received TV signal power and transmitted WSD power. Formally, this ratio was defined as

$$\mathrm{PR}(\mathrm{f_{PRI}},\mathrm{f_{WSD}}) = \frac{\mathrm{P_r}(\mathrm{f_{PRI}})}{\mathrm{P_{t,\,max}}(\mathrm{f_{WSD}})}$$

that is, the ratio of the received TV signal power centered at frequency $\mathrm{f_{PRI}}$ to the maximum allowable power transmitted by the WSD at center frequency $\mathrm{f_{WSD}}$. The maximum WSD transmit power was determined as the limit where error-free TV reception was still possible according to a subjective criterion corresponding approximately to ESR5 [8].

These measurement campaigns were performed with DVB-T and DVB-T2 signal configurations for various DTT receiver locations, signal strengths, and magnitudes of centre frequency separation between the WSD and DTT signals. Detailed descriptions of the measurement setup, scenarios, and numerical results are given in [9–11]. Complementary case studies on distributed sensing algorithms and their effect on secondary system throughput and intersystem interference were conducted using simulation model of the TVWS test network; results of these studies have been presented in [12, 13].

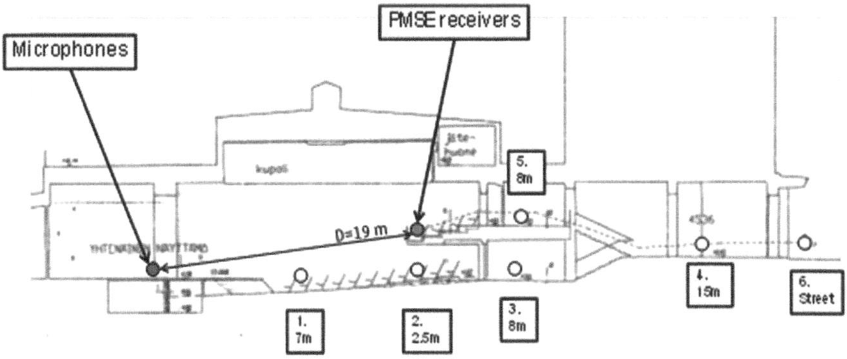

Fig. 6.4 Locations of the PMSE equipment and WSD transmitters measured at the Helsinki City Theatre. Adapted from [14]

6.1.2.3 PMSE Measurements

In addition to DTT receivers, PMSE equipment for example wireless microphone systems operating on the TV White Space frequencies in the lower UHF band also need to be protected from WSD interference. Geolocation databases are currently considered to be the primary method of protecting PMSE users [7]. Sensing techniques are currently not sufficient to provide reliable protection for PMSE [10] and the range of potential deployment scenarios causes large variability in the sensing thresholds [7]. In the following we outline a measurement campaign conducted to evaluate interference between WSD and PMSE devices in a real usage scenario [14].

PMSE equipment was installed into real operating environments in the Helsinki City Theatre. Interference to the PMSE equipment was caused by a simulated WSD operating on the co- or adjacent channel. The measurements were performed with the WSD in several locations inside and outside the building. Figure 6.4 illustrates the location of the microphones and PMSE receivers and the WSD locations (yellow circles) inside the Helsinki City Theatre. Subjective evaluation was used to determine when the WSD were causing audible interference.

The measurements clearly show that a WSD operating co-channel with the PMSE causes interference on very low power levels (−15 dBm … +5 dBm) when the WSD is in the vicinity of the PMSE equipment. Thus, the co-channel operation of PMSE equipment and WSDs is not possible in the vicinity of PMSE receivers. The protection level on the adjacent channel were approximately +30dBm and on the next adjacent channels +40 dBm or more. These values would allow adjacent channel operation with reasonable power levels for the WSDs located close to PMSE equipment.

One approach to manage interference between WSD and PMSE devices is to specify an exclusion zone of certain diameter around the PMSE equipment. WSDs could not operate co-channel with the PMSE equipment when the WSDs are inside the exclusion zone.

Therefore the PMSE equipment and their locations must be registered to the geolocation database if protection is needed. In the measurements conducted at Helsinki City Theatre it was not possible to cause co-channel interference at a distance of 560 m with the maximum transmit power of approximately 10 W. More detailed studies are still required to reliably estimate the sizes of exclusion zones for PMSE devices.

6.2 CR in ISM Bands

6.2.1 Extracting Interference Information from the ISM/RLAN bands for CR Applications

John Sydor and Salim Hanna

Communications Research Centre, Ottawa, ON, Canada

6.2.1.1 Operating CR in the ISM/RLAN Bands

The 2400–2483.5 and 5725–5825 MHz ISM and RLAN bands possesses several attractive features that make them sp. good candidates for CR. The regulatory burden associated with these bands is minimal and there are no licensing fees or licensing requirements for their use. Low-power radio equipment must only meet specific technical requirements to be licence-exempt; some very low-power equipment is even exempt from certification requirements [15, 16].

Spectrum regulators require no or little coordination amongst the different radio systems and devices operating in the ISM/RLAN bands. Therefore most radio applications operate without coordination or coexistence mechanisms, though a few (e.g., Bluetooth and Wi-Fi) are designed to coexist. Thus it is difficult for a radio equipment to perform its tasks without either falling victim or causing harmful interference to or from a multitude of ISM equipment and licence-exempt radio applications such as Wi-Fi (802.11), Bluetooth (802.15.1), ZigBee (802.15.4), video senders, cordless phones, RFID tags, Radars, or medical devices for example.

To make these bands truly attractive for CR, it is imperative that the CR Networks (CRN) working in these bands ascertain the quality of the spectrum and adapt accordingly to mitigate the deleterious effects of uncontrolled an highly variable interference. Real-time and historical knowledge of the radio environment is needed by CR processes such as Dynamic Spectrum Assignment (DSA) which rely on figures of merit and data that must be extracted from the raw sensed information that the CRN collects. The ISM/RLAN bands are rich in a number of interference metrics, of which a subset can be chosen by the CRN designers in order to create data bases or drive cognitive and adaptive algorithms.

These metrics can be collected in a number of ways, for example: by co-operative/synchronized sensing across the CRN; by individual terminals of the CRN sensing during their quiet (no-transmission) periods, or by designated terminals acting as sensors embedded throughout the CRN. For ISM and RLAN sensing the choice of metrics is dependent on the requirements of the CR algorithms and the bandwidth requirements of moving sensor information through the CRN to its CR processors and data bases. Two categories of sensor information are available to us for ISM band CR applications: spectrum analysis/RSSI measurements which give power or energy densities over a scanned spectrum band, or demodulated or "sniffed" information which is provided by specialized interference receivers responding to the largest category of interference in the ISM and RLAN bands, i.e.: Wi-Fi. In addition to sensed information, terminal performance metrics such as retransmission rate, packet error rate, and data uplink and downlink throughput/goodput, can be included and provided to the CR processors to support cognitive operation.

6.2.1.2 Power Detection Approaches

Power Spectral Density: Spectrum analysis (SA) has conventionally been used to ascertain radio activity in a band. The process involves measurement of radio signal power within a specific resolution bandwidth (RB), resulting in the creation of a Power Spectrum Density metric. Typically a band is scanned at the RB, a process that usually cannot be carried out at a rate where very rapid (50–500 μs) Wi-Fi packet events are fully captured, thus necessitating multiple scans and integration of readings.

However the process can be simply and cheaply implemented using a common RF radio receive chain, usually with the data demodulation and spectrum analysis processes operating separately because of the requirement of the SA to scan off-channel while the demodulation stays on-channel. The average received power spectral density *PSD*(b, s, T) from Wi-Fi and non-Wi-Fi signals is measured in bins having (b) (*RB*) bandwidths over a particular geometric space (s) and a specific period of time (T):

$$PSD(b, s, T) = N^{-1} \sum\nolimits_{n=1}^{N} psd_n(b, s, T) \qquad (6.2.1)$$

where N is the number of measurements over the monitoring period, and $psd_n(b, s, T)$ represents the total received power in the frequency bandwidth b (e.g.; 1 MHz) from DC to the Nyquist sampling rate as captured by the spectrum analyser.

It is noted that this measurement is applicable to the space, s, determined by the capture area of the antenna and that T should be of a sufficiently long integration time to appropriately sample the broadband but transient packet transmissions of the interference. Narrowband PSD analysis is useful primarily for detecting interference sources such as microwave ovens, radars, and video transmitters, which are deleterious sources of interference, necessitating channel changes.

RSSI: The measurement can be obtained by implementing an energy detector of bandwidth *b*. For detection of Wi-Fi, *b* is typically 17 MHz. Measurement of the average received signal strength indicator RSSI(*b*, *s*, *T*), of a frequency channel with a bandwidth *b* over a particular geometric space *s* and a specific period of time, *T*, is given by:

$$RSSI(b,s,T) = N^{-1}\sum\nolimits_{n=1}^{N} rssi_n(b,s,T) \tag{6.2.2}$$

where *N* is the number of packets captured per channel over the observation period *T*, and $rssi_n(b,s,T)$ is the received signal strength acquired during receiving an 802.11 packet.

This metric provides an indication of the severity of interference due to the interference on a channel and is useful in qualitatively assessing channel choices; the technique is often used in Wi-Fi routers to provide a channel assessment but provides no information on the frequency or duration of the interference. Readings can be skewed by the occurrence a relatively few and benign high power interference readings over the sampling period, *T*.

Band RSSI Occupancy: An improved RSSI metric considers occupancy of a band (*b*) by radio signals over a period of time, allowing a cumulative distribution function (CDF) for the RSSI to be determined. Such a metric can be determined b measurement of the RSSI in a band (*b*) over a period of time, and determining the intervals at which the RSSI was at or above specific levels. Such data then allows compilation of occurrence (probability) statistics for the RSSI, thus giving cumulative distribution function of the RSSI variable *X* as the function given by:

$$F_X(x) = P(X \leq x) \tag{6.2.3}$$

where the right-hand side represents the probability that the random variable *X* takes on a value less than or equal to *x*. The probability that *X* lies in the semi-closed interval (*a*, *b*), where $a < b$, is then:

$$P(a < X \leq b) = F_X(b) - F_X(a) \tag{6.2.4}$$

The RSSI CDF is a useful channel quality metric as it allows unambiguous comparison of the interference between channels over a common period of time. The implementation of a CDF analyser is slightly more complicated than a RSSI detector, but essentially involves the use of fixed bandwidth RSSI detectors having the standardised Wi-Fi and RLAN channels' width (17 MHz). Such detectors are sampled at rates significantly shorter than the smallest expected Wi-Fi packets to create a data base of occurrence statistics. Such detectors typically provide an assessment of the interference caused by Wi-Fi packets only, and the metric can be skewed by narrowband continuous interference (which is best detected by Narrow Band PSD). The CDF statistics shed no information of the packet dimensions or the inter-packet arrival time for the Wi-Fi interference. For such information other approaches need to be used, however they can rely on the same detector to supply their raw information.

6.2.1.3 Monitor-Mode Receivers

Wi-Fi radios with MAC layer software modifications can be highly effective ISM/RLAN band interference detectors. Operating in what is known as a 'monitoring' or 'sniffing' mode, IEEE 802.11 signals, including both desired and interference packets are received and processed to extract a number of useful channel interference metrics. Such detectors can share common RF front end receivers and operate separately from the ISM/RLAN band Wi-Fi receivers. In such a manner they can used to scan a number of channels while the data receiver remains fixed on the operating channel.

Interferer Number: Operating in monitor mode allows determination of the unique MAC source addresses of the packets and thus can identify the number of unique transmitters using the channel. The total number of interferers *I*(*b*, *s*, *T*) captured in a frequency channel with a bandwidth *b* over a particular geometric space *s* and a specific period of time *T*, is given by:

$$I(b, s, T) = M \tag{6.2.5}$$

where *M* is the total number of unique 802.11 MAC source addresses captured during the observation period *T*. The metric can be used to assess immediate channel quality and also provide insight into the long term occupancy and use of a channel if historical records are kept of MAC source addresses' occurrences on the channel with time. By such means it is possible to identify diurnal operational behaviors of specific interference sources or groups of interference sources (such as networks) and facilitate long-term dynamic channel selection approaches taken by cognitive algorithms.

Packet Duration, Mean Time between Packets: The ability to monitor interference packets allows their frequency of occurrence, duration, and mean time between packets (inter-arrival time) to be determined. Such metrics can be examined as a function of the RSSI, thus allowing determination of average levels of interference occurrence; which becomes a useful metric if the desired data link's RSSI operational threshold is known and/or is adjustable. This leads to the notion of channel and band occupancy being threshold dependent, allowing determination of the temporal qualities of the interference environment (i.e., provision of a statistic such a percentage of interference packets above a specific RSSI). Some details of how such metrics can be used to provide an enhanced assessment of interference behaviour are given in [17].

Channel Occupancy: A frequency channel with a bandwidth, *b*, is considered occupied or busy, over a particular geometric space (*s*) and a specific period of time *T*, whenever it contains radio frequency energy (due to own activity or emissions from co-channels and adjacent channels within the same system and/or other systems), *E*, that exceeds a specific threshold, T_o (the threshold is usually taken to be X dB with respect to the noise floor):

$$E(b, s, T) > T_o \tag{6.2.6}$$

Measured channel occupancy, $\varphi(b, s, T)$, can be given by a ratio representing the amount of time, t for which the channel bandwidth is sensed busy over the monitoring period T:

$$\varphi(b, s, T) = t/T \tag{6.2.7}$$

where s is the inner space of locations at which the channel is measured busy. $0 \leq \varphi(b, s, T) \leq 1$ where a value close to zero indicates low activity or channel availability while a value close to unity indicates high activity, potential channel congestion, and performance degradation.

Band Occupancy: A frequency band representing the union of bandwidths of N adjacent or overlapping channels, $B = U_{n=1}^{N} b_n$, is considered occupied, over the geometric space $S = U_{n=1}^{N} s_n$ and time period T, whenever its ensemble average of channel occupancies exceed a specific threshold T_O:

$$\Phi(B, S, T) = N^{-1} \sum\nolimits_{n=1}^{N} \phi_n(b, s, T) > T_O \tag{6.2.8}$$

where measured channel occupancies $\varphi_n(s, b, T)$ are determined for all the channels over the same monitoring period T. $0 \leq \Phi(B, S, T) \leq 1$ where a value close to zero indicates spectrum availability while a value close to unity indicates service degradation and potential band congestion.

6.2.1.4 Summary

The ISM and RLAN bands are characterized by a large variety of licensed-exempt users that create a chaotic interference environment that needs quantification if cognitive radio operation is to be supported. Traditional metrics that quantify interference alone, such as RSSI measurement and Spectrum Analysis, have limited use but can be enhanced to provide a temporal dimension to interference occurrence, as we have shown. Additionally, techniques using modified Wi-Fi receivers (sniffers) that examine Wi-Fi packet addressing can implement detectors that will provide highly useful and accurate assessments of channel occupancy, interference origin and location, and temporal behavior. Such techniques can provide a quantified, statistical assessment of the interference environment and support cognitive radio techniques such as dynamic channel selection, network power control, spectrum reuse, and collaboration by CRNs.

6.2.2 Spectrum Utilisation and Congestion of IEEE 802.11 Networks in the 2.4 GHz ISM Band

Roel Schiphorst and Jan-Willem vanBloem

University of Twente, Enschede, The Netherlands

Wi-Fi technology plays a major role in society thanks to its widespread availability, ease of use and low cost. Many new applications have emerged for such technologies, intelligent transportation systems (ITS), Dynamic Spectrum Access (DSA) systems and offloading of traffic from cellular networks. To assure its long term viability in terms of capacity and ability to share the spectrum efficiently, it is of paramount importance to study the spectrum utilisation and congestion mechanisms in live environments. In the measurements reported in this section we have focused on the crowded 2.4 GHz ISM band. The number of wireless devices (smartphones, laptops, sensors) that use this band is rapidly increasing. In many urban areas not only many WLAN networks can be found, also other systems like Bluetooth, Zigbee and wireless A/V transmission systems use this band.

On the other hand there is only a limited amount of spectrum available. So it is very likely that interference between systems in this band will occur. Due to the rapid increase of number of wireless devices, interference issue is expected to become even more important. We have addressed this issue by providing a setup to measure the service level—i.e. can all devices fulfil their communication needs—in this band with focus on WLAN standard's amendment IEEE 802.11e (the upcoming IEEE 802.11e is an extension of the 802.11 Wireless Local Area Network (WLAN) standard and is developed to enhance Quality of Service (QoS) support).

Researching spectrum utilisation and congestion in 802.11 networks is important for CR in the broad sense, as many CR implementations are based on 802.11 technologies, but translated to a different frequency band. An example being the 802.11af extension for white spaces in TV bands [18]. Identifying bottlenecks in 802.11 can be therefore used to enhance throughput and QoS which is especially relevant in the heterogeneous environments where CR is often used.

In this research, a cross-layer approach is used [19], since the service level can be measured at several levels of the protocol stack. The focus is on monitoring at both the Physical (PHY) and the Medium Access Control (MAC) link layer simultaneously by performing respectively power measurements with a spectrum analyser to assess spectrum utilization and packet sniffing to measure the congestion. Compared to traditional QoS analysis in 802.11 networks, packet sniffing allows studying the occurring congestion mechanisms more thoroughly. The measurement equipment consists of:

- Packet sniffer: to capture raw packets on MAC link layer level;
- RF monitoring equipment: spectrum analyser tuned to the 2.4 GHz ISM band.

The monitoring was applied for the following two cases. First the influence of interference between WLAN networks sharing the same radio channel was investigated in a controlled environment. It turned out that retry rate, Clear-To-Send (CTS), Request-To-Send (RTS) and (Block) Acknowledgment (ACK) frames can be used to identify congestion, whereas the spectrum analyser may be employed to identify the source of interference. Secondly, live measurements were performed at three locations to identify the type of interference in real situations.

Below we describe the results of live measurements in more detail. Figure 6.5 shows the mean number of frames per second per location. It has been split up into

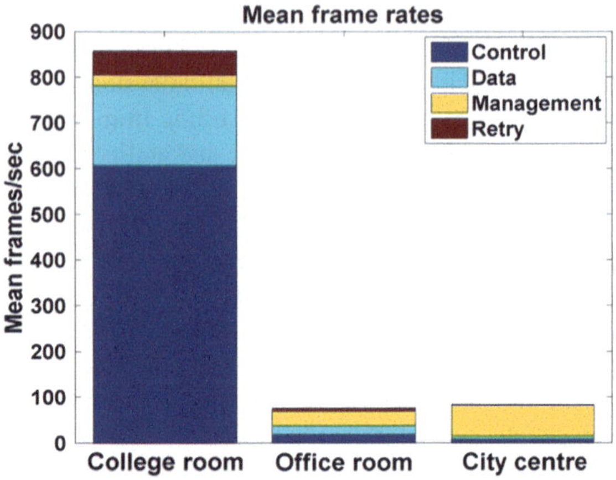

Fig. 6.5 Occurrence of different packet types

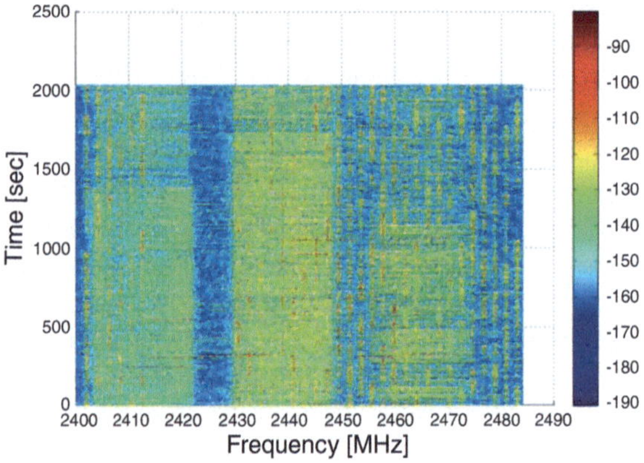

Fig. 6.6 RF spectrum in the college room

management, control, data and retry frames. First of all the results show that the college room location has most traffic which is due to measurements that were carried out at a college area with 75 to 100 students. Secondly, in this location about 70 % of the captured traffic were control frames and only about 21 % where the actual data traffic. The sub-field identifiers reveal that most of the control frames are RTS, CTS and, to a lesser extent, ACK and Block ACK packets. This is in line with the lab experiments where the same type of control frames have been shown to identify congestion.

The measurements carried out in a college are most interesting and reviewed below in more detail. Figure 6.6 shows the observed RF spectrum. From Fig. 6.7 it can be seen that the WLAN traffic is very spiky as expected. These measurements were performed during a college classes being in session until the time corresponding to 1400 s mark in scales of Figs. 6.6 and 6.7.

It may be noted that when the classes ended at 1400 clock mark, a sharp downwards transition is observed in the RF channel occupancy of Fig. 6.6. Moreover, at this particular moment the occupancy drops from values around 65–20 %.

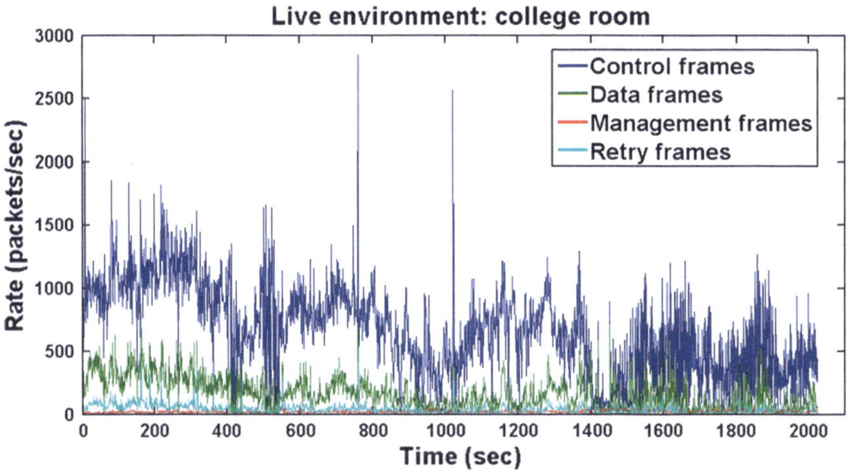

Fig. 6.7 Packet rate for the different MAC packet types versus time in the college room

6.2.2.1 Conclusion

Our measurements revealed that there is severe performance degradation in 802.11 networks in scenarios with many users, i.e. in situations where capacity is most needed. Especially the performance of the MAC layer based on CSMA/CD (Carrier Sense Multiple Access/Collision Detection) degrades in these scenarios; decreasing the throughput of data packets to less than 25 % of all packets. Similar performance degradations can be expected by CR applications based on 802.11 technology. Further research should focus on improving the performance, an example being oCSMA (optimal CSMA) [20].

6.2.3 Techno-Economic Viability of Cognitive Solutions for a Factory Scenario

Vânia Gonçalves[1] and Lieven Tytgat[2]

[1] iMinds–SMIT Vrije Universiteit Brussel, Brussels, Belgium
[2] iMinds–IBCN Ghent University, Ghent, Belgium

Wireless and mobile applications are projected to grow by orders of magnitude in the coming years. While the number of devices explodes, so do their application domains. These different application domains have specific needs, which are tackled by an ever-increasing amount of different technologies. The major concern arising is that a number of these technologies use the same limited spectrum, even though they have not been designed to coexist with one another, resulting in a severe degradation of the operation of these networks. Dynamic Spectrum Access (DSA) promises to alleviate most of the coexistence problems between different technologies. However, DSA economic and business feasibility is still to be recognised by concrete results showing economic gains. In this section we hypothesise about how valuable could DSA be in indoor settings. Towards this goal, different implementation alternatives of spectral sensing engines within an industrial context are assessed and the trade-off between the added cost and the increased usability of coexisting IEEE 802.11 and IEEE 802.15.4 networks is identified.

When considering actual deployment of DSA in real-life, there is a natural techno-economic trade-off between benefit and cost. We focus on an industrial plant where a ZigBee based wireless sensor network monitors and controls the production equipment, while WLAN provides wireless access to the data network of the plant. Within such an environment Wi-Fi and ZigBee coexist in the unlicensed ISM band. The economic benefit of implementing DSA on top of regular ZigBee and Wi-Fi is the reduction of machine failure rate and production disruption. An added cost will arise due to the actual implementation cost of the selected solution and increased cost of battery replacement due to the shortened battery lifetime.

6.2.3.1 Business Scenario

A modern electronics contract manufacturer that operates multiple Surface Mount Technology (SMT) assembly lines is the focus of our scenario. A mid-size manufacturer may operate a production floor with 15 assembly lines in parallel. Each line includes 3–4 robots and one oven, and is constantly monitored by 2 human operators on the production floor. Each robot contains 2 cameras and 6–7 different ZigBee sensors, while the ovens contain another 10 ZigBee sensors each, bringing the total number of ZigBee sensors throughout the production floor to 600. These sensors form a ZigBee wireless sensor and actuator network (WSAN). They measure the temperature and other parameters of machinery and processes on the assembly line, and transmit it periodically to a central control and monitoring system. This system alerts human operators of various types of malfunctions, e.g. component-feed problems and overheating, which typically happen multiple times every day.

The wireless LAN in the factory is composed of 100 Wi-Fi devices, including access points, laptops, portable terminals and smartphones. For example, each of the operators of the assembly lines has a portable terminal that he uses to control software download to the assembly machines, verify that proper material is loaded in the robots, etc. Since the sensors are located to monitor critical parameters in the assembly lines, loss of ZigBee data might lead to severe damage to machinery and significant loss of material. Two types of failure are possible. Major Failures are ones that risk damage to machinery. If, for example, a machine overheats while ZigBee packets are lost, the supervisors will not be alerted in time, which could lead to serious damage to the machine and a full stop of the assembly line until the damage is repaired. This would reduce production output, and decrease revenue as a result. Minor Failures are ones that only risk loss of material and profit. If, for example, one of the SMT component feeders gets jammed, then all products that continue to be produced before the problem is fixed are damaged, and considered lost. In our scenario each assembly line uses $700 worth of materials and produces $300 of profit per hour of uninterrupted operation. We assume that every assembly line develops conditions that, if not detected on time, will cause a Major Failure once every year. We also assume that, on average, every assembly line suffers a Minor Failure once every hour. Furthermore, we estimate that an assembly line that suffers a Major Failure will shut down for 24 h, and the total cost of repair, in labour, equipment and replacement parts, is $10,000. We also estimate that if a Minor Failure occurs while ZigBee packets are lost, it will take additional 30 s to detect the failure and stop production.

Due to the substantial opportunity costs and repair costs, it is clear that the factory owner is interested in reducing interference to an acceptable minimum. Therefore, we propose the solution of adding cognitive elements to the wireless devices. These come however at an investment cost that must be balanced with the performance gains they promise to deliver. We compare over a 5-year period a reference Scenario 1 of a factory with standard Wi-Fi and ZigBee networks and no cognitive solutions, to three alternative set-ups: Scenario 2 consists of deploying a sensing engine on ZigBee devices; Scenario 3 with sensing engines deployed in Wi-Fi devices; and Scenario 4 with sensing engines deployed in both ZigBee and Wi-Fi devices.

6.2.3.2 Potential Benefits

Sensing reduces the interference between the ZigBee and Wi-Fi networks. In fact, sensing therefore reduces the amount of machinery and assembly line failures, which are caused by late alerting due to interference. The economic gains of sensing are thus derived from the amount of failures (along with their costs and losses) that can be avoided. These failures can, as mentioned in the scenario description, be divided into two groups. In the absence of any monitoring sensors, Major failures would occur on average once a year on each line. With a total cost of $1.290.000 over 5 years, Major failures represent a very large potential loss for the factory. Again, in the absence of any monitoring sensors, Minor failures would occur on average once an hour on each line and would account to $5.475.000 over 5 years, representing even a larger potential loss than Major failures. In summary, the potential total cost of failures in 5-years-time amounts up to $6.765.000. This significant figure is the reason monitoring sensors are indeed deployed in assembly lines and other industrial plants.

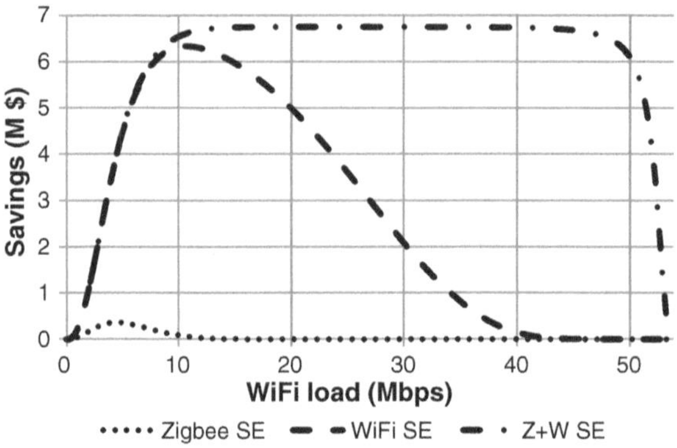

Fig. 6.8 Savings due to implementation of sensing as function of average Wi-Fi load over a 5-year period

6.2.3.3 Potential Costs

The additional investment cost comes down to the extra price of a node that is equipped with a sensing engine. The core of this engine is an Application Specific Integrated Circuit (ASIC), of which the cost is estimated at $1. Within a Wi-Fi device, no additional components need to be added and therefore we estimate the cost of one sensing engine for a Wi-Fi device at $1. For ZigBee sensors it is necessary to add additional components. We estimate the total cost of this sensing engine at $10. Because there are 600 ZigBee nodes and 100 Wi-Fi devices throughout the factory, we estimate the total additional investment in Scenario 2 at $6.000, in Scenario 3 at only $100 and in Scenario 4 at the sum of these 2 cases, $6.100.

6.2.3.4 Conclusions

Figure 6.8 presents the total expected savings due to the implementation of sensing engines compared to no sensing. Scenario 2 (ZigBee SE) presents little savings in a very small Wi-Fi load range. Scenario 3 (Wi-Fi SE) shows high savings under most common Wi-Fi loads, while Scenario 4 (Wi-Fi and ZigBee SE) presents high savings under all real-world Wi-Fi loads.

The costs due to machine failure in this setting are limited to only $1K at 28.6 Mbps, $10K at 38.5 Mbps and $100K at 46 Mbps average Wi-Fi load during a 5-year period. By effectively limiting the costs, the savings are thus the highest for a scenario in which the sensing engine is implemented in both Wi-Fi and ZigBee. This is a consequence of the high reliability, for low as well as high Wi-Fi loads. Although this analysis is within a certain factory scenario, its results can be extrapolated towards different settings where reliability of the ZigBee network is needed. Additional insights about this work can be found in [21], [22], [23].

Acknowledgments: The research leading to these results has received funding from the European Union's Seventh Framework Programme FP7/2007-2013 under grant agreements n° 257542 (CONSERN project) and n° 258301 (CREW project). It has also received funding from IWT under projects ESSENCES and WI-NETS2020. Contributions from colleagues and project partners are here acknowledged, with special contribution from Matthias Barrie.

6.3 CR in Medical Environments

Raúl Chavéz-Santiago[1], Norberto Barroca[2], Fernando J. Velez[2], Ilangko Balasingham[1], Henrique M. Saraiva[2], Paulo T. Gouveia[2], Jorge Tavares[2] and Luis M. Borges[2]

[1] Oslo University Hospital Norwegian University of Science and Technology, Trondheim, Norway
[2] Instituto de Telecomunicações–DEM Universidade da Beira Interior, Covilha, Portugal

6.3.1 Introduction

Wireless communications have the potential to impact beneficially medical practice through the development of ubiquitous health monitoring solutions [24, 25]. This can be achieved through the use of biomedical sensors combined with small wireless intercommunicating radio transceivers for measuring, transmitting, and storing different physiological signals in real time. The interconnection of these wearable and implantable devices constitutes a body area network (BAN), which is the core of modern telemedicine systems (Fig. 6.9).

A dedicated frequency band in 2360-2400 MHz for BAN use on a secondary basis has recently been designated in the United States. Nevertheless, it is anticipated that a large number of wireless biomedical sensors will operate in unlicensed frequency bands too. In fact, the BAN IEEE 802.15.6 standard [26] has recommended the unlicensed 2.4 GHz industrial, scientific, and medical (ISM) frequency band in 2400–2500 MHz and ultra wideband (UWB) in 3.1–10.6 GHz as alternative spectrum for wireless biomedical sensors. Small transceivers compliant with the IEEE 802.15.4 (ZigBee) standard are commercially available for operation in unlicensed ISM bands and have been found suitable for health and fitness monitoring in confined indoor areas [27]. In addition, a large number of wireless local area networks (WLAN) based on the family of IEEE 802.11 standards also share the 2.4 GHz ISM band, making the coexistence of these different wireless devices challenging [28]. Similar interference scenarios may be expected in the UWB band [29]. Techniques to avoid mutual interference must be applied in such cases. In hospital scenarios, the coexistence problem is more critical in small areas like intensive care units (ICUs) and operating rooms (ORs) because electromagnetic interference (EMI) from wireless devices can disrupt the performance of non-communication medical equipment that is routinely present in such

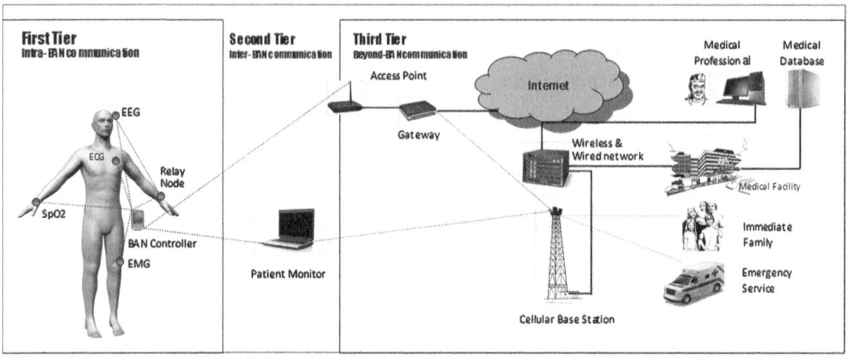

Fig. 6.9 Three-tier architecture of a telemedicine system

premises. CR is a promising technology that can ease the coexistence of wireless devices while protecting the electronic medical equipment. Despite the recognized potential benefits of CR for BANs [30], this application has not been extensively investigated and just a few solutions have been proposed.

Future improvements in radio frequency (RF) energy harvesting technology will facilitate the creation of a network with no need of dedicated transmitters, as a reliable source of wireless energy power [31]. This can be accomplished by enabling the capture of electromagnetic energy from multiple available ambient RF energy sources, such as mobile base stations, TV and radio transmitters, microwave radios, and mobile phones. Moreover, since wireless body area network (WBAN) nodes are battery operated, energy recharging is a possibility, avoiding the need of battery replacement. However, the service lifetime of the electronic components could be a major concern if there is no possibility to collect enough energy to generate the voltage needed to drive the sensor node. Medium access control (MAC) and routing protocols also play an important role in the network performance [32]. As a consequence, choosing the best opportunities poses a high effect on the overall network performance, as well as on the energy consumption.

The WBAN MAC protocols are responsible for providing the mechanisms for scheduling and allocation of the shared wireless channel. Compared to conventional WBANs, the MAC layer of WBAN nodes with CR capabilities must handle additional challenges, such as silent spectrum sensing periods and the need for high priority access mechanisms, for the distribution of spectrum sensing and decision results [33]. Therefore, the new innovative MAC protocols must be design regarding energy efficiency.

IEEE 802.15.4 has become the de facto standard for Wireless Sensor Networks (WSNs) being used in a wide range of scenarios and applications. The associated MAC protocol is responsible for triggering the current transmission allowing for multiple sensor nodes to share the same communication medium as well as to determine and change the operation mode of the radio transceivers whilst saving energy.

6.3.2 Hospital Scenarios

CR has been identified as the enabling technology to tackle spectrum scarcity and interference in healthcare and medical telemetry by enabling dynamic utilization of the wireless medical telemetry services (WMTS) frequency band, which comprises different parts of the spectrum, namely 608–614, 1395–1400, and 1427–1432 MHz [34]. A CR request-to-send/clear-to-send (RTS/CTS) protocol for e-health applications was proposed in [35], which adapts the transmit power of wireless devices operating in 2.4 GHz according to standardized electromagnetic immunity (EMI) constraints. The protocol effectively handles two different types of medical application traffic with different priorities. Through computer simulations it was demonstrated that this EMI-aware RTS/CTS protocol can reduce significantly the interference to protected non-communication medical devices in comparison to other medium access control (MAC) protocols like the specified by IEEE 802.15.4. Finally, a CR solution for BAN based on ultra wideband (UWB) technology was proposed in [36]. UWB signals offer many advantages to BANs, and some features of this technology can be exploited for effective implementation of CR. Below, the two latter approaches, namely CR for BAN in 2.4 GHz and UWB bands, are described with more detail.

6.3.2.1 Cognitive Radio Solution in 2.4 GHz

In the hospital case study in [35], two different types of traffic from two wireless e-heath applications were considered to be handled by the CR system:

(1) Real-time non-critical telemedicine, which is used to transmit data that are not delay/loss-sensitive, e.g., remote consultation, patient record transfers, and remote diagnosis.

(2) Hospital information system, which collects patient, technical, and facility data that are intended for better clinical decisions and to prevent patient complications. This system collects information with the aid of BANs and other wireless sensor networks (WSNs) located in the hospital.

In the CR context, the telemedicine system is treated as primary user (PU) and the hospital information system as secondary user (SU).

The CR system consists of three components, namely an inventory system, a CR controller (CRC), and CR clients. The inventory system is a database containing information about all the medical devices in the hospital premises. Information like location, activity status, and EMI immunity levels are stored in the database. The CRC is a computer that controls the transmission parameters of the CR clients, i.e., PUs and SUs. For this sake, the CRC uses the information in the inventory system to compute the appropriate transmit power for each CR client in order to avoid interference that exceeds the EMI immunity levels of non-communication medical devices located in the vicinity. The CR system operates using a dedicated control channel (DCC) and a data channel (DATC). Both channels are in unlicensed spectrum, e.g., the 2.4 GHz ISM band. Every CR client transmits its data through the CRC. The CRC can transmit/receive data from both channels simultaneously, whereas the CR clients can transmit just in one of the two channels at a time. The DCC is used to broadcast information to all the CR clients about their corresponding maximum power, P_{ctrl}, for transmitting RTS messages. Each CR client has a different P_{ctrl} value depending on its location, and it is calculated as

$$P_{\text{ctrl}} = \min\left\{\min_n(P_{\text{NLS}}(n)), \min_m(\text{P}_{\text{LS}}(m))\right\} \tag{6.3.1}$$

where $P_{\text{NLS}}(n)$ and $P_{\text{LS}}(m)$ are the upper bounds on transmit power for non-life-supporting (NLS) medical device n and life-supporting (LS) medical device m, respectively. For a frequency range of 800–2500 MHz, these transmit power values can be computed as:

$$P_{\text{NLS}}(n) = \left(\frac{D_{\text{NLS}}(n)E_{\text{NLS}}(n)}{7}\right)^2 \tag{6.3.2}$$

and

$$P_{\text{LS}}(m) = \left(\frac{D_{\text{LS}}(m)E_{\text{LS}}(m)}{23}\right)^2 \tag{6.3.3}$$

where $D_{\text{NLS}}(n)$ and $D_{\text{LS}}(m)$ are the distance from the CR client to the NLS device n and LS device m, respectively; $E_{\text{NLS}}(n)$ and $E_{\text{LS}}(m)$ are the EMI immunity levels for the NLS and LS medical devices n and m, respectively, the values of which are stored in the inventory system.

Since the protected non-communication medical devices can be turned ON or OFF and the locations of the CR clients change dynamically, P_{ctrl} must be computed and broadcasted every t_p slots on the DCC. All the transmissions from CR clients are paused during broadcasting of P_{ctrl} to synchronize with the CRC.

In order to access the channel, a CR client transmits a RTS message to the CRC on the DCC. If a collision occurs, the colliding CR clients wait for a random time based on a constant back-off window for PUs and exponential back-off window for SUs. The CR clients can retransmit the RTS message with probability α_1 for PUs and α_2 for SUs. A limited number of SUs can be in a queue, referred to as the imaginary orbit, waiting for retransmission, whereas this number is unlimited for PUs. When a RTS is successfully received by the CRC, the maximum transmit power on the DATC, P_{data}, is computed in the same way as P_{ctrl}. If the CRC cannot find a suitable transmit power that satisfies the minimum quality of service (QoS) of the CR client without violating the EMI constraints given by either (6.2.2) or (6.2.3), the request for data transmission is dropped. In addition, the CRC randomly drops RTS messages with probability P_{d1} for PUs and P_{d2} for SUs in order to avoid network congestion. If the CR client's RTS is dropped, a negative-CTS message is sent by the CRC; after a random number of time slots the CR client can

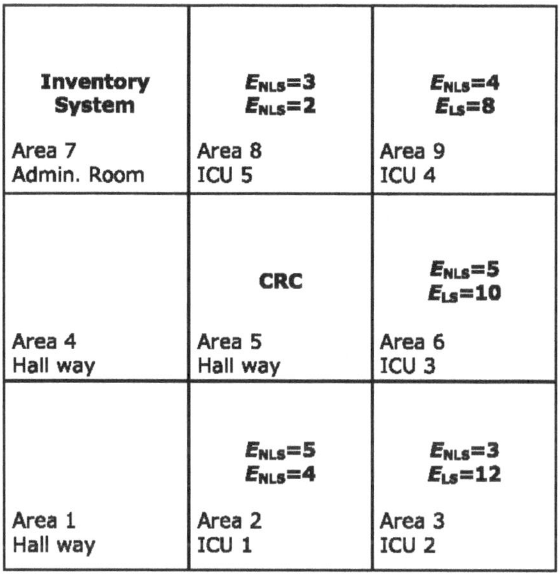

Fig. 6.10 Layout of hospital scenario used for simulations

attempt to transmit again. If the CR client is not dropped, then a CTS message is sent. After the CTS message is successfully received, the CR client will wait in a transmission queue of finite size, which means the CR client's request will be dropped if the queue is full. PUs and SUs wait in separated queues, and PUs always have priority to transmit on the DATC. The number of time slots for data transmission is geometrically distributed with parameters β_1 and β_2 for PUs and SUs, respectively.

6.3.2.2 Performance Evaluation

The EMI-aware protocol described above can be evaluated through numerical simulations in terms of interference probability and outage probability [35]. For performance comparison with a traditional MAC protocol, a scenario consisting of hospital premises over 27 m^2 arranged in nine areas of equal size comprising a hall way, an administration room, and five ICUs were used, as shown in Fig. 6.10.

In this case study, the CRC was located at the center of area 5. Ten NLS and LS non-communication medical devices were located in the ICUs, and their corresponding EMI immunity levels are also given in Fig. 6.10. The locations of the NLS and LS medical devices and the CRC were fixed, whereas the CR clients were mobile and uniformly distributed over the area. A random pedestrian mobility model mimics the wandering of the CR clients. In order to compute (6.3.2) and (6.3.3), the following indoor path loss (PL) formula as a function of the distance in meters, d ($d > 1$), was applied:

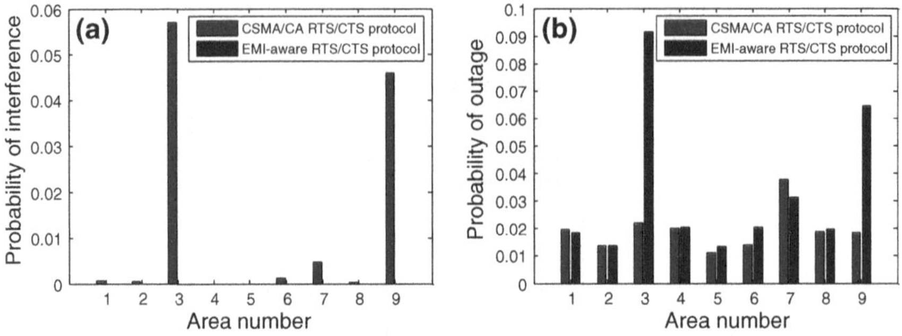

Fig. 6.11 **a** Interference probability and **b** outage probability over the nine areas of the hospital scenario

$$PL_{\text{total}} = 37.7 + 3.3\log_{10}(d) + 16.2n \tag{6.3.4}$$

where n is the number of floors (or walls) the radio signal has to traverse and PL is given in decibels (dB).

The interference probability was defined as the chance that a wireless CR client causes interference to the non-communication medical devices by violating (6.3.2) or (6.3.3). In the EMI-aware RTS/CTS protocol, interference occurs when the ON status of the medical devices is reported wrongly to the CRC. This probability of misdetection was set to 0.01.

The interference probabilities of each area for the EMI-aware RTS/CTS protocol and a traditional carrier sense multiple access with collision avoidance (CSMA/CA) RTS/CTS protocol are shown in Fig. 6.11a.

As may be observed, the EMI-aware RTS/CTS protocol successfully protected NLS and LS medical devices. The outage probability, i.e., the probability that the received signal strength at the CRC is less than −65 dBm, is depicted in Fig. 6.11b for each of the nine areas. Evidently, the interference reduction comes at the expense of high outage probability for the wireless devices in areas where protected medical equipment are located.

6.3.2.3 Enhancement Through the Use of an Additional Channel

The EMI-aware RTS/CTS protocol in [35] can be enhanced by including dual-band operation [37]. The use of an additional "emergency" channel (AEC) in a different frequency band that can serve as a control/data channel for potential interferers can reduce the outage probability. The recently allocated 2360-2400 MHz BAN frequency band and the 900 MHz ISM band (902–928 MHz) are suitable for allocation of the AEC. Through computer simulations, the performance of this MAC scheme was evaluated in terms of the outage probability, and the comparison with the EMI-aware RTS/CTS protocol in [35] is depicted in Fig. 6.12.

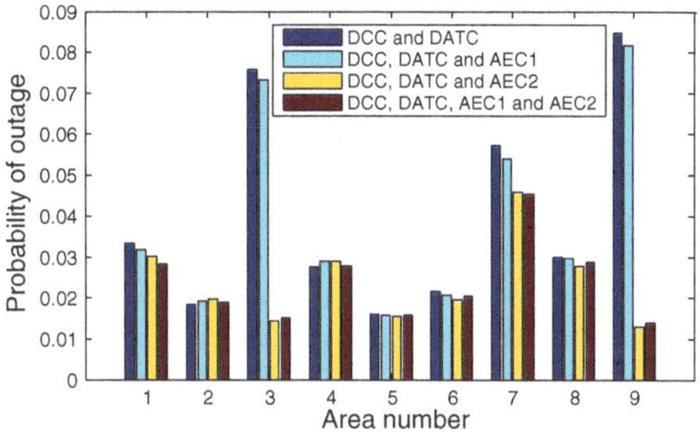

Fig. 6.12 Outage probability over the nine areas of the hospital scenario for different multiband EMI-aware RTS/CTS protocols

Clearly, the use of an AEC reduced the outage probability, but the improvement is determined by the center frequency of the AEC. Marginal improvement is obtained with an AEC in 2360–2400 MHz (AEC1), whereas significant improvement can be obtained by using an AEC in the 900 MHz ISM band (AEC2). The simultaneous use of AEC1 and AEC2 does not provide further improvement.

6.3.2.4 Cognitive Radio Solution in the UWB Band

Besides large bandwidth that enables high data rate transmission, UWB technology has other attractive characteristics for BAN implementation. UWB signals have an inherent noise-like behavior due to their extremely low maximum effective isotropically radiated power (EIRP) spectral density of –41.3 dBm/MHz. This makes UWB difficult to detect and increases its robustness against jamming, potentially rescinding the need for complex encryption algorithms in small, low-cost transceivers. In addition, compared to single-band orthogonal frequency division multiplexing (OFDM) where symbols are continually sent on one frequency band, for multiband OFDM (MB-OFDM), symbols are interleaved over multiple sub-bands across both time and frequency. By interleaving the OFDM symbols across sub-bands in this manner, multiband UWB can maintain the power level associated with a single-band OFDM transmission yet the data throughput can be significantly increased. MB-OFDM UWB technology can currently achieve rates ranging from 53.3 to 480 Mbps over distances up to 10 m.

Additionally, UWB signals do not represent a threat to patients' safety and are not significant sources of interference to other medical devices. Impulse radio (IR) UWB transceivers have a simple structure and very low power consumption characteristics. These features facilitate their miniaturization for wearable biomedical sensors.

6.3.2.5 Architecture of a Cognitive Radio Controller for BAN

Characteristics of UWB technology can be exploited to turn a BAN controller into a CR controller (CRC). The CRC is a central unit that controls the transmission parameters of the CR clients (i.e., wearable sensors) for wireless access. The wearable sensors must be low cost, small in size, and with low power dissipation; this can be achieved through the use of an IR-UWB radio interface for communications in the first communications tier (see Fig. 6.9). Hence, the communication links between the sensors and the BNC should be implemented with IR-UWB as specified by IEEE 802.15.6. On the other hand, for the connection of the BNC with the second communication tier, the MB-OFDM-based ECMA-368 interface [38] should be used. Besides supporting high data rates, MB-OFDM technology also provides a relatively straightforward way to implement detect-and-avoid (DAA) and CR features.

Figure 6.13 shows the architecture for a UWB-BAN CRC, consisting of two transceivers, an IR-UWB transceiver with on–off keying (OOK) modulation and a MB-OFDM UWB transceiver. The division of the 3.1–10.6 GHz UWB spectrum into 14 sub-bands of 528 MHz as suggested by the ECMA-368 standard can be adopted. The lower UWB frequency band (3.1–4.8 GHz) is then covered by three sub-bands with center frequencies at 3232, 3960, and 4488 MHz, respectively. Due to better propagation characteristics, these three sub-bands should be preferred for first-tier communications. The second-tier communications can be implemented in higher frequencies.

The fast Fourier transform (FFT) engine monitors the UWB spectrum and broadcasts information on available sub-bands for data transmission. It has been demonstrated that the 128-bin FFT engine of the OFDM transceiver can act as a rudimentary spectrum analyzer with a sampling frequency of 528 MHz resulting in a frequency resolution of 4.125 MHz (528 MHz/128 bins) [39]. This enables the estimation of spectrum occupancy for opportunistic spectrum usage purposes with minimal additional implementation costs. A MB-OFDM transceiver also facilitates the protection of sensitive receivers or protected medical devices from UWB interference by using frequency-domain spectrum shaping. This can be achieved by zeroing a range of subcarriers that overlap with the frequency band one intends to protect during transmission.

The broadcast channel is one of the three sub-bands of the lower UWB frequency band, which is chosen based on spectrum usage statistics. The least congested frequency sub-band is the most convenient choice for the broadcast channel and this information is stored in the memory of each of the sensor nodes and the CNC. When a wearable sensor is turned on, the device is informed (through the broadcast channel) of which sub-band it can use for data transmission. Then, it adapts accordingly the matched filter (MF) of the correlator in the receiver and the pulse generator of the transmitter. The same is done in the IR-UWB transceiver of the CNC. This can be implemented through a simple look-up table for the MF and pulse generator.

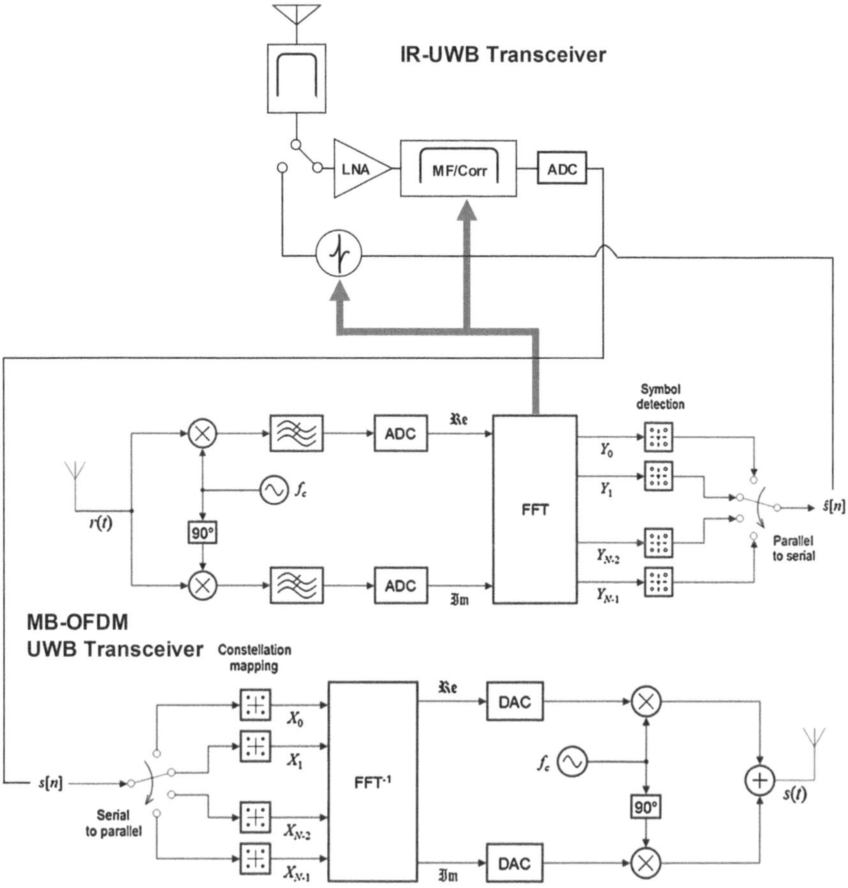

Fig. 6.13 Architecture of a BNC performing the role of a CRC for MBAN by using UWB technology. The *bold arrows* indicate the flow of information on spectrum availability from the MB-OFDM UWB transceiver to the IR-UWB transceiver

6.3.2.6 Cognitive Radio and Energy Consumption

Energy consumption is an important issue for BANs and elements therein, particularly for communications at the first tier of operation (Fig. 6.9). Energy efficiency facilitates operation for the user through reducing frequency of battery charges whereby the user may find it difficult to carry out such tasks due to his or her current condition. It also improves reliability through reducing the probability of the network or the sensors therein being rendered inoperable due to energy depletion.

BANs employ a range of means to reduce communications energy consumption [40]. CR is also highly pertinent to energy saving in BANs through at least two key means. The first is better dynamic selection of spectrum, avoiding interference and therefore reducing necessary received power hence transmitted power. MB-OFDM-based UWB leaves scope to tailor transmissions towards the lowest interference spectrum.

The second energy-saving advantage of using CR involves the use of cognition to achieve better awareness of context, including higher-layer communication requirements, the changing general environment, and the battery energy level. This awareness can be used to improve timing of transmissions and dynamic MAC and physical (PHY) positioning given the current and projected interference conditions, variations in required data rates and constraints such as delay, and variations in general channel conditions and sensor energy availability. For example, BAN sensor readings such as blood pressure, oxygen saturation, and blood sugar levels might be sampled at changing intervals and rates, depending on the condition of the patient and time of day. In order to dynamically maintain the appropriate configuration of the BAN communication with minimized outage due to battery energy depletion while still aiming to communicate all data with appropriate delay constraints, learning and cognition can play a key part in estimating the future in terms of how requirements for these readings and battery levels are likely to vary. This can be matched to sensor transmission timing, rate, and MAC/PHY characteristics dependent on battery conditions. Moreover, battery conditions might also be driven by energy harvesting as it is described later in this section. Learning and prediction of such increases is also a key use for cognition, particularly given that the temporal characteristics of such incoming charge are highly dependent on the specific patient's nature.

6.3.3 Wireless Body Area Networks

6.3.3.1 Introduction

In the context of WBANs, electromagnetic RF energy harvesting is valuable and could be accomplished by using wearable antennas that allow for power supplying the sensor nodes [41]. Ubiquitously available RF sources, operating at different bands, are therefore exploited for RF electromagnetic energy harvesting purposes.

In this section, the opportunistic radio frequency bands for RF energy harvesting have been identified. Moreover, based on power density measurements, we have been able to identify the best spectrum opportunities that may be considered in order to conceive multiband antennas for electromagnetic energy harvesting. As an example of energy harvesting implementation, the readers are directed to familiarise with the RF energy harvesting system developed in the context of the PROENERGY-WSN project [42], which consists of an impedance matching circuit, rectifier and the energy storage sub-system.

Fig. 6.14 Locations of the measurements in Covilhã, Portugal

6.3.3.2 Indoor and Outdoor RF Energy Harvesting Opportunities

In order to seek the best spectrum opportunities for RF energy harvesting, several field trial measurements have been conducted in Covilhã, Portugal, by using the NARDA-SMR spectrum analyser with a measuring antenna, in both indoor and outdoor environments.

(A) Average Received Power

By analysing the power density measurements in 36 different locations, we intend to find the best frequencies for RF energy harvesting. Besides, the identified spectrum opportunities are being considered to conceive multi-band antennas. The location for the measurements is shown in Fig. 6.14. To determine the received power, P_r, of the spectrum analyser, we multiply the power density, P_d, by the effective receiving area of the antenna, A_e, and gain, $G = 1$, as follows:

$$p_d\left[W/m^2\right] = \left|E^2\right|/(120.\pi) \tag{6.3.5}$$

$$\overline{P_r[dBm]} = 10 \cdot \log\left(P_d \frac{\lambda^2 \cdot G}{4\pi}\right) + 30 \tag{6.3.6}$$

where E is the electric field and λ is the wavelength.

To choose the best frequency bands for electromagnetic energy harvesting, we have determined the average of each of the n values of the received power, P_{ri} [W] in linear units, in five different locations, where n is the number of measurements taken, for each frequency. The average received power, in dBm, was then calculated as follows:

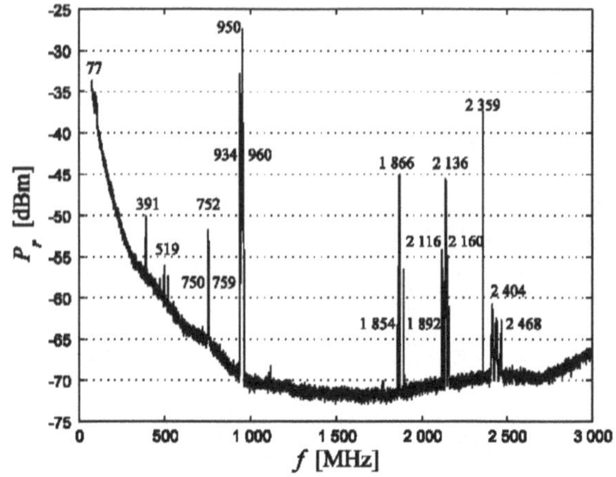

Fig. 6.15 Average received power as a function of the frequency for the university scenario (indoor)

$$\overline{P_r[dBm]} = 10 \cdot \log\left(\frac{\sum_{i=1}^{n} P_{ri}[W]}{n}\right) + 30 \quad (6.3.7)$$

(B) Indoor Opportunities

Figure 6.15 presents the indoor spectrum opportunities as observed at the higher education institution in Covilhã.

The set of frequencies with high energy available for harvesting comprises the range from 934 to 960 MHz (GSM900), 1854 to 1892 MHz (GSM1800), 2116 to 2160 MHz (UMTS), 2359 MHz (amateur, SAP/SAB applications, video), and 2404 to 2468 MHz (Wi-Fi).

(C) Outdoor Opportunities

The location of public places in the outdoor scenario for the field trial results were identified in Fig. 6.14 as locations numbered 8, 9, 12, 13, 14, 21 and 22. The corresponding values of the average received power are shown in Fig. 6.16.

The set of frequencies with more energy available for harvesting are in the range from 79 to 96 MHz (mobile/radio broadcast stations), 391 MHz (emergency broadcast stations), 750 to 759 MHz (digital television broadcast stations), 935 to 960 MHz (GSM 900 broadcast stations), 1854 to 1870 MHz (GSM 1800 broadcast stations) and 2115 to 2160 MHz (UMTS broadcast stations).

It may be concluded from the above reported measurements that the GSM900/1800 frequency bands appear to be the most promising bands for RF energy harvesting.

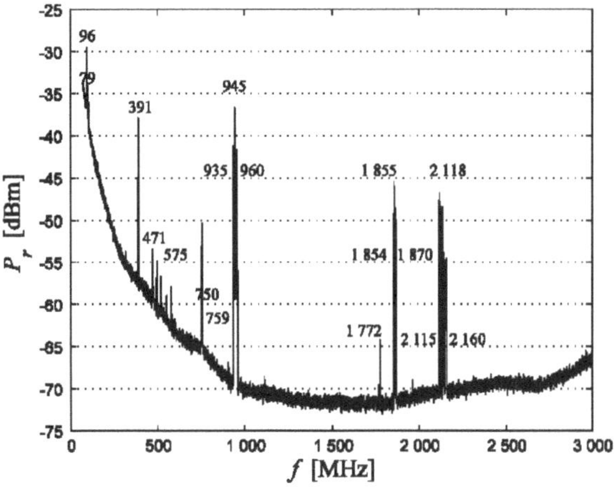

Fig. 6.16 Average received power for the outdoor scenario

6.3.3.3 Innovative MAC Protocols

In [43] the authors have shown that one fundamental reason for IEEE 802.15.4/4a MAC inefficiency is overhead, where the use of ACK control packets can decrease the bandwidth efficiency about 10 %. In this work, we propose and analyse two innovative mechanisms to reduce overhead in IEEE 802.15.4: (1) concatenation and (2) piggyback [44]. The main idea is to improve channel efficiency by aggregating several acknowledgment (ACK) responses into one single transmission (i.e., one single packet) like in the IEEE 802.11e standard. This aggregation of ACKs aims at reducing the overhead by transmitting less ACK control packets and by decreasing the time periods the transceivers should switch between different states.

We aim at increasing the throughput as well as decreasing the end-to-end delay, whilst providing a feedback mechanism for the receiver to inform the sender about how many transmitted (TX) packets were successfully received (RX). Our proposal also considers the use of the Request-To-Send/Clear-To-Send (RTS/CTS) mechanism, in order to avoid the hidden terminal problem.

In IEEE 802.15.4 the protocol overhead impacts on end to-end delay and throughput. In order to reduce end-to-end delay and increase throughput, we propose a new innovative MAC protocol that solves the above problems, along with the elimination of the backoff period repetitions, the Sensor Block Acknowledgment (SBACK)-MAC protocol [45]. The main difference compared to IEEE 802.15.4 is related to the way that SBACK-MAC treats the ACK control packets. The SBACK-MAC allows the aggregation of several ACK responses into one special packet. The *BACK Response* will be responsible to confirm a set of data packets successfully delivered to the destination. This packet has the same length as an ACK packet in IEEE 802.15.4. Hence, an ACK control packet will not be received in response to every data packet sent/received. By decreasing the number of control packets exchanged in a wireless medium, it is possible to decrease not only the number of collisions but also the number of backoff periods

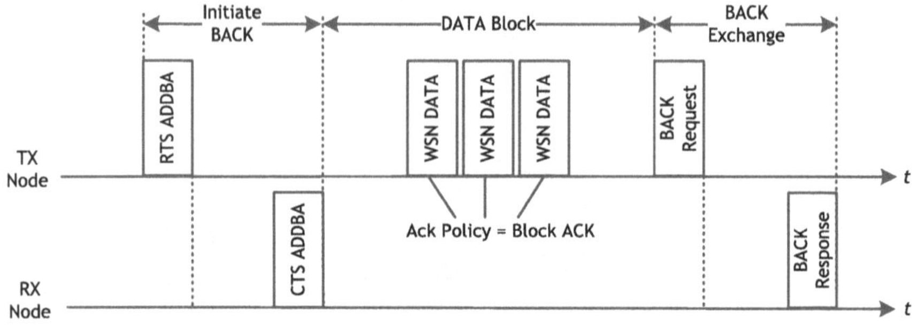

Fig. 6.17 SBACK-MAC protocol—block acknowledgment mechanism with *BACK request*

(the time a node must wait before attempting to transmit/retransmit the packet) on each node. Moreover, in WSNs the length of control packets can be of the order of magnitude of the data packets. Since nodes are battery operated, the transmission of such packets leads to energy decrease, whilst reducing the number of data packets that will be transmitted containing useful information (i.e., goodput).

The SBACK-MAC also considers the *ccaTime*. This way, during CCA nodes are able to determine the channel state (i.e., busy or idle), which allows for providing statistical information for the MAC sub-layer and upper layers. Moreover, since the CCA result is based on the obtained Received Signal Strength Indicator (RSSI), transmission power control techniques could be used to estimate the minimum transmission power for sending each packet to a neighbouring node.

(A) Block Acknowledgment Mechanism with BACK Request

The version of the proposed SBACK-MAC protocol with *BACK Request* considers the exchange of two special packets, i.e., *RTS ADDBA* and *CTS ADDBA*, where *ADDBA* stands for "Add Block Acknowledgement". After this successfully exchange, data packets are transmitted from the transmitter to the receiver (e.g., 100 packets are sent during the active periods). Afterwards, by using the BACK Request primitive, the transmitter inquires the receiver about the total number of data packets that successfully reach the destination. In response, the receiver will send a special data packet, called BACK Response identifying the packets that require retransmission, and the BACK mechanism finishes.

Figure 6.17 presents the message sequence chart for the BACK mechanism.

The exchange of two types of special control packets used at beginning and end of the BACK mechanism allows for mitigating the hidden-terminal and exposed-terminal problems like in IEEE 802.11e. The BACK mechanism also aims at reducing the power consumption by transmitting less ACK control packets and by decreasing the time periods the transceivers should switch between different states. By using the BACK there is no need to receive an ACK for every DATA packet sent, as presented in Fig. 6.18. Besides, during the data transmission there is no way to know how many packets have successfully reached the destination, except at the end of communication by using the *BACK Request/BACK Response* (we set, *BE*, equal to 0), as if there is no congestion.

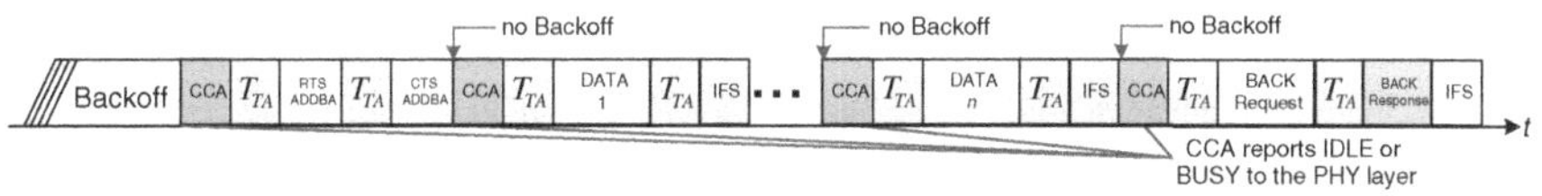

Fig. 6.18 Frame sequence for the SBACK-MAC protocol with BACK request (concatenation)

Table 6.1 IEEE 802.15.4 and SBACK-MAC typical parameters and values

Description	Symbol	Value
Backoff period duration	T_{BO}	320 μs
PHY SHR duration	T_{SHR}	160 μs
CCA detection time	T_{CCA}	128 μs
TX/RX or RX/TX switching time	T_{TA}	192 μs
Short interframe spacing (SIFS) time	T_{SIFS}	192 μs
Long interframe spacing (LIFS) time	T_{LIFS}	640 μs
ACK transmission time	T_{ACK}	352 μs
RTS transmission time	T_{RTS}	352 μs
CTS transmission time	T_{CTS}	352 μs
RTS ADDBA transmission time	T_{RTS_ADDBA}	352 μs
CTS ADDBA transmission time	T_{CTS_ADDBA}	352 μs
BACK request transmission time	$T_{BRequest}$	352 μs
BACK response transmission time	$T_{BResponse}$	352 μs
ACK wait duration time	T_{AW}	560 μs
DATA transmission time	T_{DATA}	576 μs
Time to setup radio to RX or TX states	rxSetupTime	1792 μs
PHY length overhead	L_{H_PHY}	6 bytes
MAC overhead	L_{H_MAC}	9 bytes
DATA payload	L_{H_DATA}	3 bytes
DATA frame length	L_{FL}	18 bytes
ACK/RTS/CTS frame length	L_{ACK}	11 bytes
Number of TX frames	n	1 to 100
Data rate	R	250 kb/s

When the SBACK-MAC with *BACK Request* is considered, the minimum average delay, $D_{\min}$, in seconds, is given by:

$$D_{\min\_BACK} = \left(\overline{CW} + ccaTime + T_{RTS_{ADDBA}} + H_1\right)/n \qquad (6.3.8)$$

where $H_1 = T_{TA} + T_{CTS\_ADDBA} + n \times (ccaTime + T_{TA} + T_{DATA} + T_{TA} + T_{IFS}) + ccaTime + T_{TA} + T_{BRequest} + T_{TA} + T_{BResponse} + T_{IFS}$.

The maximum average throughput, $S_{\max}$, in bits per second, is given by:

$$S_{\max\_BACK} = 8L_{DATA}/D_{\min\_BACK} \qquad (6.3.9)$$

The meaning of the parameters is defined in Table 6.1.

By analysing Eqs. (6.3.8) and (6.3.9), we conclude that by using the *BACK Request* primitive we allow several MAC Protocol Data Units (MPDUs) to be acknowledged by a single *BACK Request* packet. Therefore, compared with IEEE

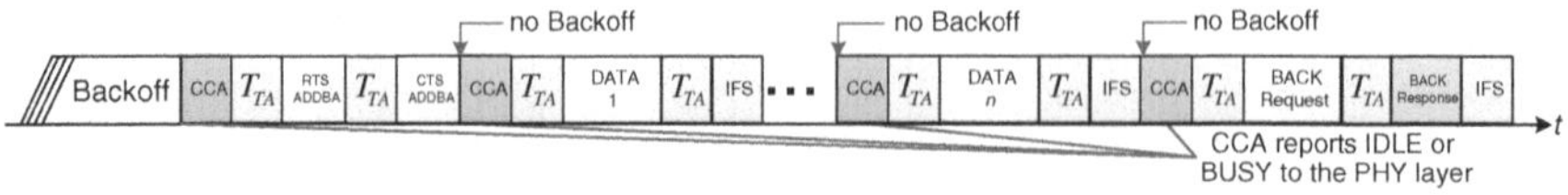

Fig. 6.19 Frame sequence for the SBACK-MAC protocol with no *BACK request* (piggyback)

802.15.4 in the basic access mode, there is no need to consider the use of individual ACK control packets. We also assume that the back-off period is equal to 0, as if there is no congestion (no activity in the shared medium). This is explained by the fact that a CTS can be only sent if there is no congestion at the receiver. The *BACK Response* contains the information about the reception of corresponding MPDUs by using a specific bitmap that is transmitted as an answer to an explicit transmitter request. This request is performed by the new *BACK Request* control frame. Both the *BACK Request* and *BACK Response* are transmitted at the same data rate (i.e., 250 kb/s).

(B) Proposed scheme with no BACK Request

The version of the SBACK-MAC protocol with no *BACK Request* ("piggyback mechanism") also considers the exchange of the *RTS ADDBA* and *CTS ADDBA* packets at the beginning of the communication. However, at the end of the communication the BACK Request primitive is not transmitted. Therefore, the last aggregated data frame must include the information about the packets previously transmitted, as shown in Fig. 6.19.

The SBACK-MAC version with "piggyback" does not consider the use of the BACK Request primitive, as shown in Fig. 6.19. As a consequence the control overhead and the end-to-end delay are reduced whilst increasing the throughput, by piggybacking the BACK information into the last data fragment. However, with this scheme, the system becomes less robust. If the last aggregated frame (DATA frame *n*) is lost, the destination does not know that an ACK needs to be sent back.

When the SBACK-MAC with no *BACK Request* is considered, the minimum average delay, $D_{\text{min\_Piggy}}$, in seconds, is given by:

$$D_{\text{min\_Piggy}} = \left(\overline{CW} + ccaTime + T_{RTS_{ADDBA}} + H_2\right)/n \tag{6.3.10}$$

where $H_2 = T_{TA} + T_{CTS\_ADDBA} + (n-1) \times (ccaTime + T_{TA} + T_{DATA} + T_{TA} + T_{IFS}) + ccaTime + T_{TA} + T_{DATA} + T_{TA} + T_{B\text{Response}} + T_{IFS}$

The maximum average throughput, $S_{\text{max\_Piggy}}$, in bits per second, is given by:

$$S_{\text{max\_Piggy}} = 8L_{DATA}/D_{\text{min\_Piggy}} \tag{6.3.11}$$

The meaning of the parameters is defined in Table 6.1.

By analysing Eqs. (6.3.10) and (6.3.11), we conclude that in the SBACK-MAC with no BACK Request, (*n*-1) frames are transmitted (which corresponds to less one IFS) like in the SBACK-MAC with *BACK Request*. However, since the last data packet includes the information about the total number of packets previously transmitted, the *BACK Response* is transmitted immediately after the reception of the last data frame. Therefore, there is no need to transmit the *BACK Request* packet.

(C) Modelling and simulation results

The SBACK-MAC was evaluated by using the MiXiM simulation framework [39] from the OMNeT++ simulator. SBACKMAC throughput and end-to-end delay with and with no *BACK Request* have been compared against IEEE 802.15.4, by considering a 95 % confidence interval, however, as it is too small, we decided not to plot it in the figures. Table 6.1 presents the MAC parameters considered for the network in our simulations. The performance analysis of the proposed schemes is conducted for the best-case scenario. Therefore, we are assuming that the channel is an ideal channel, with no transmission errors. During the active period, there is only one node that always has a frame to be sent. The other stations can only accept frames and provide acknowledgments.

Figure 6.20 presents the maximum average throughput and the minimum average delay versus the payload size for the SBACK-MAC protocol with and with no *BACK Request*. The discontinuity around 18 bytes is due to the use of SIFS and LIFS (i.e., MPDU less of equal than 18 bytes must be followed by a SIFS, whilst MPDU longer than 18 bytes must be followed by a LIFS). The number of transmitted frames, n, is 10 (i.e., for the SBACK-MAC, the frames are aggregated and transmitted in a burst).

It is observed that, by increasing the payload size, S_{max} also increases. This conclusion is valid for all the three presented mechanisms. For small packet sizes (i.e., data payload less or equal than 18 bytes) by comparing the IEEE 802.15.4 with the SBACK-MAC protocol, with and with no *BACK Request*, S_{max} increases 17 % and 25 %, respectively. Moreover, by using the IEEE 802.15.4 basic access mode with DATA/ACK, the maximum achievable throughput is approximately 108.7 kb/s whereas, by using the SBACK-MAC with and with no *BACK Request*, the maximum achievable throughput is 118.1 and 123.2 kb/s, respectively. Results for D_{min} as a function of the payload size show that, by using SBACK-MAC with and with no *BACK Request* for small packets sizes (i.e., data payload less or equal to 18 bytes), D_{min} decreases 17 and 25 %, respectively. For larger packet sizes, by considering SBACK-MAC with and with no *BACK Request*, D_{min} decreases by 8 and 13 %, respectively.

Figure 6.21 presents S_{max} and D_{min} as a function of the number of TX packets. A fixed payload size of 3 bytes (i.e., LDAT A = 3 bytes) is considered, since it is one of the values in the range from 1 to 18 bytes presented in Fig. 6.20, by considering the worst throughput performance, when taking into account the BACK mechanism. Even for the shortest payload sizes, it is possible to improve the network performance by using the proposed BACK mechanisms.

It may be observed that when the number of TX packets is less than 4, the IEEE 802.15.4 standard through the basic access mode, achieves higher throughput in comparison to SBACK-MAC (either with or with no *BACK Request*). Moreover, by considering the IEEE 802.15.4 standard in the basic access mode, S_{max} does not depend on the number of TX packets, and achieve the maximum value of 5.2 kb/s.

Fig. 6.20 Maximum throughput and minimum delay versus payload size

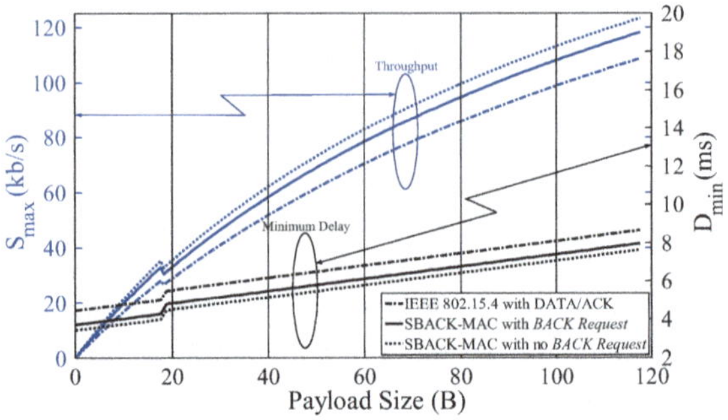

Fig. 6.21 Maximum throughput and minimum delay versus number of TX packets

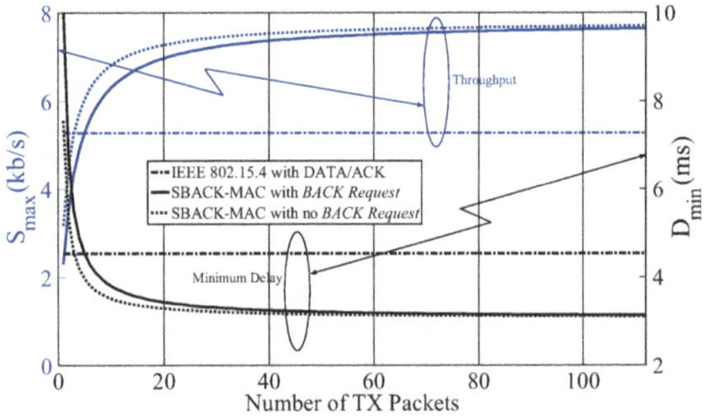

In SBACK-MAC with and with no *BACK Request* (i.e., concatenation and piggyback), S_{max} increases by increasing the number of TX packets (i.e., the number of aggregated packets). For a number of TX packets equal to 18, by considering the SBACK-MAC with *BACK Request* (i.e., concatenation version) S_{max} is about 6.3 kb/s. This value corresponds to an increase of 21 % in the throughput in comparison to the MAC protocol from the IEEE 802.15.4 standard in the basic access mode, whereas by considering the SBACK-MAC with no *BACK Request* (i.e., piggyback version), the achievable throughput is 6.8 kb/s, an increase of 30 %. However, the difference on the throughput between the SBACK-MAC with and with no *BACK Request* tends to decrease by increasing the total number of TX packets (i.e., by aggregating more packets). We also conclude that, for more than 4 TX packets, SBACKMAC (with and with no *BACK Request*) delay is significantly shorter than for IEEE 802.15.4 in the basic access mode. The difference is mitigated by increasing the total number of TX packets (i.e., by aggregating more packets).

The previous results have shown that the use of the proposed BACK mechanisms enables to improve the network performance whilst increasing the channel use optimization. Therefore in the context of cognitive radio wireless sensor networks (CRWSN) the proposed BACK mechanisms can be applied in a scenario where the secondary user's flow alternates between sensing (i.e., CCA mechanism) and sending rapidly at short intervals [46], see Fig. 6.22.

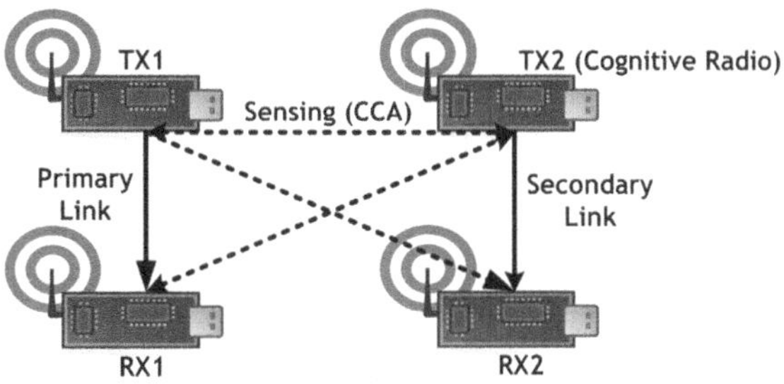

Fig. 6.22 Signal and interference paths in a cognitive radio sensor network

By considering that the primary user data flow (from TX_1 to RX_1) is bursty, there will be inactive periods between transmissions. These inactive periods are used by the secondary users to send their own traffic. Since, by using the proposed BACK mechanism we decrease the end-to-end delay whilst increasing throughput we intend to exploit the temporal opportunities (i.e., temporal "white spaces") [46] more efficiently, by decreasing the transmissions times of secondary users. Moreover, CRWSN nodes must perform sensing and make a decision about the channel state (i.e., busy or idle). This spectrum decision time can also be seen has overhead, leading to an increase of the energy consumption. Therefore, in the case of re-transmissions, the SBACK-MAC protocol with and with no *BACK Request*, is able to decrease the time periods before making CCA again, since there is no back-off phase between two consecutive data packets, which allows for decreasing the total overhead.

6.3.3.4 Summary

We have identified the spectrum opportunities for RF energy harvesting to power supply the wireless sensor nodes in real indoor/outdoor scenarios. The set of indoor/outdoor most promising frequency bands are 79–96 MHz (mobile/radio broadcast stations), 391 MHz (emergency broadcast stations), 750–758 MHz (digital television broadcast stations), 935–960 MHz (GSM 900 broadcast stations), 1855–1868 MHz (GSM 1800 broadcast stations) and 2115–2160 MHz (UMTS broadcast stations). A scenario of using RF harvesting with dedicated devices had been analysed in the PROENERGY-WSN [42]. It enables future WBAN to operate without the need of replacing batteries. Moreover, in order to reduce the overhead that affects WBANs communications we have proposed SBACK-MAC—a new contention-based MAC protocol that uses a BACK mechanism to improve channel efficiency by aggregating several ACK into one special packet, the *BACK Response*. By using a concatenation mechanism the total overhead is decreased, since there is no back-off between two consecutive data packets. Two innovative solutions have been proposed to improve the IEEE 802.15.4 performance. The first one considers the SBACK-MAC protocol in the presence of *BACK Request* (concatenation mechanism), while the second considers the SBACK-MAC in the absence of BACK Request (piggyback mechanism).

6.3.4 Conclusions

The growing use of radio communication technology for healthcare is triggering the deployment of BANs in medical premises mostly on a license-exempt secondary basis. This fact motivates further studies on interference mitigation techniques for medical communication environments. CR offers viable cost-effective and future-proof solutions addressing both scalability and coexistence issues.

The next generation of CR networks will be supplied by renewable energy from natural resources, where RF energy harvesting plays an important role. In a near future, RF energy will enable to power supply all the nodes without the need of replacement of the primary source of energy (i.e., batteries).

In order to improve the efficiency of WBANs, a new innovative Sensor Block Acknowledgment (SBACK)-MAC protocol which aggregates several ACK responses into one special packet has been proposed. The results presented here show that the proposed concatenation mechanisms considerably improve network performance in terms of throughput and end-to-end delay. In the context of CR wireless sensor networks enhancements will be achieved because of the reduction of queue lengths of secondary users as well as the channel utilisation.

Dynamic spectrum access can be used to mitigate spectrum scarcity. This is accomplished by enabling an unlicensed user (i.e., secondary user) to adaptively adjust its operating parameters and exploit the spectrum which is unused by licensed users (i.e., primary users), in an opportunistic manner.

References

1. ITU Final Acts of the Regional Radiocommunication Conference for planning of the digital terrestrial broadcasting service in parts of Regions 1 and 3, in the frequency bands 174–230 MHz and 470–862 MHz (RRC-06), International Telecommunication Union, Geneva, 2006
2. Więcek, D.: Methodology of White Space estimation in TV bands based on the ITU GE06 technical conditions, COST IC0905 TERRA 3rd Workshop, Brussels, 21st June 2011
3. ECC Report 185, Complementary Report to ECC Report 159. Further definition of technical and operational requirements for the operation of white space devices in the band 470–790 MHz, ECC, 2013
4. ITU-R, Recommendation P.1546-4 Method for point-to-area predictions for terrestrial services in the frequency range 30–300 MHz. International Telecommunication Union, Geneva, October 2009
5. WISE, http://wise.turkuamk.fi, cited 31.5.2013
6. Tekes Trial Programme, http://www.tekes.fi/programmes/Trial, cited 31.5.2013

7. ECC report 159, Technical and operational requirements for the possible operation of CR systems in the 'white spaces' of the frequency band 470–790 MHz, January 2011
8. ITU-R Document 6E/64-E, The ESR5 criterion for the Assessment of DVB-T transmission Quality, April 2004
9. Talmola, P., Kalliovaara, J., Paavola, J., Ekman, R., Kokkinen, H., Heiska, K., Wichman, R., Poikonen, J.: Field measurements of WSD-DTT protection ratios over outdoor and indoor reference geometries. In: Proceedings of International Conference on Cognitive Radio Oriented Wireless Networks CROWNCOM 2012, Stockholm, Sweden, 2012
10. CEPT/ECC SE 43(11)36 WISE Project Measurement Report, WSD maximum Power Measurements in Turku Test Network, 10th SE43 meeting, Bologna, Italy, July 2011
11. CEPT/ECC SE43(11)81 WISE Project Measurement Report, WSD maximum Power Indoor Measurements in Turku Test Network, 12th SE43 meeting, Cambridge, UK, December 2011
12. Ojaniemi, J., Poikonen, J., Wichman, R.: Effect of geolocation database update algorithms to the use of TV white spaces. In: Proceedings of International Conference on Cognitive Radio Oriented Wireless Networks CROWNCOM 2012, Stockholm, Sweden, 2012
13. Ojaniemi, J., Kalliovaara, J., Alam, A., Poikonen, J., Wichman, R.: Optimal field measurement design for radio environment mapping. In: Proceedings of 47th Annual Conference on Information Sciences and Systems (CISS 2013)
14. CEPT/ECC SE43(11)82 WISE Project Measurement Report, PMSE protection measurements in Helsinki City Theatre. 12th SE43 meeting, Cambridge, UK, December 2011
15. Radio Standards Specifications (RSS-Gen, RSS-210, RSS-310), Procedures and regulations. http://www.ic.gc.ca/eic/site/smt-gst.nsf/eng/h_sf01841.html
16. FCC Part-15 and FCC Part-18 rules, http://www.fcc.gov
17. Enhancing the Security of Corporate Wi-Fi Networks Using DAIR. Paramvir Bahl et al. MobiSys'06, June 19–22, 2006, Uppsala, Sweden
18. Flores, A.B., Guerra, R.E., Knightly, E.W., Ecclesine, P., Pandey, S.: IEEE 802.11af: a standard for TV white space spectrum sharing. IEEE Commun. Mag. 51(10), 92–100 (2013)
19. van Bloem, J.W.H., Schiphorst, R., Kluwer, T., Slump, C.H.: Spectrum utilization and congestion of IEEE 802.11 networks in the 2.4 GHz ISM band. J. Green Eng. 2(4), 401–430 (2012). ISSN 1904-4720
20. Nardelli, B., Lee, J.,Lee, K., Yi, Y., Chong, S., Knightly, E.W., Chiang, M.: Experimental evaluation of optimal CSMA. In: INFOCOM, 2011 Proceedings IEEE, pp. 1188–1196, 10–15 April 2011
21. Tytgat, L., Barrie, M., Gonçalves, V., Yaron, O.Y., Moerman, I., Demeester, P., Pollin, S., Ballon, P., Delaere, S.: Techno-economical viability of cognitive solutions for a factory scenario. In: IEEE International Dynamic Spectrum Access Networks (DySPAN) Symposium (2011)
22. Barrie, M., Tytgat, L, Gonçalves, V., Yaron, O.Y., Moerman, I., Demeester, P., Pollin, S., Ballon, P., Delaere, S.: Techno-Economic Evaluation of Cognitive Radio in a Factory Scenario. Networking 2011 Workshop on Performance Evaluation of Cognitive Radio Networks: from Theory to Reality PE-CRN 2011, Springer LNCS book (2011)
23. Tytgat, L., Yaron, O., Pollin, S., Moerman, I., Demeester, P.: Avoiding collisions between IEEE 802.11 and IEEE 802.15.4 through coexistence aware clear channel assessment. EURASIP J. Wireless Commun. Netw. 2012, 137 (2012). doi:10.1186/1687-1499-2012-137
24. Alemdar, H., Ersoy, C.: Wireless sensor networks for healthcare: a survey. Comput. Netw. **54**(15), 2688–2710 (2010)
25. Jovanov, E., Milenkovic, A.: Body area networks for ubiquitous healthcare applications: opportunities and challenges. J. Med. Syst. **35**(5), 1245–1254 (2011)
26. IEEE Standard for Local and Metropolitan Area Networks–Part 15.6: Wireless Body Area Networks, IEEE Standard 802.15.6-2012, 2012
27. Cao, H., Leung, V., Chow, C., Chan, H.: Enabling technologies for wireless body area networks: a survey and outlook. IEEE Commun. Mag. **47**(12), 84–93 (2009)
28. Hauer, J.-H., Handziski, V., Wolisz, A.: Experimental study of the impact of WLAN interference on IEEE 802.15.4 body area networks. Wireless Sens. Netw. **5432**, 17–32 (2009)

29. Chávez-Santiago, R., Khaleghi, A., Balasingham, I., Ramstad, T.A.: Architecture of an ultra wideband wireless body area network for medical applications. In: Proceedings of 2nd International Symposium on Applied Sciences in Biomedical and Communication Technologies (ISABEL), Bratislava, Slovak Republic, November 24–27, 2009, pp. 1–6
30. Wang, J., Gosh, M., Chan, H.: Emerging cognitive radio applications: a survey. IEEE Commun. Mag. **49**(3), 74–81 (2011)
31. Jabbar, H., Song, Y.S., Jeong, T.T.: RF energy harvesting system and circuits for charging of mobile devices. IEEE Trans. Consum. Electron. **56**(1), 247–253 (2010)
32. Barroca, N., Ferro, J.M., Borges, L.M., Tavares, J., Velez, F.J.: Electromagnetic energy harvesting for wireless body area networks with cognitive radio capabilities. In: Proceedings of URSI Seminar of the Portuguese Committee, Lisbon, Portugal, November 2012
33. Cotton, S.L., Scanlon, W.G.: Characterization and modeling of the indoor radio channel at 868 MHz for a mobile bodyworn wireless personal area network. IEEE Antennas Wireless Propag. Lett. **6**(1), 51–55 (2007)
34. Doost-Mohammady, R., Chowdhury, K.R.: Transforming healthcare and medical telemetry through cognitive radio networks. IEEE Wireless Commun. **19**(4), 67–73 (2012)
35. Phunchongharn, P., Hossain, E., Niyato, D., Carmolinga, S.: A cognitive radio system for e-health applications in a hospital environment. IEEE Wireless Commun. **17**(1), 20–28 (2010)
36. Chávez-Santiago, R., et al.: Cognitive radio for medical body area networks using ultra wideband. IEEE Wireless Commun. **19**(4), 74–81 (2012)
37. Chávez-Santiago, R., Jankūnas, D., Fomin, V.V., Balasingham, I.: A dual-band MAC protocol for indoor cognitive radio networks: an e-health case study. In: Proceedings of 8th International Conference on Body Area Networks (BodyNets), Boston, MA, September 30–October 2, 2013
38. High Rate Ultra Wideband PHY and MAC Standard, ECMA-368 Standard, December 2008
39. Batra, A., Lingam, S., Balakrishman, J.: Multi-band OFDM: a cognitive radio for UWB. In: Proceedings of IEEE International Symposium on Circuits Systems (ISCAS), Island of Kos, Greece, May 21–24, 2006, pp. 4094–4097
40. Chen, M., Gonzalez, S., Vasilakos, A., Cao, H., Leung, V.C.M.: Body area networks: a survey. Mob. Netw. Appl. J. **16**(2), 171–193 (2011)
41. Bellon, J.S., Cabedo-Fabres, M., Antonino-Daviu, E., Ferrando-Bataller, M., Penaran-da-Foix, F.: Textile MIMO antenna for wireless body area networks. In: Proceedings of the 5th European Conference on Antennas and Propagation (EUCAP), Rome, Italy, April 2011, pp. 428–432
42. PROENERGY-WSN, Prototypes for Efficient Energy Self-sustainable Wireless Sensor Networks. http://www.e-projects.ubi.pt/proenergy-wsn, Apr. 2013
43. Chowdhury, M.S., Ullah, N., Kabir, Md.H., Khan, P., Kwak, K.S.: Throughput, Delay and Bandwidth Efficiency of IEEE 802.15.4a Using CSS PHY. In: Proceedings of the International Conference on Information and Communication Technology Convergence (ICTC), Jeju Island, Korea 2010
44. IEEE Standard for Local and Metropolitan Area Networks–Part 15.4: Wireless Body Area Networks, IEEE Standard 802.15.4, 2011
45. Barroca, N., Velez, F.J., Chatzimisios, P.: Block acknowledgment mechanisms for the optimization of channel use in wireless sensor networks. In: Proceedings of the 24th Annual IEEE International Symposium on Personal, Indoor and Mobile Radio Communications (PIMRC 2013), London, UK, September 2013
46. Dash, D., Sabharwal, A.: Secondary transmission profile for a single-band cognitive interference channel. In: Proceedings of 42nd Asilomar Conference on Signals, Systems and Computers, Pacific Grove, California, USA, October 2008, pp. 1547–1551

Chapter 7
Policy Suggestions for the Way Forward for CR

Arturas Medeisis and Oliver Holland

Abstract The key message that emerges from this book is of an extremely complex and intricate picture of regulatory considerations inevitably surrounding the introduction and development of CR and DSA technologies. Yet, it is also a hopeful picture, given that one may already clearly see the emergence of some key paths along which that development could take place. This chapter therefore wraps up the discussion of CR policy and regulation in preceding chapters by taking a forward-looking stance and proposing some realistic and concrete examples towards building the elements of a regulatory policy framework that would be conducive to the broader deployment of CR technologies in general and DSA applications in particular. The chapter opens with Sect. 7.1 that presents a novel approach to licensing of primary services, which might become a fair, flexible, and viable means to introduce hierarchical spectrum sharing. This vision is backed by a techno-economic case study that proves a financial advantage of such a concept and thus builds a compelling case for the regulators to consider this novel tool as part of their licensing toolbox. Section 7.2 contains joint contribution from a group of authors who discuss the idea of ISM-Advanced band—the concept of using CR to make better use of unlicensed commons, such as 2.4 GHz Wi-Fi band. They outline a number of possible changes to the existing set of rules that govern access to ISM bands, which would allow making those bands the hot-bed of CR innovation and, most importantly, bringing real benefits by means of increasing the efficiency of using those vital bands. Section 7.3 builds on the thread of an actor-centric approach that was started in previous chapters in order to explore the notion of an CR "sweet spot", i.e. finding a suitable combination of regulatory arrangements, an incubator, which would establish favourable conditions for

A. Medeisis (✉)
Department of Telecommunication Engineering, Vilnius Gediminas Technical University, Vilnius, Lithuania
e-mail: arturas.medeisis@vgtu.lt

O. Holland
Centre for Telecommunications Research, King's College London, London, UK
e-mail: oliver.holland@kcl.ac.uk

A. Medeisis and O. Holland (eds.), *Cognitive Radio Policy and Regulation*, Signals and Communication Technology, DOI: 10.1007/978-3-319-04022-6_7,

exploration of CR technologies. Section 7.4 presents analysis of the lessons learned over the decade of efforts to introduce CR devices in TV White Spaces. It is shown how one of the most profound developments of that age—the Geolocation Database concept—may be used as a foundation for further advances in TVWS and beyond. And finally, the ultimate Sect. 7.5 offers some prescient concluding remarks as regards the role of spectrum as key resource facilitating growth of wireless services and CR technologies that could unleash the next big wave of wireless innovation.

7.1 Pluralistic Licensing Concept

Oliver Holland[1] and Arturo Basaure[2]

[1]King's College London, London, UK
[2]Aalto University, Espoo, Finland

7.1.1 Background

Spectrum management is a difficult balancing act, presenting the challenge of realizing market and capacity demands, economic potential, technical innovation and fairness in spectrum usage, while taking into account the concerns and requirements of licensed users.

Since the inception of radio spectrum regulation, management of it has primarily been through "command and control" means [1]. Licenses awarded through such means allow reliable services, some examples of which include near-ubiquitous voice connectivity in public land mobile, reliable and highly innovative data-capable services such as UMTS and LTE, and high-quality broadcast services, among many others. However, the efficiency of such means is questionable, leading to the frequency bands in which they are applied being sparsely used at many times and locations.

At the other extreme is license-exempt spectrum [1], which has proven to be a catalyst for innovation as well as an entry point for "free" access to spectrum. However, such bands are built on solely the caveat of limited (low) transmit power, on the premise of limited propagation (especially at higher frequencies). This hampers the scope/range of wireless services in license-exempt bands, and although it is partially effective as a rudimentary method of interference mitigation, it does not effectively handle interference in busy areas.

Advanced spectrum sharing techniques essentially involve a blended approach that fits between these extremes. Light-licensing, for example, can be viewed as a coordinated sharing mechanism whereby (under a common light licensing model) a per-transmitter license fee might be set in order to indirectly influence the interference density (transmitter density) in an area. As another example, Authorised Shared Access (ASA) is an approach proposed by Qualcomm and Nokia [2] and initially driven by industry. The concept involves license holders authorizing secondary usage of spare spectrum within their licensed bands but under tight controls to prevent any disruption. It was originally intended to support business cases for the build-out of mobile broadband network infrastructure, where it is both economically and technically feasible. The Licensed Shared Access (LSA) variant on ASA, which regulators have pushed, brings licensing upfront in the sense of secondary users being granted a license to utilize under-used spectrum without interfering with the incumbent user. It is there argued that LSA allows such secondary licensees to maintain a certain level of quality of service [2].

7.1.2 Our Concept

As a highly flexible approach which takes in account recent tendencies to use financial and other means to leverage better spectrum usage, we propose Pluralistic Licensing (PL). We describe PL as "the award of licenses under the assumption that opportunistic secondary spectrum access will be allowed, and that interference may be caused to the primary with parameters and rules that are known to the primary at the point of obtaining the license" [3]. A key aspect of the concept is that the primary will choose from a range of offered "pluralistic licenses" each with associated fees, and each specifying alternative opportunistic access rules for secondary users operating in the same spectrum as the primary, with associated interference characteristics caused to the primary. The objective is to incentivize the primary to obtain this type of license through means such as a reduced license fee, whereby the opportunistic secondary spectrum access will use "cognitive radio" mechanisms to keep interference to the primary within known parameters [3].

We see that the primary's ability to select from numerous possible licenses as it would require under our scheme, embodied in a well-defined/automated mechanism in addition to the incentive scaling to primary based on implied interference, as being a novel. Further, whereas LSA bringing licensing up front for ASA, and therefore implies that the "secondary" is licensed, PL implies no license for the secondary and is therefore distinguished from LSA going forward. Hence, we see PL as simpler and quicker to implement, particularly for the secondary, being a complementary approach to LSA and increasing the range of tools that the regulator has in order to manage and promote enhanced spectrum usage.

In order to implement such as concept, we employ the following assumptions. First, a primary user must meet the following criteria:

- Its technical characteristics and geographic location (or location bounds) are available from an authoritative register or database of radiocommunications licences.

- It may or may not be required to adopt a technique for secondary awareness assistance such as transmission of beacons, or to support other users in the coexistence process.
- It operates under the assumption that all other users will put in place mechanisms for detecting its presence with a minimum required detection performance, whereby the definition of such a minimum performance, and likely also precise detection metrics required by the secondary to achieve that detection performance, will be specified in the license that the primary user has obtained.

A secondary user must meet the following criteria:

- It uses devices equipped with adequate means to achieve awareness of and guarantee coexistence with the primary user, at least being capable of meeting the minimum required primary detection performance, and likely achieving better than that dependent on incentivisation.
- It operates with no guarantee of quality of service. Secondary users might nevertheless use probabilistic or other methods to understand exactly which levels of service they might be achieve with which probabilities, but this is as far as they can go regarding "guaranteeing" service.
- Depending on implementation, cross-checking with a GDB may be required to check the form of PL operating in a given band and location before transmitting (see below).

Primary users, who might obtain a license for a very short-term or longer-term duration or perhaps even geographically (e.g., per-transmitter) limited in line with light licensing concepts, are allowed to access the spectrum at will. A method for coordinating among primary assignments by the regulator would nevertheless be useful in cases where there are multiple primaries coexisting. Secondary users must use a "cognitive" mechanism ensure the band's availability before access, whereby the detail of the "cognitive" mechanism (e.g., the use and configuration of spectrum sensing, a GDB, or other method, etc.), as well as the secondary's radio characteristics, are dependent on the context within which the band is chosen to operate. Such context could include the protection of primary services(s) in the band for which the primary has obtained a license to serve, an assessment of an appropriate "burden" on the primaries in terms of acceptance of a slightly higher probability or net amount of interference, the possible use of another form of burden for the primary such as the primary deploying a beacon mechanism to ward off secondaries, and the degree to which the primary and secondary should/will negotiate, among other aspects. Of course, these considerations define the extent to which the secondary must avoid interfering with the primary, hence the technical requirements placed upon the secondary for primary avoidance. It is noted, importantly, that the primary will be effectively making a decision on such aspects when it obtains its PL with associated technical (e.g., secondary opportunistic access) characteristics from the regulator, and therefore predict the effects of the secondary access on its service in advance before obtaining the license, e.g., through the modelling of those effects.

Given all of the above, it is likely that such a concept, depending on how it is implemented, could lead to a very spatially and temporally dynamic situation in terms of the types of PLs operating in different areas, and associated requirements on the secondary users for detecting and avoiding interference with the primary. Depending on the implementation specifics of the PL concept, the secondary may therefore additionally be required to cross-reference a GDB which will return detail on the PL that is being used in a given band and location, before it can be allowed to transmit.

7.1.3 Analysis and Discussion

A first aspect of analysis and discussion of our PL concept is the weighing of the benefits and drawbacks of it. First assessing the benefits of the approach, we see as:

- Able to propel more robust or better design of devices/systems. Namely, for primary devices it might lead to the ability to cope with an increased degree or risk of interference, among other benefits, through resilience to such effects being built into the primary. For secondary devices it might also lead to the ability to better share spectrum opportunities with other secondaries, if such a concept was also applied to the pricing of awareness mechanisms for secondary—secondary coexistence as might be done at the point of secondary device certification, for example. Both of these aspects greatly improve spectrum usage efficiency and fairness.
- Highly flexible dependent on the case-by-case deployment context (e.g., dependent on the intended primary service), even so far as allowing solutions such as spectrum sensing and primary beacon transmissions which would otherwise not be practical or desirable, while defaulting to safe mechanisms such as a geolocation database in cases where a lower interference variance is required.
- Reducing the need for the secondary to cope with the inefficiencies of legacy primary systems as might exist in some bands. Indeed, in obtaining the license, primary users implicitly accept the rules of the band hence will be designed and manufactured with better technical capabilities that are able to cope with those rules. Through such a link, the primary might for example incorporate better rejection of adjacent channel interference.
- Highly scalable to progressive deployment in more spectrum bands. Rules for each deployed band (e.g., the use of spectrum sensing, the geolocation database, or potentially even a beaconing mechanism being employed by the primary) being determined based on the intended context for that band (e.g., the expected primary services(s), expected burden on the primary, etc.), as typically selected by the primary in obtaining its license.
- Lending well to progressive deployment of primary transmitters as a network expands, for example, using a licensing regime similar to light-licensing.

- Incentivising the primary to make more efficient use of spectrum, and to make unused spectrum available.

Conversely, drawbacks of the approach we see are:

- Complexity: instead of a simple license and pre-defined rules, this concept implies a far more dynamic mechanism potentially both temporally and spatially, requiring technology for its management (e.g., a GDB, with more advanced capabilities than current proposed database implementations), and possibly will require enhanced assessment (pre-deployment) and monitoring (post-deployment) of its operation. PL can nevertheless operate in far simpler forms, ultimately to a level that is no more complicated than currently proposed TV white space rules. Hence, the extent of this drawback is very much dependent on the implementation specifics as implied by the primary's choice of PL configuration.
- A further drawback is that does not implicitly guarantee a given level of service for secondary users. Such a lack of guarantee is nevertheless typically a characteristic of "secondary" status per se. Probabilistic and other modelling/simulation methods can nevertheless be used by the secondary to understand what will be available at certain times and locations and therefore understand its implied service quality.

Based on the weighting of the benefits of the concept versus drawbacks, we see PL as a method that presents a compelling case for inclusion in the arsenal of solutions available to the regulator.

A further aspect of consideration is the configuration of the concept based on the context, e.g., the type of service that the primary is aiming to protect in obtaining its license, potentially also in consideration of the types of secondary systems that might access the band. Figure 7.1 gives an example of pricing considerations under the concept, in this case applied to two alternative combinations of primary and secondary services: (i) dual-priority (primary and secondary) mobile broadband, and (ii) primary broadcast with secondary mobile broadband. In the dual-priority broadband case, the primary (which is also broadband) will have a return channel so be able to feedback information on channel quality to the transmitter for link adaptation purposes, and will be able to trigger retransmissions of lost data It will therefore be able to accept a higher variance of interference, noting that it is also more likely that the primary users will be running higher-layer applications that are tolerant of a variation in service quality such as web browsing for example. The primary's acceptance of a higher degree or variance of interference in this case therefore means it should pay a lower licensing fee. Further, a more lenient mechanism for primary protection may be used by the secondary, such as spectrum sensing, thereby improving spectrum opportunity for the secondary.

On the other hand, for the primary broadcast case in Fig. 7.1, the primary will usually not have a return channel, so will not be able to adapt and will therefore be

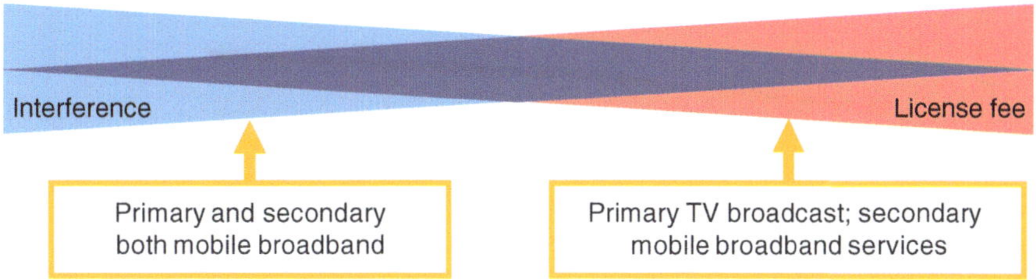

Fig. 7.1 Interference/license pricing for two different examples of primary and secondary services

able to accept a much lower variance of interference. Hence, in this case the primary must pay a higher fee in return for the use of a more robust protection mechanism from secondaries in the band. Nevertheless, conditions can be created for the primary broadcast to accept a higher level of interference through increased coding at a cost to its channel usage efficiency; this would be one way of implementing a reduced fee for that primary broadcast service. It would, however, impact on the rate that the primary can achieve.

Such pricing configurations are further investigated through simulation as follows. Here we simulate a primary broadband network, and secondary point-to-point connections separated and under a similar methodology and rules to the FCC's TV White Space rules. However, we vary the default distance that the secondary user is allowed to travel from the point at which the database was last consulter (hereafter referred to as the "leasing distance") from 100 m, up to 200 m and 300 m, thereby creating (in theory), the situation where some secondary users may get closer to the primary coverage area while still transmitting and interfering than would usually be the case. As we increase the leasing distance, we observe that the average SINR achieved by the primaries increases marginally, as would be expected. The standard deviation in the SINR received by the primaries further increases. Moreover, we assume that there is a fixed rate (SINR) requirement at each primary receiver above which the primary operator will receive payment from the user, whereby below this threshold the user is regarded as being in outage and no payment is received.

In this scenario, the dynamic of profit against the varying leasing distance requirement for the primary operator, a secondary (best effort) service and the sum of these profits, are given in Fig. 7.2.

It can be seen that there is a small impact on the primary profit caused by the greater interference uncertainty if the leasing distance increases. However, there is a bigger increase in profit for the secondary, and for the sum profit of the primary and secondary. Hence, in this case, if the primary were given 68 USD compensation (plus a small incentive) to compensate for a reduction in profit from 261 USDs (100 m leasing distance) to 193 USDs (300 m leasing distance), the net

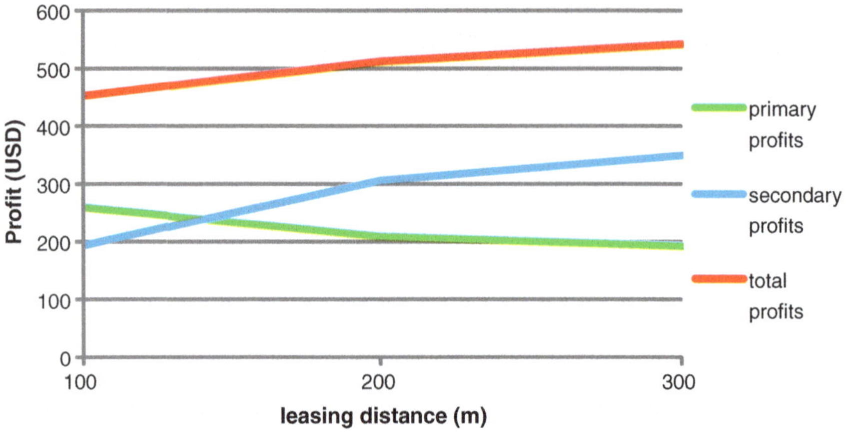

Fig. 7.2 Change in profit (USD) against "leasing distance", for the primary, secondary and the sum of the primary and secondary profits (Key simulation parameters: Primary base station transmission power: 47 dBm; primary terminal transmission power: 21 dBm; secondary terminal transmission power: 20 dBm; primary SINR outage threshold: 10 dB)

profit for the secondary could increase from 193 USDs to 350 USDs, a gain of 157 USDs (89 USD more than the compensation given to the primary). Hence, if the secondary were to compensate the primary directly or indirectly, both parties could gain from the situation. Moreover, the sum profit would increase from 454 to 543 USDs, making a compelling case for the regulator to allow such a sharing solution or to otherwise compensate the primary itself in the form of a reduced license fee for the overall good of the economy. This greater flexibility in leasing distance (i.e., flexibility in the allowance of opportunistic secondary spectrum access), and the associated compensation for the overall societal or financial good is a good example for our PL concept in practice.

7.1.4 Conclusions

In this section, we have introduced the concept of Pluralistic Licensing, which we see as a fair, flexible, reliable and viable means to introduce hierarchical spectrum sharing. We have discussed the specifics of the concept, including some implications and requirements for primary and secondary users. Moreover, we have analysed aspects of the concept, which we believe presents a compelling case for incorporation by the regulator to its range of licensing solutions.

Acknowledgements This work has been supported by COST Action IC0905 "TERRA" and the ICT-ACROPOLIS Network of Excellence, FP7 project no. 257626, www.ict-acropolis.eu.

7.2 ISM-Advanced Band Concept

Arturas Medeisis[1], John Sydor[2], Ligia Cremene[3], Dariusz Więcek[4], Yoram Haddad[5], Oliver Holland[6], Aurimas Anskaitis[1] and Tomas Cuzanauskas[1]

[1]Vilnius Gediminas Technical University, Vilnius, Lithuania
[2]Communications Research Centre, Ottawa, Canada
[3]Technical University of Cluj-Napoca, Cluj-Napoca, Romania
[4]National Institute of Telecommunications, Wrocław, Poland
[5]Jerusalem College of Technology, Jerusalem, Israel
[6]King's College London, London, UK

7.2.1 Introduction

In this section we describe the concept of "ISM Advanced" band, i.e. possible conversion of one (or a part of) traditional ISM bands, such as well-known 2.4 GHz band, into a CR-friendly spectrum access framework that would retain the original flexibility of ISM band while giving it new possibilities as a development ground for more powerful and intelligent wireless technologies.

Such an ambitious task is warranted at this time since the ISM/RLAN bands, and Wi-Fi technology in particular, are poised at becoming the dominant means of wireless access around the globe. Much of continued Wi-Fi proliferation and traffic growth are expected, but there are also challenges, and these are discussed in this section. Wi-Fi is spectrally inefficient and its variable quality of service makes its use in many scenarios questionable. Furthermore the ISM/RLAN bands are treated in a free-for-all manner, consequently their utility is constrained by uncontrolled interference. These limitations make it questionable whether Wi-Fi, with its license-exempt status and decades-old media access techniques, and the ISM/RLAN bands, will be able to evolve to meet the demands of the small cell, heterogeneous adaptive networks that are being touted for next generation wireless.

Today the co-existence in ISM bands, such as 2.4 GHz used for Wi-Fi and many other SRDs, relies heavily on setting a low ceiling for Effective Isotropically Radiated Power (EIRP, in Europe limited to 100 mW for Wi-Fi in 2.4 GHz) and low Duty Cycle (i.e. less than 0.1…1 %) as primary mitigation factors to contain interference on the local level. Both of these methods constitute severe inhibitors that dramatically limit the communication range (or link quality) and effective throughput of wireless applications in the ISM and other unlicensed bands. Such a paradigm for constructing license-exempt spectrum access rules is now decades old, therefore this paper aims to consider whether the prospect of using modern CR capabilities may allow proposing a novel way of sharing unlicensed bands with more efficiency and less limitations for deployment of innovative systems with higher bandwidth and service quality.

The original spectrum regulations that led to the proliferation of Wi-Fi and gave it its novel socio-economic/wireless niche and unprecedented growth also result in its poor reliability and hinder its use. The emissions mask of Wi-Fi devices, channelization plan and media access approach are, for the most part, left unaddressed by regulations although some standardisation of these issues might be beneficial to improve efficient utilisation of spectrum, as is the case in many other (licensed) bands. Thus the regulations, which lead to the development of wireless systems that must unresponsively tolerate all interference, have created a situation where the spectrum and devices' utility is constrained by uncontrolled and unpredictable interference. These limitations make it questionable whether Wi-Fi, as it is presently constructed, will be able to evolve and meet the demands of small cell heterogeneous adaptive networks that are now being touted for the next generation wireless applications. Commensurate with this is the concern that the public bands, despite their growing encumbrances of highly suitable spectrum, will satisfy the expanding socio-economic expectations of a society increasingly dependent on wireless [4].

Technical or regulatory changes to Wi-Fi and the ISM bands are complicated by the necessity to maintain backward compatibility with the huge number of legacy Wi-Fi based technologies that are deployed globally. Consequently, if the Wi-Fi standard's bodies work to conform to new spectrum regulations, they will likely need to follow an evolutionary path that supports the core attributes of Wi-Fi, principally at its PHY layer, but allows modifications to its MAC attributes, augmenting them in a manner that allows implementation of functions common to intelligent networks. The challenge to new spectrum regulations and policy supporting introduction of CR technologies in ISM band evolution is that it must remain technologically neutral and not force future Wi-Fi evolutions or other wireless applications toward a specific technique. New regulations need to stimulate the standards process in a way that engages researchers in as wide a technical discussion as possible without skewing the choice of CR networking solutions or limiting options.

7.2.2 Problems of Existing Wi-Fi Solution

Some of the most pronounced performance limitations with IEEE 802.11-based Wi-Fi devices is due to their OOB emission spectrum. The poor suppression of RF energy outside of the OFDM modulation bandwidth (having a 17 MHz wide spectrum) results in poor adjacent channel interference ratios (ACIR). Consequently, co-located Wi-Fi networks operating on adjacent channels see combined performance degradation that decreases only when there is increased physical separation between networks. Wi-Fi congestion problems in one channel consequently affect adjacent channels, especially when inter-terminal distances are reduced. Such problems give rise to poor spectrum efficiency and spectrum re-use, and contribute to the poor performance often noted within the ISM/RLAN bands.

Part of this problem can be attributed to the current globally implemented license-exempt ISM and RLAN band regulations, which do not provide channelization plans nor stipulate in-band emission and reception requirements on devices using the bands, other than transmitted power. This lightly regulated approach was taken when the unlicensed spectrum was underused and the number of devices was low, and low ACIR criteria were acceptable and kept at easily achieved levels; consequently ACIR as low as −1 dB for Wi-Fi is acceptable. In comparison, recently developed industry standards for licensed spectrum, such as the LTE-EUTRA, have ACIR ratios in the order of 44 dB and specify much sharper filtering characteristics on the OFDM modulation.

To overcome some of the Wi-Fi bandwidth limitations due to ACIR, the IEEE 802.11 standards process implemented channel bonding which essentially increases the OFDM modulation bandwidth of the Wi-Fi signal. IEEE 802.11n and 802.11ac, for example, have modulation bandwidths of 40–160 MHz in the 2.4 and 5.8 GHz bands. Though in theory channel bonding (combined with MIMO) significantly increases spectrum efficiency and utilization, practical deployments seem to indicate otherwise. It is noted that channel bonded Wi-Fi devices are severely impacted by interference from other Wi-Fi devices operating on a co- and adjacent channel basis; resulting in a performance anomaly where the higher rate modulation schemes become degraded by lower rate schemes. To overcome this problem, requires knowledge about both the desired and interfering links transmissions, their channel distances, and their physical rates, and use of techniques based on power control, explicit link scheduling, and resolution of hidden and exposed nodes (terminals), a problem which increases significantly with ACI and overlapping and adjacent channels.

Whereas the use of RTS-CTS or other MAC layer techniques succeed in solving most of the hidden node scenario instances, the exposed node issue remains in most of the cases unsolved (some solutions require strong assumption such as node synchronization to solve the issue or require change in the MAC protocol).

In the exposed node scenario, a node is prevented to perform a concurrent transmission because it senses the activity of another node, even though the concurrent transmission could have been successful, due to the fact that their respective receivers are far enough from the interfering transmitter. Due to its decentralized nature, the CSMA/CA mechanism doesn't allow the identification of whose concurrent transmission could take place which yields to a meaningful decrease in the overall system capacity. This not so new problem is emphasized if we consider the possibility to increase the EIRP. The number of nodes that will be "exposed" will increase leading to a decrease in the achievable throughput. Therefore the benefits of higher EIRP would be shifted off by this issue. The use of some kind of databases that gather which station transmit at what time could not only avoid this problem but even leverage the opportunities of possible multiple concurrent transmissions.

As regards channel access modes, the IEEE 802.11 standard encompasses two MAC mechanisms namely: DCF (Distributed Coordination Function) and PCF (Point Coordination Function). The DCF mechanism is the one that has been mostly deployed in all the Wi-Fi compatible devices because it is the only mechanism certified by the Wi-Fi alliance. DCF rely on the famous CSMA/CA mechanism which turns out to be the most preferred one, by the manufacturers, due to its fully decentralized nature. However, this latter mechanism exhibits poor performance in a highly interfered environment. Therefore, a priori, unlimited EIRP should be the last thing to do in this context. However, as we will see in the following, PCF could be the right answer for this case.

PCF is a polling mechanism. In this mechanism a coordinator initiates what is referred to as a "contention free period (CFP)" (to be distinguished from the contention period when DCF mechanism is used). The coordinator is generally the AP. During the CFP, no collision occurs between the users served by the same AP which saves all the wasted time spent by the devices during a collision or the recover from a collision. The coordinator polls in a round robin manner all the users to send the data they buffered. Originally, PCF was intended mostly for delay sensitive traffic. A fixed period of successive CFP followed by a contention phase for background traffic, is scheduled. Interestingly this approach did not attract too much attention from manufacturers as mentioned and was not really implemented. Some of the reasons can be due to the polling overhead or the requirement that a station that has nothing to send must anyway send a null frame.

Nevertheless, the increasing amount of deployed Wi-Fi networks with a so small number of non-overlapping channels available generates such an interfered environment that the using of PCF has been recently re-considered. Therefore using of PCF should be considered as one possibility to exploit more efficiently the ISM spectrum. It worth to mention that even in the ad hoc mode of Wi-Fi where no AP Is deployed and which requires a distributed mechanism such as DCF, it is still possible to use PCF. Recently some modifications were proposed in order to adapt PCF to a distributed environment without an AP where the point coordinator is chosen among all the participants and become the master of the cluster

The measurements carried out in a Canadian urban area [5] showed that the 2.4 GHz ISM band in an urban *outdoor* environment is not significantly occupied. At detection levels as low as −82 dBm, about 5–15 % of the band is occupied by IEEE 802.11. Its spectrum use efficiency is poor, since most of its bandwidth is spent transferring management packets (mostly beacons) to maintain Wi-Fi's Infrastructure Mode Operation.

Furthermore, the measurements noted that the distribution of Wi-Fi interference is highly irregular and non-homogeneous. Despite the large number of transceivers that may be observed in the ISM band in urban environments, the number of dominant interferers was shown to be quite small, and thus might be addressed and managed efficiently. Another important circumstance is that detectable Wi-Fi interfering signal packets contain MAC addresses making them traceable to a particular device. Such interference knowledge leads us to believe that it should be reasonably straightforward and easy to implement CR in the ISM bands and thus take advantage of the observed interference in order to improve spectrum utilization efficiency. Furthermore, Wi-Fi itself can achieve significantly improved spectrum efficiency if appropriate MAC and PHY changes are undertaken.

7.2.3 State-of-the-Art in Wi-Fi Technology

The pressing nature of the above discussed problems is evidenced by the fact, that industry had been actively testing various proprietary solutions for improving the efficiency of Wi-Fi operations in the ISM band. Most of these recent technological developments show a clear trend towards the CR paradigm, albeit falling short of complete realisation without a supportive regulatory framework.

The most typical road followed by many manufacturers of Wi-Fi APs is imbuing them with certain autonomous sensing that in turn provides input for dynamic frequency selection (DFS mechanism). Typical examples of such an approach include Cisco Aironet, Infinet, and Ruckus' products.

Cisco *CleanAIR* technology with proprietary Prime Network Control System is designed to aggregate information from multiple APs in order to build a multi-faceted picture of radio environment. A similar approach is followed by Infinet with their *iDFS* feature. It uses a dedicated secondary radio to scan the environment which feeds the data to DFS functioning. The collected data includes signal levels and amount of traffic on each channel, which allows performance grading of channels according their interference potential.

Yet another example of DFS implementation may be found in Ruckus' *ChannelFly* solution. Its specific feature is the use of smart antenna array, which collects radio environmental data from different directions. Importantly, the channel selection is carried out based on capacity averages across all channels. An optimal channel is selected based on gathered historical data.

Another good example of the fully fledged CR-ready Wi-Fi technological architecture may be found by looking at the multi-functional platform from Canadian Communications Research Centre (CRC) named *CORAL* [6]. It can work as centrally managed and fully re-programmable network of Wi-Fi APs, see Fig. 7.3. The radio environment data is collected by network nodes, which can perform continued monitoring without interrupting data transmissions.

The heart of CORAL is the Cognitive Radio Network Management System (CRNMS), which performs all the necessary evaluation of received data both from network nodes as well as end-users. CRNMS includes such features as controlling network nodes by changing operating modes: slave/master, operating frequency, transmit power, data rate, and packet transmission schedules. The spectrum monitoring data is accumulated from Radio Environment Memory Map (REAM) database, which allows analysing it from multiple angles in order to take a decision on most optimal spectrum access configuration.

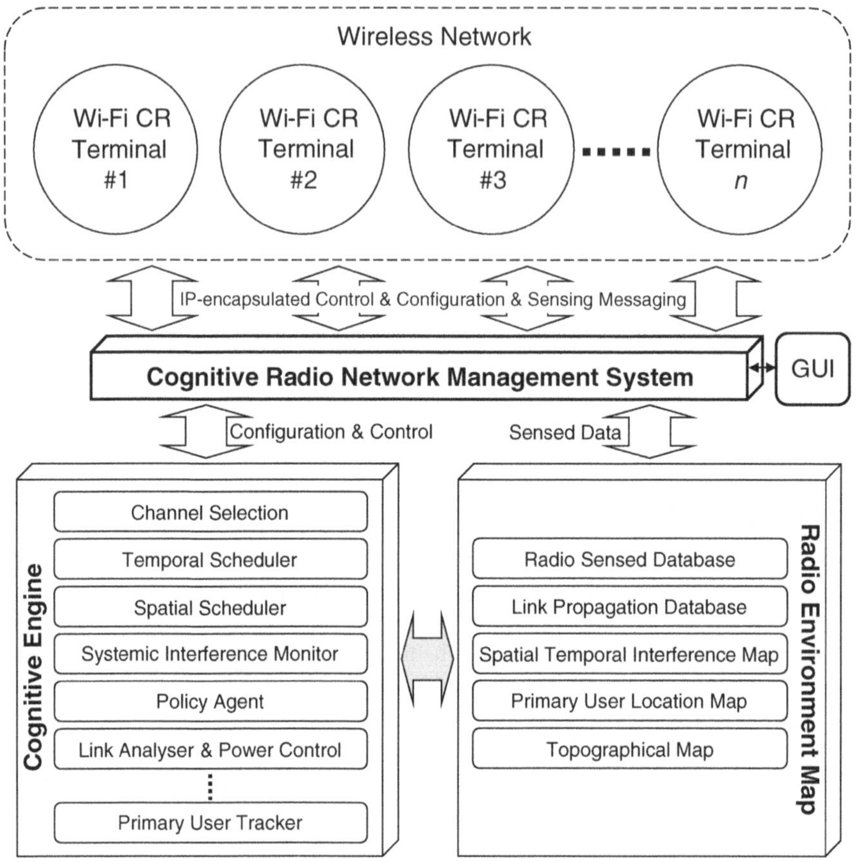

Fig. 7.3 Structure of CRC CORAL system

The above examples demonstrate that industry is trying to address the inefficiencies of current ISM use and that technology that is being developed is fully capable of taking full advantage of CR modes of operation, including both local and distributed environmental sensing, extensive analysis of real-time and historic data, and instant re-configuration of radio access parameters to address the optimisation challenge. Therefore there seems to be no obvious reasons why the regulation governing the access to ISM (and other licence-exempt bands) could not be updated so that the rules correspond to current technological capabilities of modern consumer devices.

7.2.4 Proposed Organisation of the ISM-A Band and Channel Access Modes

As derived from the above discussion, the considerations for modernising the spectrum access framework for ISM bands such as 2.4 GHz band may be directed along two broad avenues:

- Revisiting the rules on general organisation of the band (i.e. channelling options), spectrum emissions (i.e. OOB limits) of devices and channel access rules (MAC);
- Instituting provisions for using CR technologies as the means for improving quality of service to users while maximising spectrum use efficiency.

Accordingly, the rest of this section will review some fundamental principles that might be considered in instituting the ISM-Advanced regulatory framework.

7.2.4.1 Improved Organisation of the Band

With regards to the first identified direction, the prime objective would be to consider stipulating more stringent OOB emission requirements, which would support the move toward channel orthogonality. Such a stipulation could be proposed with a fixed channel plan, preferably one based on the de-facto 20 MHz bandwidth occupied by the vast majority of legacy Wi-Fi systems. Such a plan could propose to use the current Wi-Fi channels of 1, 5, 9 and 13, and would allow 4 orthogonal channels to fit within the 2.400–2.483 GHz ISM band.

A related issue would be to consider a dynamic channel bandwidth allocation, based on fluctuating traffic requirements. Today most of the Wi-Fi devices operate with fixed channel bandwidth, even though the majority of APs have the capability to work with both 20 and 40 MHz channels (since 802.11n stipulates such default capability in mixed mode). Therefore an adaptive channel bandwidth based on real-time throughput requirement would be an improvement in order to increase the spectrum capacity for multiple devices [7]. This is supported by findings [8] proving that even though a 40 MHz channel bandwidth may provide the peak throughput value, the 20 MHz channel bandwidth is better suited to real world environment where signal level might not be high.

The adaptive channel bandwidth selection mechanism would need to evaluate such link parameters as SNR, antenna mode (SISO, MIMO), current RF modulation and the required throughput. If throughput requirements from the served client are not high, then an AP would issue request-to-send using the smallest possible channel bandwidth that is sufficient to achieve the required throughput. With such approach, the access points would always roll back to the minimum channel bandwidth without degrading link performance, or in fact possibly even improving it, as a consequence of higher power density (and less interference) in the narrower channels.

A group of devices with distributed sensing and adaptive channel bandwidth mechanism would be able to share spectrum most efficiently based on their throughput requirements, due to the availability of better mapping of the environment. Whenever channel bandwidth adaptation would no longer provide the sufficient throughput, then the DFS mechanism would be able to kick in by choosing a more optimal channel.

Another aspect is that while 802.11ac enhanced bandwidth management to a new level with an improved RTS/CTS mechanism, the airtime occupation levels are still a main issue due to dominance of management packets, which are sent between devices. By modifying the MAC layer and decreasing the amount of management packets, it's possible to dedicate more airtime for traffic packets.

7.2.4.2 Deriving Benefit from CR-Technologies

As regards to the provisions for using of CR technologies, the previous discussion showed that the DFS mechanism is already being progressively implemented by the industry and therefore does not require any additional regulatory intervention in order to promote it further. Therefore it is proposed to focus the additional consideration of potential benefits of CR on the question of power (EIRP) limit that often significantly restricts the range and/or link quality of wireless access devices. It is rightful to ponder whether "intelligent" devices really need to be told as to what maximum power they should adhere to. We posit that with appropriately designed rules, the CR-enabled devices should be perfectly capable of choosing most appropriate transmit power while seeking the optimum compromise between link range/quality, ambient interference level, and its own energy consumption.

In modelling this optimisation task, Game Theory was shown to be a suitable tool to model interaction of devices within the domain of CR [9]. An example simulation of a power allocation game was reported, using 2–20 links co-located randomly within 100 × 100 m area. The results of these simulations are presented below in Fig. 7.4.

It can be clearly seen from these results that in this dense deployment scenario, removing the 100 mW limit and allowing users to operate at up to 1 W EIRP does have positive consequences on the overall capacity of the links sustained in a given bandwidth. This is especially striking noting the rather small scale of the simulated scenario of just 100 × 100 m, where the higher power would not be considered necessary based on purely link distance considerations. This shows that allowing higher power would lead the devices to use the additional power margin in order to increase the SNR and thus improve the quality of the link. At the same time it is shown, that given the clear rules, the system would converge and no excessive over-exploitation of power would occur.

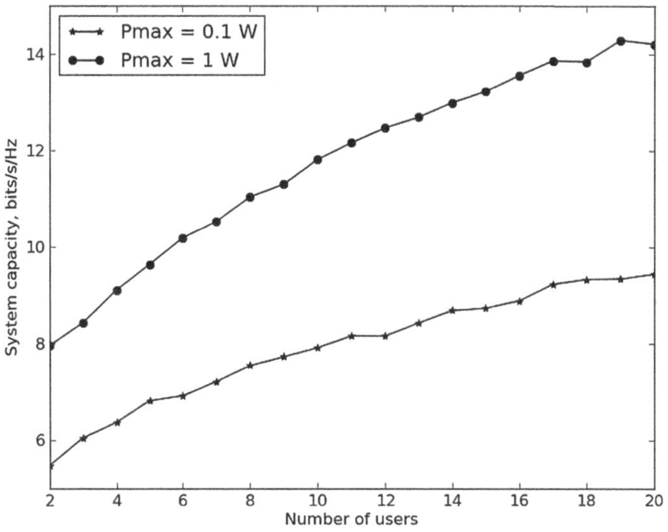

Fig. 7.4 Simulation results of a distributed interference-aware power control game as total capacity of links deployed within 100 × 100 m area [21]

On the other hand, simulations also showed that the issue of fairness in power allocation is highly relevant in today's Wi-Fi networks and would remain challenging if the power limit is lifted. This is illustrated by the following Fig. 7.5 that shows an average capacity per user, which deteriorates proportionally with the growth of number of supported users in the band.

It may be observed from the figure that the individual links would suffer spectrum congestion regardless of the power limit. It is especially illustrative that the 100 mW EIRP limit is not really a solution to address that problem. Yet even in a very congested situation, having the higher power limit would allow higher link capacities. Furthermore, having the devices with CR-capabilities might allow addressing the fairness more "intelligently", i.e. by appropriate adjustments to power allocation and MAC algorithms. The details of such improvements would need additional consideration.

Yet another possible avenue for increasing EIRP would be through setting the limit on transmitter output power and instead requiring that any additional EIRP growth would be obtained through antenna gain, i.e. the use of more directional antennas. This would correspond well with the logic of containing any excessive interference within an increasingly restricted geo-spatial beam. It was shown by statistical simulations [9] that such an approach helps reduce the probability of harmful interference significantly, by around 5–10 times compared with the reference base case using unidirectional antennas.

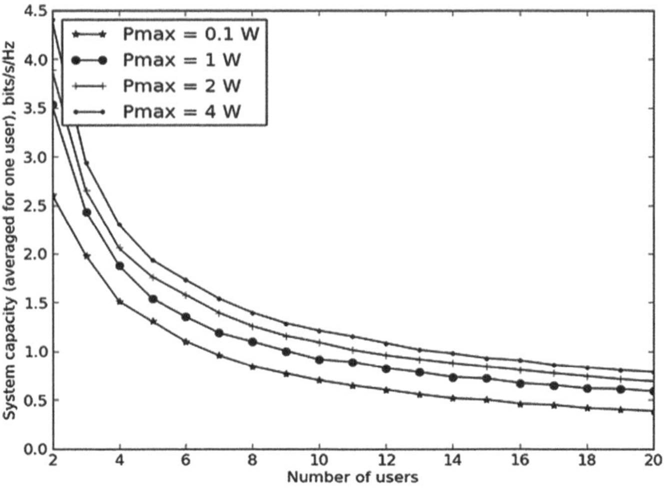

Fig. 7.5 Average link capacity in the simulated system as function of total number of users [21]

7.2.5 Summary and Proposals

It may be posited that the low EIRP limit represents an embodiment of barriers to innovation in the ISM and other license-exempt bands. If this might be overcome, then by itself the more liberal spectrum access paradigm for unlicensed band will unleash a wave of new uses of the band and will allow testing and validating in real life of the main premises of CR technology.

Along the way, the new concept and the occasion of reviewing the overall regulatory package governing the use of subject band, would also allow addressing and resolving some of the other above described critical problems that are plaguing the efficiency of using Wi-Fi and ISM bands in general, such as:

- Optimization of PHY/MAC layers, by further improving CSMA/CA mechanism that is widely used today and has limitations due to absence of any enforceable coordination features (e.g. the PCF method envisaged in Wi-Fi yet rarely used in reality);
- Poor Out-Of-Band (OOB) emissions limits;
- Quality of Service experienced by the user;
- Energy efficiency of transceivers, i.e. with a view on reducing the battery drain of portable devices.

Ideally, any solution governing the use of ISM bands should be aimed at autonomously deployed centrally uncoordinated devices, which would be more in line with the spontaneous nature of a commons band. However, when justified for improving overall efficiency (and thus avoiding the "tragedy of commons"), it may be reasonable to assume that some form of cooperation and coordination between devices might be necessary. And these are the areas where CR technology might be also helpful and become a deciding factor to dramatically improve the utilisation of ISM bands.

7.3 Finding a Sweet Spot for CR

Peter Anker

Ministry of Economic Affairs, Delft University of Technology, Delft, The Netherlands

7.3.1 Analyzing the Case of White Space Access in the Television Broadcasting Band

Having established in Sect. 5.2 the concept of actor-centric perspective as a valuable instrument in achieving alignment between technology and institutions in the past, we will now apply this perspective to a case of which its resolution lies in the future. It concerns the introduction of CR technology in the so-called white spaces in TV-bands.

The first application for CR that was put forward was the use of white spaces in the TV broadcasting bands. The US Federal Communication Commission (FCC) made these white spaces available for unlicensed broadband Internet. Its intended use is, above all, to provide more affordable broadband deployment in rural areas [10].

In this case CR technology is intended to share the TV-band with the legitimate primary users, the TV broadcasting stations and low power auxiliary service stations (notably wireless microphones). Given the latter, it is understandable that the FCC removed sensing from the original requirements and took alternative measures to guarantee access to spectrum for wireless microphones and to prevent wireless microphones from being subjected to interference from CR devices. First of all, at the current state of technology sensing is not sufficiently reliable. More importantly, to prevent interference to the primary user, the output power of the CR device should be low relative to the primary users.[1] These primary users are not only TV broadcasting stations but also these low power wireless microphones. Restriction of the output power of CR devices to a level that is low compared to the wireless microphones would have been detrimental for the business case of rural broadband access.

In taking the perspective of the private actor, the first question to be asked is: Why is there no service provided at the moment? There certainly is no scarcity of

[1] See discussion in Sect. 5.2.

radio spectrum; the 2G/3G bands are under-utilized in these rural areas. The main reason appears to be that the costs to provide the service are too high in relation to the willingness to pay for the provided service.

The second question to be asked is: how will the business case for CR improve the situation? For the business case to become viable either the willingness to pay for the CR enabled services has to become higher or the cost reduction needs to be greater than the additional costs associated with the new (more capable and sophisticated) cognitive technology. Combined they need to bridge the gap between the provision of services based on the current technology and the current willingness to pay.

Under the FCC white space ruling, rural broadband access is made more feasible due to the fact that a lower frequency range is made available, which extends the coverage area of a base station, compared to the existing alternatives to provide the service. However, existing mobile networks operate at frequencies that are just above the television band. This means that potential gains of using TV band white spaces by reducing the number of base stations needed to serve a given area are very limited.[2] Therefore the business case for deployment of a wide area network in rural areas based on white space access remains highly questionable. It is much more likely that white space access will be used to provide localized access to the Internet at specific backbone nodes. This is a business case that is comparable to Wi-Fi hot space access, although over larger distances.

The next question is whether the capacity that can be supported by white space access is high enough to support the demand from users. In areas where the required demand for capacity is bigger, the coverage area of the base station may have to be made smaller. This conflicts with the reasoning to make these lower frequencies available. This means that the business case will be restricted to areas with a population density below a certain limit. This limit will be lower if the demand per customer is higher. It remains to be seen whether the assigned band will have enough white space capacity available for the intended application—broadband Internet access—to support a successful business case.

Studies performed on the use of the UHF broadcasting bands for CR in Europe showed that the amount of white space is limited, because of the tight digital broadcast planning. Moreover, the TV band is already heavily used "opportunistically" for Program Making and Special Event services, especially wireless microphones. Furthermore, the upper part of the band has been made available as a harmonized sub-band for mobile use [11]. Hence, the amount of available spectrum for white space devices is far less than in the United States [11, 12]. This amount may be even further reduced in Europe through the decision of the World

[2] As the use of white spaces is considered to be free of charge, this normally represents a benefit compared to the business case for existing 2G/3G deployments, which may be subject to the recovery of a hefty spectrum licensing/auction fee. However, in serving the rural areas, economists will consider the auction fee as sunk costs and will calculate the business case on marginal costs, thus the comparative economic advantage of "free spectrum" is also reduced.

Radio Conference 2012 to extend the possibility of the use of the TV band for mobile services to the 694–790 MHz band.

To conclude, the white space access regulations appear to be a technological-fit instead of a business-case-fit, driven by the regulator to realize a social goal. Whether the business case is viable remains highly questionable. It would explain why the intended service providers are relatively absent in the standardization activities and other discussions around white space access in the TV band. Moreover, it may explain why there is, as yet, no viable business model for the commercial operation of a database in support of sharing the spectrum with wireless microphones.

7.3.2 Next Steps: Finding a Sweet Spot for CR

Although there are possibilities to use CR under the current radio spectrum management regime, there is still no compelling business case. To assure development and deployment of CR technologies, it is worthwhile to review potential product-market combinations where CR functionality provides a 'value add' and determine whether these cases are attractive enough to be taken up by the industry as first applications of CR, as first steps on the road toward broader deployment of CR technologies.

The government can facilitate this process through the initiation of a platform in which the equipment industry, the service providers and the government itself closely cooperate with the aim to find a "sweet spot". This sweet spot may serve as a catalyst to both the private sector and the government; for the private sector to develop products and services based on cognitive technology and for the government to realize the ultimate goal of more efficient use of spectrum.

The RSPG (Radio Spectrum Policy Group) has already recommended creating a platform to allow researchers, academia, manufacturers, operators, service providers and regulators to coordinate research activities. According to the RSPG, this platform could build upon already existing platforms with comparable purposes, notably COST-TERRA [13]. This notion of the RSPG on COST-TERRA is quite relevant. The discussions within COST-TERRA are very fruitful, but are rather academic in nature.

As the discussion within COST-TERRA is too academic, discussions will benefit from a new extended platform that serves as a Community of Practice that involves all stakeholders. In order to do so, participation should be widened in two directions. Firstly, participation should be extended to service providers and users of spectrum. This may strengthen the discussions on the incentives for primary users and possible business cases for the primary and secondary users. Secondly, participation should be widened to industry players to incorporate the ideas and solutions in the development of new technology and technology standards.

In this platform all participants should work together with the national spectrum regulators to find and enable a sweet spot. A sweet spot needs a fit between a specific CR technology, an initial business opportunity and an associated regulatory regime. The regulators can enable this sweet spot on a European level by specifying the necessary and specific regulatory regime in a European decision and/or European recommendation.

This requires participation at the working level. Intended participation should include the broad and balanced representation from regulatory bodies in different countries (as organised in CEPT/ECC), industry players (as organised in ETSI) and academia (as organised in COST-TERRA and other/future similar research projects).

There is already some experience with a Community of Practice related to CR in the Netherlands (CRplatform.NL[3]). This Community of Practice aims to identify the uncertainties surrounding potential deployment areas of CR and through discussion among stakeholders to find ways and means of addressing and reducing these uncertainties; thereby facilitating the successful deployment of CR-based products and services. This initiative evolved from the regular interaction between representatives of the Dutch Ministry of Economic Affairs, responsible for radio spectrum policy and the industry.

In addressing uncertainties and finding ways towards resolution, the Community of Practices organizes workshops to explore potential application areas of CR, so-called Use Cases. The following application areas were among the Use Cases as discussed during the past two years: container terminals in the Rotterdam harbour; coverage of special events by broadcasting organisations; public safety communications by the police force; high intensity communications at airports; and CR facilitating Home Automation. Each workshop brought together potential users, potential suppliers, policy makers and regulators, as well as academic researchers.

In these explorations, one of the first questions to be asked is: what are the gains from the use of this new technology, and are these gains high enough to cover the increased cost of the use of this technology compared to the alternatives? The Use Cases as discussed suggest that CR functionality adds most value in situations that are typically niche applications or are a small segment of the overall market for wireless technologies. One of the reasons is the fact that CR technology is basically a technology to (more efficiently) share the radio spectrum. As CR provides additional functionality compared to current radio technology this will come at increased costs, at least initially. Situations of high intensity demand are expected to provide the highest willingness-to-pay by the end-users.

Each use case discussed so far addressed a specific market segment, or even a market niche. Hence, potential market volumes are (relatively) low to moderate, which impacts the viability of the CR business case. Nonetheless, the use cases also show similarities, in particular if CR-based solutions are considered as variants of a more generic CR-platform solution. Especially the combined business case of the communication needs of the public safety services in case of an emergency and the registration of this emergency and other news gathering seems

[3] See online at: http://crplatform.nl/ (Accessed 28-Oct-2013).

to be logical and promising. This became apparent during the Use Case Workshop on Special Events, as during (ad-hoc) events the needs of public safety and broadcasting converge at the same place and time. The type of communication needs show a strong parallel. Hence, pursuing solutions for one group of actors (broadcasters) should best be done cognizant of the needs of the other group of actors (public safety).

This example shows that finding a sweet spot for CR might be easier if the solutions for one group are similar to the solutions for the other group, at least on the platform level. This increases the addressable market and hence the viability of the business case. The unresolved issue is the capacity issue. How much capacity is available for CR use and is there enough capacity available to support the (combined) business case?

The use cases further show that a viable business case for CR will require economies of scale. This extends the need for coordination to the European Union level, if not at the global level. Such coordination may still be left to be organized by the industry actors. However, the use case experience suggests that lacking a very compelling business case the likelihood that industry actors will take the lead is expected to be low. This ties in with the fact that discussions within the Community of Practice confirmed the role of the regulator to facilitate this search for a sweet spot.

7.3.3 Conclusions and Recommendations

For successful introduction of CR, it is necessary but not enough to align the specific CR technology with the regulatory environment that is chosen. Next to it, the business opportunities that are enabled by specific choices should serve the objectives of both the entrepreneur and the government.

Exploring use cases can be a good instrument to bring all interested parties together and in an explorative modus to find and enable a "sweet spot" for the use of new technology. A "sweet spot" is enabled if the institutional arrangements and the characteristics of the new technology are aligned in such a way that an intended business opportunity can be realized.

This exploration can take place in a Community of Practice. An initial exploration of possible business cases revealed that the type of CR technology to be used and the appropriate regulatory regime to support it depend on the specifics of the intended business case and the specifics of the users with which the bands will be shared. When a viable combination is found, the spectrum regulator should set up the specific regulations to facilitate the CR deployment and thereby make an important step towards a more efficient utilization of the radio spectrum.

It is recommended to introduce this Community of Practice for CR on a European level. Such a Community could make use of, and build upon, the experiences of COST-TERRA. In order to encompass all interested stakeholders, this platform should be broader than the COST-TERRA participation. It should include representatives of service providers, user communities, industry players, academia and national regulators.

7.4 Lessons from TVWS and Anticipated Future of the GDB

Jeff Schmidt

Spectrum Bridge, Orlando, Florida, USA

7.4.1 Background

In the early days of CR, efforts were focused on the concept of RF sensing and how it could be used to enable spectrum sharing. RF sensing, or the ability to detect the use of spectrum at certain frequencies, is a means for enabling co-existence among radios and networks. A rudimentary example of this is the CSMA (Carrier Sense Multiple Access) MAC within an 802.11 radio device which attempts to avoid collisions by considering only what *itself* can sense. Fortunately, abundant internet connectivity, cloud based computing, and low cost geo-location technology have given rise to the geo-location database (GDB), a vehicle for centrally managing spectrum allocation. But because the concept of a geo-location database is rapidly evolving, this term no longer provides an adequate description. Today, a geo-location database is part spectrum manager, part policy manager and part co-existence manager. A geo-location database, in its simplest form, accepts requests for spectrum access from a radio and makes allocations in accordance with usage policies as a function of time, frequency and geography, as in TVWS. If the process is successful, the potential for harmful interference is mitigated, or at least controlled. This technique is alternatively referred to as Radio Environment Mapping (REM) in which the radio environment of CRs in multiple domains is tracked spatially in terms of geographical features, regulation, policy, radio equipment capability profile, and radio frequency (RF) emissions [14]. Although the advantages of frequency management, co-existence and interference mitigation through centralized coordination and control may be obvious, there remains some debate whether the geo-location database or RF sensing is the best approach for spectrum management. The answer may be surprising—neither or both, as circumstances may dictate which approach is better. But when the elements of both approaches are combined, the best solution often emerges. Consequently, changing the perception of the geo-location database is important as we consider future spectrum sharing, as TV White Space applications are only the beginning.

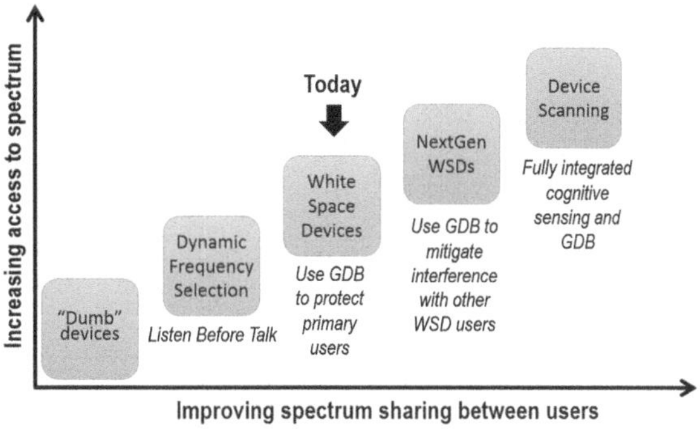

Fig. 7.6 Potential roadmap for the evolution of wireless devices accessing shared spectrum resources, adapted from [15]

7.4.2 Future Developments and Trends

The advent of cloud based computing, low cost geo-location capability and the growing availability of internet connectivity has afforded new opportunities in spectrum management and CR. It is no longer necessary (or efficient) to embed the majority of cognitive intelligence in a radio and consider spectrum management as an afterthought. When a radio or network is able to augment self-derived sensing data with RF environment information gained from a database, spectrum use efficiency can be increased drastically. Individual devices share sensing data with the database to enable this cooperative model. This method of cooperation is similar to our modern day air traffic control system. Although aircraft have relied on the use of sophisticated avionics to enable all weather operation for many years, it is virtually impossible to guarantee safe operation between multiple aircraft without the use of a central coordinating authority. Air traffic control systems not only ensure safe travel (similar to a radio's interference free operation), they can drastically increase efficiency by packing more planes (or radios if you wish) into tighter spaces! Upgrades to systems can also be made without the requirement to individually re-program the many millions of devices that are expected to be deployed, further enabling continuous innovation without the fear of obsolescence. This progression is clearly indicated in a recent Ofcom Consultation [15], contemplating future spectrum management methodologies, as illustrated in Fig. 7.6.

Arguably, the TVWS rules adopted by regulatory bodies to date are rudimentary and certainly leave room for improvement. Fortunately, in bureaucracies where perfection is too often the enemy of good enough, the rules used to create today's conservative and simplistic policies can be modified easily, through the use of a cloud-based policy engine. The effectiveness of policies can also be measured using the same system, enabling straightforward rule adaptations and optimizations as shifts in policy occur. Some aspects suitable for further development include the migration from granular to more flexible policy models, dynamic protection and allocation mechanisms, business models that incorporate consumption metrics and enabling better quality of access and co-existence mechanisms.

Fig. 7.7 GIS shapefile (A geospatial vector data format developed by ESRI) mapping approach depicting TVWS availability in the eastern US ("Spectrum Bridge Inc. I Products and Services I TV White Spaces Solutions I US Interactive Map I View TV White Spaces Nationwide." Spectrum Bridge Inc. I Products & Services I TV White Spaces Solutions I US Interactive Map I View TV White Spaces Nationwide. Web. 18 Oct 2013)

7.4.2.1 Data Resolution

One area of implementation well suited for further development is rooted in granularity or resolution. In the US, broadcast licenses are individually owned and the protected (service) area for each TV station is solely determined by a propagation model known as an R-6602 [16]. Per FCC rules, WSDs are not permitted to operate co-channel within a protected contour. As shown in Fig. 7.7 this method results in large and perhaps overprotected areas defined by binary exclusion criteria, with no opportunity for secondary users (WSDs) to co-exist by using power control. This condition is exacerbated by the insular out of band emission requirements requiring every device to meet the same stringent specification. Although the impact of these rules is limiting in some respects, they are elegant in their simplicity.

In contrast, the UK television distribution and transmission network is commonly owned and the service areas for viewers are defined by a combination of high resolution propagation models and empirical data (see Fig. 7.8). As such, the UK regulator, Ofcom, adopted a different approach to defining protection for incumbents. Ofcom relies on pixel based maps with 100 m resolution that explicitly define the maximum permissible transmit power as a function of channel, device class, and out of band emissions characteristics. This represents approximately 300 GB of data that must be parsed for each channel request at

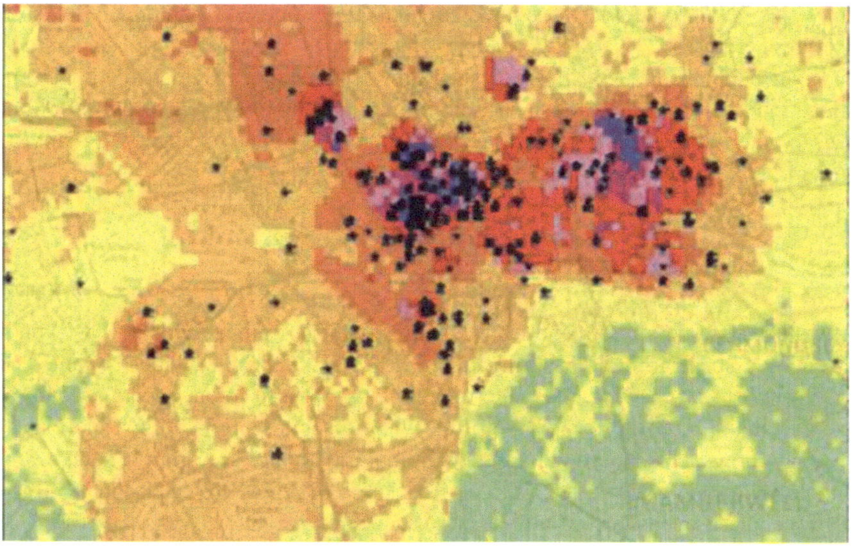

Fig. 7.8 Pixel based mapping approach projecting TVWS availability near London ("Geographic Maps Showing TVWS Availability." Ofcom. Web. 18 Oct 2013)

intervals of several minutes. Fortunately, a well-designed cloud based geo-location database is well suited for this task. As a result, the Ofcom approach allows for variable power, per channel, rather than the fixed maximums defined by the FCC. This provides more flexibility but is limited in that the additional white space afforded by this method is often not usable due to a higher induced noise floor.

So which approach is better? In reality, there are tradeoffs, but both are effective. The advantage of the FCC or US model lies primarily within the simplicity in defining contours as closed shapes, although this approach may err on the conservative side when considering how much TVWS is made available for secondary use. Alternatively, the Ofcom or UK model, derived from ETSI 301 598 [17], is significantly more complex, but provides a higher degree of resolution when defining protection for incumbent operations. The Ofcom or ETSI model also allows an authority to fine tune or adjust the protections. It is expected that these two distinctly different approaches will evolve and may converge over time as the benefits of each become clearer.

7.4.2.2 Dynamic Adaptation

Because GDBs are designed to operate in real-time, it is entirely possible to conceive of a spectrum allocation system that relies on historical usage patterns and data to predict what allocation(s) might be optimal in the future. A good example of dynamic behaviour is a large sports event in which many people converge in an otherwise unpopulated area and impose a significant load on existing networks. The usage models for these types of events is well understood and can be managed by allowing spectrum holders to engage in the spectrum allocation process and establish specific rules and policies for temporary spectrum use. In other words, it is not necessary to assume that policy managers eternally be a single government or regulatory body.

7.4.2.3 Business Models

It would also be quite beneficial to implement billing solutions that would incentivize spectrum holders to make spectrum available to a wider audience of spectrum users. Likewise, the same model could be used to encourage secondary users to maintain a right-sized footprint, while discouraging hoarding and excessive emissions. This would require the commoditization of spectrum in such a way that consumption could be definitively measured and fees collected that are proportional to usage. This would positively incentivize the holders of otherwise fallow spectrum to make it available. Obviously one of the great challenges in this model is to ensure a critical mass of compatible spectrum is made available to promote capital investment in radio hardware and infrastructure through continuous access to spectrum. However, as wider band, frequency agile, software defined radios become more available this problem becomes less acute. The use of taxed based systems that discourage secondary spectrum users through tariffs proportional to the emissions (in band and out of band) have also been proposed. This approach would likewise incentivize spectrum holders to make spectrum more available and secondary users to be efficient when consuming spectrum resources.

7.4.2.4 Quality of Access and Co-Existence

Rudimentary geo-location database systems have already been used to increase the network capacity of large commercial networks. An example is the neighbourhood hotspot initiatives launched by large wireless carriers, in which Wi-Fi routers are configured to create two parallel Wi-Fi access paths. The first path is private while the second is open. However, there is no coordination of frequencies based on location in these systems. Perhaps this is less critical when using low transmit power and high (>2 GHz) frequencies. A wireless network of this sort could certainly benefit from centralized frequency management in dense applications. Another example is the SON (self-organizing network) technology being adopted by the purveyors of 3G/4G femtocells. Unfortunately the SON technology adopted for femtocell networks is by definition 'self'-serving, as the objective is to minimize any risk of interference to the inter-network macro cellular overlay. This is facilitated through sensing and location technology, which are two essential components of GDB enabled CR systems, but no means exists to share information between dissimilar networks. The net result is room-size femtocell coverage with no ability to opportunistically co-mingle non-coherent networks. Despite the apparent limitations in these examples, they clearly illustrate the utility and state of dynamic spectrum management technology. Non-coherent or dissimilar networks could be intermingled if different GDBs could communicate with each other. This is one objective of the co-existence functions described in 802.19 [18], but little progress has been made in practice.

Recently, crowd sourcing has also become an effective means for managing common network resources accessed and shared by many. In this way, a centrally managed spectrum management platform can enable co-existence and quality of access between an eclectic mix of users, such as in a typical unlicensed band. When a spectrum management database acquires usage and interference data (feedback) through the sensing capability of individual radios this is a means of crowd-sourcing. This is much like the traffic applications in our smartphones that we rely on to navigate crowded roads. This is perhaps the most important distinction between existing methods of spectrum management and future implementations, as the augmentation of centralized coordination and feedback will bridge co-existence across dissimilar networks in which coordination would not ordinarily be possible. Channel guidance will eventually be provided as active (command) or passive (recommendation) and operate in real-time. In a world that is seeing relentless wireless broadband growth, dynamic spectrum allocation that relies on feedback really is the future of database driven cognitive networks and essential to the future of wireless networks.

7.4.3 Applicability in Other Bands

While the primary objective of geo-location databases has been to manage TV White Space, it is important to emphasize the applicability of this technology in other bands. Significant effort has been invested by commercial enterprises to develop technology for managing unfettered access to other unlicensed and licensed exempt spectrum using some type of central management scheme. But as previously mentioned, these efforts are generally self-serving (e.g. self-organizing networks), and little effort has been focused on co-existence between dissimilar networks. A fundamental difference between management of TVWS and successful real-time spectrum management in other bands is the need for feedback. Fortunately, the process of commoditizing spectrum in terms of time, frequency and geography are similar, while a well-designed policy engine can be configured to handle the unique requirements of protection ratios and exclusivity. This leaves the problem of normalizing and sharing operational feedback across dissimilar network domains and coordinating spectrum allocation between spectrum managers. Without feedback and collaboration, co-existence *between* individual self-organizing networks is futile. Fortunately, this is a technical problem that many companies are working to solve and significant progress is being made. The

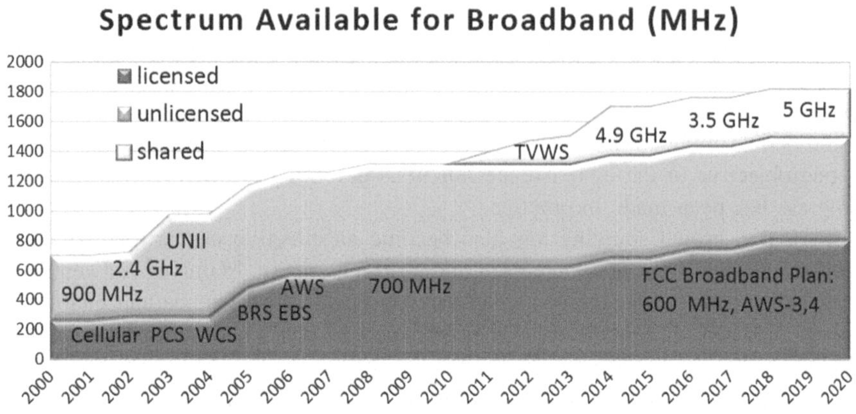

Fig. 7.9 Historical and projected spectrum allocations by band in the United States

remaining need is to create regulatory frameworks that are more conducive for adoption of dynamic spectrum management and as a consequence stimulate more widespread use. There is increasing belief in that the only way to gain access to significantly more spectrum will be through sharing. Spectrum is a finite resource and as the practical limits of spectral efficiency are approached—through methods such as MIMO, advanced modulation and MAC layer protocols, and exclusive (brute force) spectrum allocation—dynamic spectrum management becomes a very attractive option and perhaps essential.

Several spectrum bands have been recently identified in the US [19], which are underutilized by the government and military. The use of these spectrum bands is much more varied and dynamic than broadcasting spectrum, and will require more sophisticated management solutions than TVWS. The biggest difference between a geo-location database used to support TVWS and other dynamic spectrum access applications will be the dynamic behavior of protected entities, so much that central coordination aided by sensing mechanisms will be essential. Other bands in which GDBs are being contemplated are 3.5 GHz (lightly licensed), 4.9 GHz (public safety) and 5 GHz UN-II (unlicensed). It is expected that as spectrum sharing and access models are proven in unlicensed bands, growth will continue in licensed and government bands.

Sections 2.5 and 2.6 discussed the relentless appetite for mobile broadband services and the resulting growth of mobile traffic. Figure 7.9 shows projections of potentially available spectrum in the US that might be shared using geo-location based spectrum sharing mechanisms.

When comparing this linear growth of allocated spectrum bandwidth against the exponential explosion of data traffic (see Sect. 2.5), it is clear that without new means for increasing spectrum usage efficiency, growth is not sustainable, even when considering new allocations of spectrum. Today, 'shared' and 'unlicensed' are often used interchangeably, but there is a distinction that should be made in that 'shared' implies the use of a central management component and can be applied in licensed and unlicensed bands.

In applications beyond TVWS, there exists the notion that access could be managed on a Priority, Quality of Service or Quality of Access basis, rather than the 'all welcome' scheme adopted for secondary users in TVWS. This type of access may even be extended to incorporate managed payment schemes, such that spectrum is rented on an ephemeral basis, rather than bought or allocated permanently. Pay per use schemes require a spectrum management platform to track use more quantitatively and must include clearing house functions to facilitate the financial transactions associated with usage. While many of the concepts used to manage scarce spectrum resources are similar to those in other markets, it will be a new challenge to invent the provisions necessary to manage spectrum trading in terms of interference rights.

Other limited sharing schemes have also been proposed by cellular operators. These approaches are known collectively by terms such as Authorized Shared Access (ASA) or Licensed Shared Access (LSA) [20] and infer the possibility of exclusive access by secondary users (see discussion of LSA in Sect. 2.6). These schemes also contemplate a geo-location database to coordinate co-existence between primary and secondary users while providing clearing house functions to support consumption tracking and billing. Perhaps the biggest fear by licensed spectrum holders is the possibility of enabling a competitor. This is hardly a technical issue and indeed not new, as the concept of roaming and leasing network capacity rely on similar paradigms and hopefully serve as examples to alleviate these concerns. Several existing frameworks for spectrum sharing are capable of supporting these use cases from a technical perspective. It may be argued that some spectrum management frameworks offer more or less capacity for precision or flexibility, but the essential technology exists, awaiting favorable regulatory policies as a catalyst for adoption.

7.5 Concluding Remarks

Jeff Schmidt

Spectrum Bridge, Orlando, Florida, USA

It is evident that modern societies' appetite for broadband connectivity will remain unabated and consume the full availability of wireless services available. There remains debate, however, whether consumption will be stimulated by increased demand for wireless services, increased capacity or a combination of both.

This classic debate is well described in economist John Maynard Keynes', *The General Theory of Employment, Interest and Money*,[4] which summarizes economist John Say's law[5] as, "supply creates its own demand", and argued that "demand creates its own supply". Regardless of which side of the debate one chooses, spectrum is perhaps the most fundamental and finite resource that enables the growth of wireless services. This is significant because spectrum is a peculiar commodity that lends itself to incredible efficiencies in the operation of wireless systems and often measured in bits/second/area/hertz, i.e. throughput per area per spectral bandwidth.

Consequently, the largest untapped potential for increasing efficiency in spectrum consumption lies in the ability of a radio device or wireless network to acquire situational awareness and consume only what spectrum is necessary while making the best use of otherwise fallow spectrum assets. The ability to achieve this is totally within the capability of today's technology, enabled through a confluence of software defined radios, cloud based computing platforms, and widespread internet connectivity. Together these technologies are combined to form various embodiments of CR, including those enabled by the geo-location databases. Ironically, the single biggest challenge is not technology, but the evolution of regulatory policies and business models needed to galvanize the landscape.

References

1. Cave, M., Doyle, C., Webb, W.: Essentials of Modern Spectrum Management, 1 edn. Cambridge University Press, USA (2007) ISBN-13: 978-0521876698
2. Radio Spectrum Policy Group (RSPG11-392) Collective use of spectrum (CUS) and other spectrum sharing approaches. Report on November 2011. At http://rspg.groups.eu.int/rspg_opinions/index_en.htm Accessed May 2012
3. Holland, O et al.: Pluralistic licensing. In: IEEE DySPAN 2012, Bellevue, WA, USA, October 2012
4. Study on impact of traffic off-loading and related technological trends on the demand for wireless broadband spectrum. Report for the European Commission, 2013. Available in the EU Bookshop at: http://dx.doi.org/10.2759/19531
5. Hanna, S., Sydor, J.: Distributed sensing of spectrum occupancy and interference in outdoor 2.4 GHz Wi-Fi networks. In: Global Communications Conference (GLOBECOM) IEEE, Anaheim, California, pp. 1453–1459 (2012)
6. CRC Coral Manual v1.2. Available online at: http://www.crc.gc.ca/en/html/crc/home/Wi-Fi_cr/knowledge_exchange/coral_inventory
7. Pediaditaki, S., Marina, M. K., Tyrode, D.: Traffic-aware channel width adaptation in long-distance 802.11 mesh networks. In: Proceedings of the 15th ACM International Conference on Modeling Analysis and Simulation of Wireless and Mobile Systems, pp. 261–270, 2012
8. Gast, M.: 802.11 ac: A Survival Guide. O'Reilly Media Inc., Sebastopol 2013

[4] Keynes, John Maynard. The General Theory of Employment, Interest and Money. New York: Harcourt, Brace, 1936. Print. Reviewed online at: http://en.wikipedia.org/wiki/The_General_Theory_of_Employment,_Interest_and_Money.

[5] See discussion online at: http://en.wikipedia.org/wiki/Say%27s_law.

9. Medeisis, A., Sydor, J., Cremene, L., Wiecek, D., Haddad, Y., Holland, O., Anskaitis, A., Cuzanauskas, T.: ISM-advanced: Improved access rules for unlicensed spectrum (to appear)
10. FCC 10-174 second memorandum opinion and order in the matter of unlicensed operation in the TV broadcast BANDS (ET Docket No. 04-186) and additional spectrum for unlicensed devices below 900 MHz and in the 3 GHz band (ET Docket No. 02-380). In: Federal Communications Commission, Washington 2010
11. CEPT Report 24: A preliminary assessment of the feasibility of fitting new/future applications/services into non-harmonised spectrum of the digital dividend (namely the so-called "white spaces" between allotments). In: Electronic Communications Committee (ECC) within the European Conference of Postal and Telecommunications Administrations (CEPT), 2008
12. UHF white space in Europe—a quantitative study into the potential of the 470–790 MHz band. In: IEEE International Symposium on Dynamic Spectrum Access Networks (DySPAN), Aachen, 2011
13. RSPG10-348 Final RSPG Opinion on Cognitive Technologies. Brussels, 2011
14. Fette, Bruce: Cognitive Radio Technology. Newness, Boston (2006)
15. Spectrum management strategy: Ofcom's approach to and priorities for spectrum management over the next ten years. Consultation. Ofcom, London 2013
16. Damelin, J., Daniel, W. A., Fine, H., Waldo, G. V.: Development of VHF and UHF propagation curves for TV and FM broadcasting. Tech. no. R-6602 Washington, FCC (1966)
17. White Space Devices (WSD); Wireless access systems operating in the 470 MHz to 790 MHz frequency band. European Standard EN 301 598. Sophia-Antipolis, France: ETSI, 2013
18. TV White Space Coexistence Methods. IEEE Standard 802.19.1 [DRAFT], 2013
19. Realizing the full potential of government-held spectrum to spur economic growth. Report Washington DC: Executive Office of the President President's Council of Advisors on Science and Technology, July 2012
20. GSMA public policy position on licensed shared access (LSA) and authorised shared access (ASA): Issue brief. GSMA, London, February 2013
21. Python language programme used for simulation of power control game. Available online in the document download area at: http://www.cost-terra.org/our-publications

Index

A. Medeisis and O. Holland (eds.), *Cognitive Radio Policy and Regulation*, Signals and Communication Technology, DOI: 10.1007/978-3-319-04022-6,

The manufacturer's authorised representative in the EU is Springer Nature Customer Service Centre GmbH, Europaplatz 3, 69115 Heidelberg, Germany. If you have any concerns regarding our products, please contact ProductSafety@springernature.com

Printed and bound by CPI Group (UK) Ltd, Croydon, CR0 4YY
30/09/2026
02225569-0015